Marine Ecosystem Dynamics Models

Construction, Application and Development

Marine Ecosystem Dynamics Models

Construction, Application and Development

Honghua Shi
First Institute of Oceanography, MNR, China

Chengcheng Shen
Second Institute of Oceanology, MNR, China

Yongzhi Liu
First Institute of Oceanography, MNR, China

经典中国国际出版工程 China Classics International · China Ocean Press · World Scientific

Published by

World Scientific Publishing Co. Pte. Ltd.
5 Toh Tuck Link, Singapore 596224
USA office: 27 Warren Street, Suite 401-402, Hackensack, NJ 07601
UK office: 57 Shelton Street, Covent Garden, London WC2H 9HE

Library of Congress Cataloging-in-Publication Data
Names: Shi, Honghua, 1978– author.
Title: Marine ecosystem dynamics models : construction, application and development / Honghua Shi, First Institute of Oceanography, MNR, China, Chengcheng Shen, Second Institute of Oceanology, MNR, China, Yongzhi Liu, First Institute of Oceanography, MNR, China.
Description: Hackensack, NJ : World SCientific, [2024] | Originally published in Chinese by China Ocean Press Ltd., 2018. | Includes bibliographical references and index.
Identifiers: LCCN 2023013864 | ISBN 9789811264207 (hardcover) | ISBN 9789811264214 (ebook for institutions) | ISBN 9789811264221 (ebook for individuals)
Subjects: LCSH: Marine ecosystem management. | Marine ecology.
Classification: LCC QH541.5.S3 S46 2024 | DDC 577.7--dc23/eng/20230516
LC record available at https://lccn.loc.gov/2023013864

British Library Cataloguing-in-Publication Data
A catalogue record for this book is available from the British Library.

典型海洋生态系统动力学模型构建、应用及发展
Originally published in Chinese by China Ocean Press Ltd.

For any available supplementary material, please visit
https://www.worldscientific.com/worldscibooks/10.1142/13080#t=suppl

Desk Editors: Aanand Jayaraman/Amanda Yun

Typeset by Stallion Press
Email: enquiries@stallionpress.com

Foreword

There are strong land-ocean interactions in coastal zones. Coastal ecosystems are subject to dual interferences from climate change and human activities, and thus they are typically complex and giant open systems with complicated change dynamics. A large number of land-based pollutants, especially large amounts of industrial, agricultural, and domestic pollutants, are discharged into coastal waters via rivers, which, along with global climate change, overfishing, and many industrial, urban, and tourism-related activities, have changed coastlines and imposed serious threats to the health and biodiversity of coastal ecosystems. Exploring and utilizing marine resources and protecting marine ecosystem health based on land–sea coordination are inevitable requirements for constructing the marine ecological civilization.

In recent years, we have undertaken China's National Programs for Science and Technology Development, including the National Key Technologies R & D Program of China during the 9th Five-Year Plan Period entitled *Key Technologies for the Utilization of Coastal Resources and Environment* project, and the *Eco-Environmental Effects of Pollution Inputs into the Coastal Ocean* subproject, which is part of the "973" project entitled *Typical Estuary Land-Sea Interactions and Eco-Environmental Effects in China*, and the *Bioremediation Technology* subproject of the "863" project entitled *Typical Coastal Zone Environmental Restoration Technology for Bohai Sea.* With the support of the National Natural Science Foundation of China, the Ministry of Science and Technology, the State Oceanic Administration, and the Chinese Academy of Engineering, I and several of my students have been working on related strategic and technical research and development, with a focus on coastal zone system science and regulation technologies pertaining to the relationship between humans and the ocean. We have gradually realized that the coastal zone is a typical artificial ecosystem and the coastal ecosystem is affected by the synergies of multiple human activities. Terrestrial pollution, land reclamation, and marine aquaculture have coexisted for a long time, and all of them directly affect the health of coastal ecosystems. The developments in the study of artificial coastal ecosystems are critical for the implementation of ecosystem-based coastal zone management, and the regulation of the human–ocean relationship.

We have also noticed that the change process of a coastal ecosystem is extremely complicated. The deduction of the overall states of a marine ecosystem based on sample surveys and the insufficiency of case study conclusions have limited the development of research in marine ecology and its applications, which makes the study of the change, restoration, and regulation of ecosystems difficult. This is due to the lack of technical methods for inverting the whole from the part and abstracting the whole from the sample. In addition, the research method of marine ecosystem digital twin modeling is still under development. Ecosystem dynamics modeling is an important tool for revealing the mechanism and complexity of the synergistic effects of multiple human activities on coastal ecosystems, and is expected to promote the development of digital twin modelling in marine ecology.

The authors of this book have focused on the theoretical frontiers and lack of key technology in marine ecosystem dynamics modeling, such as stability, reliability, prediction accuracy, and parameter optimization, and explored model construction based on principles, methodologies, and case studies. This book exhibits the systematic and innovative research achievements of Dr. Honghua Shi and his research team, in more than 10 years of marine ecosystem dynamics research, with clear research strategies, rich content, novel methods, combined theoretical analysis and application case studies, and reliable results.

This book is a rare representative work in the field of marine ecosystem dynamics research worldwide. The authors have clarified the research objective of the study of marine ecosystem dynamics, proposed a technical framework and modeling principles, analyzed the key bottlenecks affecting model applications, such as model stability and reliability, and proposed potential solutions. They studied the construction, development, and application of typical models using the population growth dynamics model of macroalgae, coupled biological-physical biogeochemical models, and complex ecosystem modeling methods constructed by differential dynamic systems, artificial neural networks, and statistical dynamics methods as the example. In all, in this book, multiple types of marine ecosystem dynamics models are constructed using differential equations, artificial neural networks, and statistical-dynamics methods, which are suitable for different application needs and research areas. These models are highly accurate and can extend the traditional research methods of marine ecosystem dynamics.

I hope that Dr. Shi Honghua will continue his work on the development of theories for coastal ecosystems and artificial ecosystems, and actively explore the theory and technology of marine ecosystem ecology via marine ecosystem dynamics modeling. I believe that Dr. Shi and his team will be able to contribute more to the sustainable development of coastal zones and marine ecological civilization construction.

Dewen Ding
Academician of Chinese Academy of Engineering

Preface

Marine ecosystem dynamics takes the marine ecosystem as the research object, quantitatively reveals the physical, chemical, and biological processes and influence of human activities and natural change on it and the related response and feedback mechanisms, and predicts changes in the marine ecosystem's dynamics to explore the structure and function of the marine ecosystem and its spatial and temporal change. We have focused on the frontier and difficult research topics in marine ecosystem dynamics studies, such as model stability, reliability, prediction accuracy, and parameter optimization, from the aspects of the model construction principle, methodology, and case studies. An *Ulva prolifera* growth model, a phytoplankton model for coastal macroalgae aquaculture, and a coupled biological-physical coastal model are demonstrated, and their application is explored using scenario simulation, sensitivity analysis, and parameter optimization. Several marine ecosystem dynamics models are constructed using artificial neural networks and cellular automata computational methods. A dimensionality reduction theorem is proposed for analyzing the stability of the solutions of a class of self-conserving marine ecosystem dynamics models, and a technical framework is presented for improving the performance of marine ecosystem dynamics models. We also analyze and review the research progress, difficulties, and challenges of end-to-end marine ecosystem dynamics modeling.

The main content and innovative results of this book are summarized as follows:

(1) **Macroalgae ecosystem dynamics modeling.** All parameters of the coastal *Ulva prolifera* model and Nutrient-phytoplankton-zooplankton-detritus-kelp (NPZDK) model for aquaculture ranches are subjected to the global sensitivity analysis method and are ranked. Global sensitivity analysis is not affected by the initial parameter values and can be used to identify the key influence parameters of the model by analyzing the interactions among model parameters. Based on the sensitivity analysis, a global optimization strategy for the marine ecosystem dynamics modeling is proposed with the effects of parameters on the multiple state variables included.

(2) **Construction and application of a coupled biological-physical coastal model.** Scenario simulation is conducted for the impacts of typical human activities on a coastal ecosystem. The response characteristics of the coastal ecosystem of Laizhou Bay to the river pollutant inputs into the ocean are analyzed by the coupled biological-physical model. The phytoplankton fluctuation and nutrient cycle of Laizhou Bay are simulated with the nutrient-phytoplankton-zooplankton-detritus (NPZD) model coupled with the tidal current field in the bay. The characteristics of state variables of the NPZD model for Laizhou Bay under different pollution scenarios are determined.

(3) **Complex system-based modeling.** The authors propose the workflow chart and principle for the artificial neural networks based on a marine ecosystem dynamics model and establish a back propagation (BP) artificial neural network model, taking Laizhou Bay, Jiaozhou Bay, Quanzhou Bay, and Guangxi Coastal Waters—to represent the typical coastal waters of Bohai Sea, Yellow Sea, East China Sea, and the South China Sea, respectively—as the study areas to simulate the primary productivity and biomass of phytoplankton. The water environment factors affecting the phytoplankton biomass in these study areas are mostly identified through sensitivity analysis to explore the dynamic mechanism of phytoplankton biomass change. A risk-assessment and early-warning model for incidents of jellyfish-related injury is constructed using cellular automata to predict the probability of a jellyfish sting and to assess the corresponding level of risk for a simulated seaside resort to provide a theoretical

reference for scientific seaside resort management. The principle of marine ecosystems modeling by statistical-dynamics model is proposed. The case studies presented suggest that the average simulation accuracy of the target state variables is higher after the model parameters are optimized. Statistical-dynamics modeling avoids the difficulties in parameter acquisition and optimization encountered by traditional methods, takes full advantage of existing observation data, and exhibits areal characteristics.

(4) **Performance optimization of marine ecosystem dynamics models.** Based on the methods and strategies used in marine ecosystem dynamics modeling, a technology framework is proposed to improve model performance. Global parameter sensitivity analysis and parameter optimization are included to improve model performance and calibrate the model. The model is validated by analyzing and evaluating the fit of the optimized model to the measured data. Finally, the degree of influence of different factors on the model's performance are evaluated to provide the model's users or decision-makers with a quantitative assessment of the reliability of the model.

(5) **Differential dynamics-based theoretical study on the stability of marine ecosystem dynamics models.** The existence and uniqueness of solutions of a class of self-conserving marine ecosystem dynamics models are demonstrated, and the defects and limitations of the linear approximation method and Lyapunov–Krasovski method in the analysis of such ecosystem dynamics models are proven. To achieve a valid stability analysis of such models, a dimensionality reduction theorem is proposed and applied to the stability analysis of the solutions of the nutrient-phytoplankton-zooplankton (NPZ) model and NPZD model. The two ecosystem models are numerically simulated, and the numerical simulation results and theoretical analysis results are mutually verified.

(6) Research objectives and approaches of marine ecosystem dynamics are proposed from ecology and dynamics perspectives. Marine ecosystem dynamics models are classified. We believe that the research framework of ecosystem dynamics can be simply described as "pressure-state-response (PSR)", where the factors that cause the change of the ecosystem and its components are the "pressure", the structure or function of ecosystems is the "state", and the changes and development trends of the ecosystem are the "response". The study of marine ecosystem dynamics should be focused on the development of mathematical models suitable for assessing a marine ecosystem's status and predicting future changes in the marine ecosystem based on an in-depth understanding of the interaction mechanisms of key processes, including physical, biological, and chemical processes, as well as human activities. The end-to-end approach that emphasizes the characterization of system integrity has gradually become the focus of current marine ecosystem dynamics modeling. We summarize the features of the marine ecosystem dynamics models established by the end-to-end approach and analyze the challenges faced by model construction, such as the integration of the models of different resolutions, and how to handle model uncertainty. We also propose classification methods of marine ecosystem dynamics models from different perspectives and classify ecosystem dynamics models into population dynamics models, biogeochemical models, coupled human-natural marine ecosystem dynamics models, and coastal ecosystem models based on the main ecological processes that the model is focused on. Finally, we propose a technical framework to carry out the global optimization of marine ecosystem dynamics models (GOMEDM) throughout the entire modeling process in order to improve the comprehensive performance of the model.

At present, it has been well known that marine ecosystem dynamics modeling is of significant importance. However, there still is room for major improvements in model expression, ecological significance, simulation accuracy, and parameter selection. In particular, the evaluation and improvement of model accuracy remain the bottlenecks of model application. Different types of marine ecosystem dynamics models can be constructed under the "pressure-state-response" framework, while the extension of modeling methods has provided more options for researchers to solve specific problems. Ecosystem dynamics is one of the main basic disciplines. The dynamic methods, especially modern mathematical methods, have become more and more powerful in model improvement. In this book, we constructed differential dynamics, artificial neural networks, and statistical dynamics models to quantitatively reveal the changes in marine ecosystems using novel strategies. These different types of models complement each other, and their application is explored from the aspects of parameter sensitivity analysis and parameter optimization. The models exhibit high accuracies and applicability and are valuable for promoting the development of marine ecosystem dynamics research methods. The relevant models are verified by case studies in typical coastal areas like Jiaozhou Bay, Sanggou Bay, Laizhou Bay, Quanzhou Bay, and the coastal waters of the Guangxi Zhuang Autonomous Region. The results and conclusions provide scientific foundations for regional marine ecosystem protection and planning. Many of the methods adopted and developed in this book are first attempts. They are novel, yet may need further improvement or are even wrong. We will continuously work to improve them in future research and applications.

This book includes my research accomplishments, as well as those of Dr. Chengcheng Shen, Prof. Long Hu, Prof. Wei Zheng, Dr. Yongzhi Liu and Ms. Li Fen, and other research group members in the past 15 years, especially those from my postdoctoral work on marine ecosystem modeling. Chapter 1 is written by Honghua Shi and Chengcheng Shen. Chapter 2 is written by Honghua Shi and Haoyu Guo. Chapter 3 is written and edited by Honghua Shi and Long Hu. Chapter 4 is written and edited by Honghua Shi and Chengcheng Shen. Chapters 5 and 6 are written by Honghua Shi, Wei Zheng, Yongzhi Liu, Fen Li and Chengcheng Shen. Chapter 7 is written by Chengcheng Shen and Honghua Shi. Chapter 8 is written and edited by Yongzhi Liu. Chapter 9 is written by Honghua Shi, Yongzhi Liu and Long Hu. Chapter 10 is written and edited by Honghua Shi and Long Hu. Chapter 11 is written by Chengcheng Shen and Honghua Shi. Chapter 12 is written by Honghua Shi and Wei Zheng. The entire book is edited by Honghua Shi.

I started my study on marine ecosystem dynamics modeling in 2005 and was encouraged, guided, and supported by my PhD supervisor, Academician Dewen Ding, and postdoctoral supervisor, Academician Guohong Fang. Prof. Fangli Qiao, Prof. Zongling Wang, Prof. Zexun Wei, Prof. Xianqing Lv, Prof. Zhiming Yu, Prof. Jingzhong Zou, Prof. Huiwang Gao, Prof. Mingyuan Zhu, Prof. Baodong Wang, Prof. Ruixiang Li, Prof. Shang Chen and Prof. Xuelei Zhang have given me critical suggestions, guidance, and assistance for my research. Mr. Xuewu Wang contributed significantly to the code writing, debugging, and drawing of model codes. Prof. Meng Gao, Dr. Xikun Song, Dr. Zhan Lian, and Dr. Xiaoyan

Chen gave many useful comments and suggestions. Ms. Liting Yin, Mr. Jingfang Lu, Ms. Yanfei Bi, Ms. Rui Jiang, Mr. Yadong Sui, Mr. Yuxian Ma, Dr. Yongzhi Wang, Dr. Deming Ma, Dr. Yuanzi Huo, Mr. Huimin Xu, Mr. Tao Xia, Dr. Yan Li, Dr. Zhen Guo, Dr. Yuan Chi, Dr. Yucheng Wang, Ms. Yuanyuan Wang, Ms. Liyuan Gao, and Mr. Mingyang Qiao worked hard on the data collection, methodological discussion, and proofreading. I also would like to thank all the researchers, experts, colleagues, students, and friends who have helped me to complete and publish this book.

This book is part of our research on the complexity of coastal ecosystems and the regulation of the human–ocean relationship, which is supported by the National Natural Science Foundation of China (No. U1806214, 41206111, 51779048), and Special Funding for Scientific Research in Marine Public Welfare Industries of China (No. 200705029, 200805080, 201005008, 201005009, 201505012), and is partially supported by the Youth Scientist Fund of the State Oceanic Administration of China (No. 2010225).

Marine ecosystem dynamics modeling, as an important means to reveal the change mechanism of marine ecosystems and predict future changes in marine ecosystems, has become one of the basic theories and powerful technologies for marine ecosystem conservation and integrated ecosystem-based coastal zone management. Considering there are rarely monographs on marine ecosystem dynamics modeling and theoretical research, the authors humbly collate and publish this work, hoping to promote the development of marine ecosystem dynamics. Many institutes and experts have provided us with their selfless help through data sharing and contributing to our model optimization. I would like to express my sincere appreciation to them. Although this book has been revised repeatedly, it is inevitably insufficient to cover every aspect. I appreciate any critical and constructive comments and suggestions.

Prof. Dr. Honghua Shi

About the Authors

Honghua Shi is a Research Professor at the First Institute of Oceanography, Ministry of Natural Resources, China, and Laoshan Laboratory. He obtained his doctorate degree from the Ocean University of China and has been working at the First Institute of Oceanography, Ministry of Natural Resources, China, since then. He is dedicated to the study of marine ecosystem dynamics, ecological evaluation and coastal ecosystem management. He and his team have been focusing on issues and problems, such as the stability, reliability, prediction accuracy and parameter optimization of marine ecosystem dynamics models, and explorations in the basic theory, methodology and typical cases of ecological model construction. Dr. Shi has also contributed to the improvement of marine ecology research to extend toward supporting the sustainable development of coastal zones and the conservation of marine ecosystem health.

Chengcheng Shen is a Research Associate at the Second Institute of Oceanography, Ministry of Natural Resources, China. She obtained her doctorate degree from the Ocean University of China. She has been focusing on marine ecosystem dynamics modeling and ecological evaluation of the impact of human activities.

Yongzhi Liu is a Senior Research Assistant at the First Institute of Oceanography, Ministry of Natural Resources, China. He obtained his doctorate degree from the Ocean University of China. He contributed to the data preprocessing, nonlinear analysis and calibration of the three-dimensional coupled physical-ecological model.

Contents

Chapter 1

Research Background of Marine Ecosystem Dynamics and Characteristics of Typical Models

1.1 Research Objective of Marine Ecosystem Dynamics

Ecology is a systems science linking biology, the environment, and human society, and conducting in-depth research and continuously expanding it can provide the basis for sustainable development (Odum, 1997). Dynamics is a branch of theoretical mechanics, which mainly studies the relationship between the force acting on an object and the motion of the object. It is also a part of Newtonian mechanics or classical mechanics. Since the 20th century, dynamics has been considered a branch of mechanics with a focus on the application of engineering technologies. Ecosystem dynamics arise from the social needs of sustainable development and the development of ecology disciplines. It is an interdisciplinary study of ecology and dynamics that mainly focuses on the interactions between biology, the environment, and human society, and the dynamic mechanism and approaches for sustainable development (Pei *et al.*, 2001; 2004).

Ecology and dynamics are disciplines based on biology and mathematics, respectively, and their internal relationship is constantly involved in ecosystem dynamics. Therefore, many advanced mathematical methods are of great significance in the study of ecosystem dynamics (Pei *et al.*, 2001). The application of mathematics in ecological studies was inadequate before the 1920s. With the development of science and technology and upgraded technical equipment, the application of mathematics in ecology studies has gradually become extensive. Like biology, mathematics is also an important scientific foundation and tool for studying ecosystem dynamics.

The research framework of ecosystem dynamics can be simply described as "pressure-state-response". Specifically, the factors that cause the change of an ecosystem and its components are the "pressure"; the structure and function of the ecosystem are the "state"; and the changes in the ecosystem and changing trends are the "response". This framework reflects a dynamic process, which is the important basis for ecological cybernetics (Pei *et al.*, 2003).

Marine ecosystem dynamics is a discipline that takes marine ecosystems as its research object. It is an important branch of ecosystem dynamics. Marine ecosystem dynamics studies the structures, functions, and their temporal and spatial changes of marine ecosystems. It also quantifies the effects of physical, chemical, and biological processes and human activities on a marine ecosystem and the related response and feedback mechanisms, and predicts the dynamic changes in the future (Shi *et al.*, 2012). As a newly emerged research topic in marine science, marine ecosystem dynamics study links various marine disciplines. Therefore, its development represents the overall research level and direction of oceanography to a certain extent (Chen, 2003). In recent years, marine ecosystem dynamics has become the forefront of marine scientific research. The study of marine ecosystem dynamics has been focused on establishing mathematical models suitable for assessing the state of a marine ecosystem, and predicting how it may change based on an in-depth understanding of the interaction mechanisms among physical, biological, and chemical processes, and the stresses of human activities and natural change.

1.2 Necessity of Marine Ecosystem Dynamics Modeling

Unlike land-based ecosystems, marine ecosystems are characterized by openness, borderlessness, mobility, and complex ecological dynamics. Coastal ecosystems are simultaneously affected by air-sea and sea-land interactions, as well as human activities. Its change mechanism is particularly complex under such serious threats. Therefore, to maintain the healthy and sustainable development of coastal ecosystems, it is necessary to reveal the change mechanism of coastal ecosystems from the perspective of ecosystem dynamics, predict their changing trends, and propose reasonable development and protection countermeasures accordingly. Marine ecosystem dynamics modeling aims to understand the change mechanism of marine ecosystems, simulate and predict their dynamic changes, and provide scientific guidance to assist in decision-making for maintaining the health of marine ecosystems and repairing damaged ecosystems (Wang, 1998; Chen *et al.*, 1999).

The scale of marine surveys continues to expand with the improvement of survey instruments and methods. However, large-scale continuous observations are time-consuming and labor-intensive and can no longer meet increasingly complex marine environmental management needs (Rose *et al.*, 2010).

The physical processes of marine ecosystem dynamics have been well studied. Universally applicable equations have been formed to simulate the movements of seawater and the changes in thermohaline circulation. Yet marine biological and chemical processes remain unclear due to obvious differences in different sea areas. Research has revealed, to a certain extent, the laws followed by the interactions between biological processes — for example, Liebig's law of the minimum and the bottom-up and top-down strategies — and corresponding empirical formulas have been developed (Sun, 2001; Shen *et al.*, 2010). In addition, the development of numerical algorithms and the continuously improved computing capabilities provide computational support for the construction and calculation of marine ecological dynamics models. Therefore, marine ecosystem dynamics modeling is another important means to quantitatively describe the dynamic processes of marine ecosystems, in addition to on-site observation and laboratory experiments.

At present, ecological problems in coastal waters are severe and the causes of these problems are complicated. The responses of coastal ecosystems to human activities are often complex and unpredictable, which makes marine ecosystem management highly uncertain. Marine ecosystem dynamics models are established based on marine survey data and experimental results, as well as the basic laws of physical, chemical, and biological processes. They can quantitatively simulate the structure, function, and temporal and spatial changes of marine ecosystems; allowing for the study of the impacts of natural and human activities on the marine ecosystem and the corresponding response and feedback mechanisms, and the prediction of the dynamic changes of the marine ecosystem (Shi *et al.*, 2012). Therefore, marine ecosystem dynamics modeling is an important tool for studying the response and adaptation of marine ecosystems to human activities.

1.3 Progresses on Marine Ecosystem Dynamics Modeling

Marine ecosystem dynamics modeling relies on the knowledge of marine ecosystems as well as the availability of data and numerical calculation. Therefore, the development of marine ecosystem modeling has been relatively slow. Marine ecosystem dynamics modeling can be traced back to the 1940s. Riley *et al.* (1949) constructed a vertical one-dimensional model based on the upper mixed layer dynamics equations to simulate the seasonal changes of plankton in the North Sea. With the increasing case studies in the North Atlantic, the bays of the United States, and Japan, and other areas since the 1990s, a variety of models of different nutrition levels and types have gradually emerged (Fennel and Neumann, 2004; Plaganyi, 2007). The studies of marine ecosystems in China gradually moved towards quantitative analysis in multiple disciplines since the late 1990s (Wu and Yu, 1996; Gao *et al.*, 2000; Shu *et al.*, 2002; Chen, 2003; Tang *et al.*, 2005; and Feng *et al.*, 2007), and the marine ecosystem dynamics models suitable for the Bohai Sea (Gao and Zhai, 1999; Shu *et al.*, 2002; Chen *et al.*, 2003; Wei *et al.*, 2004; Zhao *et al.*, 2005; Liu and Yin, 2006; Feng *et al.*, 2007), the Jiaozhou Bay and Sanggou Bay in the Yellow Sea (Wan, 1999; Chen *et al.*, 1999; Cui, 2001; Shi *et al.*, 2011; Zheng *et al.*, 2012; Shi *et al.*, 2014), the Yangtze Estuary of the East China Sea (Qiao *et al.*, 2000; Han *et al.*, 2016) and the South China Sea (Chai *et al.*, 2009; Liu and Chai, 2009) have been successively established.

Different models have been developed for different modeling purposes. The traditional models mainly focus on the low and high trophic levels of a marine ecosystem. The studies on the dynamic changes of coastal ecosystems and their responses to human activities by marine ecosystem dynamics modeling are usually conducted by altering model inputs, such as parameters, physical environments, source terms, and so on, or setting multiple scenarios to provide guidance for marine ecosystem management (Gibson *et al.*, 2011; Rombouts *et al.*, 2013; Shen *et al.*, 2016). However, good ecosystem management requires highly accurate models. Therefore, in addition to the model types and modeling methods, model performance optimization, stability, and predictability analysis are also important research focuses of marine ecosystem dynamics modeling (Shi *et al.*, 2016; Shen *et al.*, 2016).

1.4 Classification of Marine Ecosystem Dynamics Models

Marine ecosystem dynamics models can be classified by different methods from different perspectives.

Based on the mathematical methods used, marine ecological dynamics models can be classified based on differential dynamics, statistics, artificial neural networks, and cellular automata. According to the research scope of the model, marine ecosystem dynamics models can be divided into the following model types: the global or ocean basin-scale ecosystem dynamics model, the marginal sea ecosystem dynamics model, the estuary ecosystem dynamics model, and the bay ecosystem dynamics model.

Based on the characteristics of the ecosystem involved in the model, marine ecosystem dynamics models can be classified into the following dynamic ecosystem model categories: aquaculture, upwelling, mangrove, coral reef, and seagrass bed.

The models can also be classified according to the accuracy and spatial resolution of the model description; here, models are classified as zero-dimensional, one-dimensional, two-dimensional, or three-dimensional models.

If we base the criteria for classification on the research objects involved, the models can be identified as individual-based models, population dynamics models, or biomass-based models. It is worth pointing out that the number of state variables and biological processes included in the biomass-based ecosystem model is unrelated to the accuracy of the model. The model structure should be determined based on the understanding of the ecosystem dynamics.

It is obvious that there are diverse marine ecosystem dynamics models. However, marine ecosystem dynamics modeling is based on ecological processes. The model classification based on the main ecological processes can more precisely describe the basic dynamic and structural characteristics of a model. The components in a marine ecosystem are diverse and affect each other.

Understanding marine ecosystems is a continuous accumulative process. Different models have been developed in this process, with the focuses on different aspects of marine ecosystems to solve different issues in the marine ecological environment. Traditional marine ecosystem dynamics models mainly include population dynamics models and biogeochemical models (Traversetal, 2007). The former focuses on the dynamic changes of several species of interest, while the latter generally emphasizes the interactions between environmental factors and the organisms of low trophic levels, yet both describe the structural characteristics of the marine ecosystem under study. With the increasing complexity of marine environmental issues and the popularization of sustainable development concepts, the focus of marine ecosystem dynamics modeling has been gradually formed through the characterization of all components of marine ecosystems, and the mutual coupling with socioeconomic models to study the responses and adaption of marine ecosystems to climate change and human activities (Travers *et al.*, 2007; Plaganyi, 2007; Fulton, 2010). This type of model is called the coupled human-natural marine ecosystem dynamics model in this book.

A coastal zone displays the characteristics of both ocean and land ecosystems with the strongest land-ocean interactions. The coastal environment is mainly affected by socioeconomic interference. Studying the marine and land ecosystems separately would cause the separation of pressure (source) of the marine ecosystem from the carrier. Off-land coastal marine environmental management and off-sea terrestrial environmental management make the implementation of regulatory measures very difficult. Therefore, it is necessary to treat the coastal zone as a special ecosystem to construct a composite coastal ecosystem including coupled human-natural, coupled ocean-land, and coupled resource-environment-social economic subsystems (Shi *et al.*, 2012).

The coastal ecosystem is one of the most representative marine ecosystems. Its evolution, simulation, and regulation models are important means for marine ecosystem protection, restoration, and management, which bridges marine ecosystem dynamics modeling and ocean management and decision making. Therefore, it should be the focus of marine ecosystem dynamics modeling.

1.5 Typical Models and Their Characteristics

Marine ecosystem dynamics models can be classified into the marine population dynamics model, marine biogeochemical model, coupled human-natural marine ecosystem dynamics model, and coastal ecosystem dynamics model, based on ecological processes. Their characteristics are briefly introduced in the following sections.

1.5.1 *Population dynamics model*

Population dynamics models include single population dynamics models and multiple population dynamics models with a focus on the interactions between populations. A single-population dynamics model only considers the dynamic changes of one population and indirectly reflects the impacts of others by involving their population growth rate and mortality. This type of model is simple and can reveal the internal population size determination mechanism. However, the model excludes interactions with other populations in the community and is thus is not ideal for describing the characteristics and balance of biological populations in a marine ecosystem (Chen, 2003). The models considering the different characteristics of the individual components, such as zooplankton, fish, and other populations in different life cycles in marine ecosystems are generally referred to as the individual-based model (IBM). They are used to examine population changes in the community structure using age, growth stage, and individual size (Chen, 2003).

A multi-population dynamics model includes interactions, such as competition, predation, mutual benefit, symbiosis, etc., among the populations in the microbial community, which has become an important research direction of population dynamics modeling (Chen, 2003). The multi-population dynamics model of a predator-prey relationship, including multiple trophic organisms from feed species (such as zooplankton) to top predators, is often used to study the characteristics of ocean food webs and thus is also called the nutrition-based dynamics model. Multi-population dynamics models are mostly focused on the biological communities with important economic and social values, such as fish, for guiding fisheries management (Travers *et al.*, 2007). Multi-species virtual population analysis (MSVPA) (Pope, 1979; Helgason and Gislason, 1979) and an object-oriented simulator of marine ecosystem exploitation (OSMOSE) (Shin and Cury, 2001; 2004) are the most widely used multi-species dynamics models in fisheries management.

The MSVPA model was proposed in the late 1970s (Pope, 1979; Helgason and Gislason, 1979). It is constructed based on species survival and fishing formulas to simulate the dynamic changes of developed fish species. The model is relatively complex (Travers *et al.*, 2007) and requires detailed population information, such as the age structure of the population, which makes it difficult to obtain model parameters. In addition, it is also very challenging to use the existing model structures for the further study of the effects of low trophic level species or functional groups and environmental factors on food web components (Travers *et al.*, 2007).

The OSMOSE model adopts the IBM. It considers predation as an opportunistic process based on individual size matching and the spatial coexistence of predators and prey (Shin and Cury, 2001; 2004). The model characterizes low-trophic level organisms with carrying capacity parameters, and can indirectly include the impacts of environmental factors by tuning the carrying capacity parameters (Shin and Cury, 2001; 2014; Travers *et al.*, 2007).

1.5.2 *Marine biogeochemical model*

Marine biogeochemical models are mainly used to characterize the material cycling between ecosystem components. The models usually focus on low-trophic level organisms and consider the upward effects of nutrients on phytoplankton. The nutrient-phytoplankton-zooplankton (NPZ) structure (Steele, 1974; Evans and Parslow, 1985; Wroblewski *et al.*, 1988; Franks, 2002) is a typical biogeochemical model.

The NPZ model, when including the decomposition and mineralization of bacteria as debris in its observation, is a classic nutrient-phytoplankton-zooplankton-detritus (NPZD) model. The NPZD model can detail the lowest trophic level in a marine ecosystem and can indirectly characterize the predation effects of high trophic level organisms on the low trophic level organisms using zooplankton mortality as the closed term. Because the zooplankton mortality in the NPZD model includes the natural death of zooplankton and the predation of zooplankton by fish, the mortality parameter is often corrected during the modeling. Therefore, the dynamic changes of zooplankton simulated by the model are often significantly different from the measured results.

The NPZD model can be further improved by refining key ecological processes including physical processes, biological processes, and chemical processes, and their interactions. In terms of chemical processes, the combined effects of nitrate nitrogen, ammonia nitrogen, phosphate, and silicate can be included in the model to simulate the nitrogen and carbon fluxes. For the modeling of biological processes, the population composition and structure, such as the age and individual size, of plankton can be considered during the modeling. Further coupling the benthic organisms of high-trophic levels, such as fish, can further improve the model. The coupling with hydrodynamic models for physical processes can make physical environment spatially heterogeneous (Travers *et al.*, 2007; Fulton, 2010).

Among various improved NPZD models, the coupled biogeochemical-hydrodynamic model, also known as the coupled bio-physical model, is the most commonly used marine ecosystem dynamics model. It combines physical environmental factors, such as seawater temperature and salinity, solar radiation, tidal current, and the depth of the mixed layer, and thus can be used to evaluate the effects of environmental or climate changes on phytoplankton. This type of coupled model is usually based on relatively complete physical processes and couples relatively simple biological processes to gradually form the population dynamics models of the organisms of higher trophic levels in the attempt to characterize all components of the ecosystem (Travers *et al.*, 2007).

The commonly used highly complex biogeochemical models are mainly based on the North Pacific Ecosystem Model for Understanding Regional Oceanography (NEMURO). The NEMURO model mainly focuses on low-trophic level organisms. It includes 11 state variables, e.g., nitrate nitrogen, ammonium nitrogen, silicic acid, small phytoplankton, large phytoplankton, small zooplankton, large zooplankton, predatory zooplankton, particulate organic nitrogen, dissolved organic nitrogen, and particulate silicon to detail the characteristics of the planktonic community (Kishi *et al.*, 2007).

1.5.3 *Coupled human-natural marine ecosystem dynamics model*

The coupled human-natural marine ecosystem dynamics model uses human activity as the inherent parameter to simulate the dynamic interaction between human activity and the marine ecosystem. It can also be applied to ecosystem management by combining assessment and management to simulate each part of the adaptive management cycle (Fulton, 2010; Kaplan and Marshall, 2016). The Atlantis ecosystem model and the individual-based *in vitro* model are the representative coupled models.

The Atlantis model also includes multiple processes, such as industrial production, monitoring, evaluation, management, and socioeconomics activities, in addition to various biophysical components, such as the components of physical and biogeochemical systems, and high-trophic-level organisms. It is suitable to study each phase of the adaptive management cycle (Fulton, 2010), and the interactive effects of human activities, such as marine fisheries, changes of land use, nutrient pollution, and wind farms on marine ecosystems. The Atlantis model is considered to be the most effective decision support tool for ecosystem-based management (Plaganyi, 2007). The *in vitro* model is an agent-based model that couples human activity and its socioeconomic drivers as a subject into the system and thus is considered as an assessment and management tool for evaluating the dynamic characterization of human activities and related activities.

The *in vitro* model involves multiple types of human activities, such as commercial fishing, recreational fishing, tourism, shipping, oil and gas exploration, waste dumping and sewage, and the coastal development, protection, mining, fortifications, ports, regional economies, and infrastructure construction (Plaganyi, 2007).

The coupled human-natural marine ecosystem dynamics model often adopts an end-to-end approach to form a complete system and a two-way coupled model framework. The model can be simplified based on the modeling purpose by taking operational efficiency into account. The coupled human-natural system is an open and complex giant system. The interdisciplinary research of social sciences and natural sciences is still slowly developing and the modeling method to couple the human factor is relatively immature. Therefore, the coupled human-natural marine ecosystem model is still underdeveloped and immature. It is often more difficult to characterize the dynamics of the human factor than to characterize the biophysical components of an ecological system. Currently, modelers usually set up multi-scenario simulations based on different behaviors, patterns, and intensities of human activities, which is also a feasible research method (Fulton, 2010).

1.5.4 *Coastal ecosystem model*

Coastal zones are located in the land-ocean interlaced zone with extensive human activities, high socioeconomic development, and strong land-ocean interactions. The rapid development of any coastal zone and marine economy imposes great pressure on marine resources and the environment. Such coastal zones contain the most concentrated and contradictory relationships between humans and the ocean; they are the key area for the comprehensive development of land and ocean as well as for the protection, restoration, and building of the ecological civilization of the marine ecosystem.

Marine ecosystem management has progressed from focusing on natural ecosystem management to complex coastal ecosystem management strategies. The corresponding development and research of marine ecosystem dynamics models should also be adapted to this progress. Marine ecosystem dynamics modeling should focus on the development of simulation and regulation models for the changes likely to occur to complex coastal ecosystems, including the impact of human activities (type and intensity) and the pressure on the ecosystem caused by human activities, the structure and function of the coastal ecosystem, and the regulation strategies based on the continuous development of population dynamics and biogeochemical models from the concept of system ecology, especially the coastal ecosystem, to provide scientific guidance for ecosystem-based adaptive marine management. The primary characteristics and regulation strategies of coastal ecosystems are briefly introduced in the following section to provide a theoretical reference for coastal ecosystem dynamics modeling (Shi *et al.*, 2017).

A complex ecosystem considers the socioeconomic system and natural ecosystem as one, and takes advantage of both systems by adjusting the interrelationships among various components. Internationally, a complex ecosystem is often classified as the coupled human-natural system and coupled socioecological system from the natural and unnatural aspects. In the past two decades, significant progress has been made in the construction of complex coastal ecosystems, especially those combining the socioeconomic system and natural ecosystem via the evaluation of marine natural resources (Constanza *et al.*, 1997).

Researchers in China have also made great contributions to the development of complex ecosystem theories. In the early 1980s, Chinese ecologists, Dr. Shijun Ma and Dr. Rusong Wang (1984) proposed the first concept of a socioeconomic-natural complex ecosystem based on the overall, coordinated, circular and self-sustaining ecological cybernetics. They pointed out that sustainable development was the coordinated development of human life, the living and working environment, the material production environment, and sociocultural environments. Around the same time, the famous Chinese geographer, Dr. Chuanjun Wu, also published his theory that the man–earth areal system was the core of geographical study. His theory has provided important guidance for the study of other areal relationships (Lu and Guo, 1998).

The coastal zone system science and engineering research team at the Key Laboratory of Marine Eco-Environmental Science and Engineering of the State Oceanic Administration was established to conduct research on the joint regulation and control of complex coastal ecosystems and human–ocean relationships. By combining the unique geography and ecology of the coastal zone, the team proposed the conceptual framework of the "Coastal Complex Ecosystem" (Ding *et al.*, 2009; Shi *et al.*, 2012). The coastal complex ecosystem embodies the coupled human-natural system and socioeconomic-natural system in the relationship between humans and the ocean, an important part of the relationship between humans and the Earth, and the extension of the man–earth areal system theory. Further studying of the coastal ecosystem is of significant importance, both theoretically and practically, for promoting the co-development of land and the ocean, integrated ecosystem-based coastal zone management, and the eco-civilization construction of marine ecosystems.

1.5.4.1 *Characteristics of coastal ecosystem*

Systematically understanding the characteristics of a coastal ecosystem is the basis for its modeling. The primary characteristics of coastal ecosystems will be discussed based on the system components and functions in the following section (Shi *et al.*, 2017).

(1) *A coastal ecosystem is an open and complex giant system* with many different types of components. It is composed of coastal resources and environmental, social, and economic subsystems in terms of material cycling and energy exchanging processes. Each subsystem can be classified into several types. For example, the coastal resource subsystem is composed of biological and abiotic resources. The coastal environment includes the coastal water environment, sedimentary environment, and natural geographical conditions of the coast. The economy mainly refers to various industrial activities in the coastal zone, while "society" is understood as composed of the production, living, and consumption patterns of human beings in the coastal zone. The coastal ecosystem is also affected by global changes and the globalization of economic and environmental issues.
(2) *Human beings play an important role in a coastal ecosystem.* A coastal ecosystem incorporates human beings into its component structure. The trophic structure of an ecosystem consists of producers, consumers, and decomposers, in addition to the abiotic environment. Human activities can fall under all three tropic classifications; therefore, attention should be paid to the multiple attributes of human beings. Human activity contributes most to the regulatory functions of the ecosystem and human beings play a unique role as the regulator in all the energy transfer, material cycling, and information exchange processes of a coastal ecosystem. These activities are also the main cause of environmental stress in coastal ecosystems, though humans are also the solvers of and adaptors to these environmental problems. Human activities and the corresponding regulation strategies are the main driving forces which change the coastal ecosystem. Therefore,

coastal ecosystem research should be focused on the regulation of human–ocean relationships and set harmonious human–ocean relationships as the goal of marine ecological civilization building.

(3) *The coastal ecosystem is a complex living organism.* Its functional structure is composed of the immune system, behavioral system, nervous system, and metabolic system. The metabolic system is the foundation and driving force that supports life activities in the coastal zone. It is mainly manifested in the health and development of the population, the utilization and consumption of resources, and the discharge and comprehensive utilization of the "Three Wastes". The immune system is the barrier that maintains the orderly, coordinated, and safe operation of the coastal zone, which is mainly manifested in the abundance and sustainability of regional natural resources, economic risk resistance capacity, an orderly and harmonious society, etc. The nervous system refers to the regulatory center that conducts self-regulation and ensures that the ecosystem functions well. It is mainly reflected in the ability of the ecosystem to sense the changes in the internal and external environments, make scientific decisions, and implement them. The behavior system is a comprehensive system describing the production of the ecosystem and the internal characteristics and external performances of human activities. It is mainly manifested in the functions and values of the services of a regional ecosystem.

The four functional systems of the coastal ecosystem reflect the diverse service functions of the marine ecosystem. They constitute the functional structure of the coastal ecosystem, thereby maintaining the sustainable and healthy development of the ecosystem.

(4) *The functions of a coastal ecosystem are highly diverse.* They comprise cultural functions, such as recreation and entertainment, scientific research, and educational functions in the social system, in addition to its inherent functions, which include the production, consumption, and decomposition of the natural ecosystem. A coastal ecosystem possesses high levels of ecosystem service functions and values. It is also an important carrier of socioeconomic development in the coastal zone and provides important support for the development and utilization

of ocean resources. Its resource and environment-carrying capacities reflect the functions of a coastal ecosystem.

1.5.4.2 *Framework for coastal ecosystem modeling*

A coastal ecosystem is an open and complex giant system. Managing it with systems engineering techniques can provide a scientific and effective model framework. The Driving force-Pressure-State-Impact-Response (DPSIR) framework established by the European Environment Agency based on Pressure-State-Response (PSR) and Driving force-State-Response (DSR) is a systems engineering framework widely used in the environmental field. The PSR model was proposed by the International Economic Cooperation and Development Organization in the late 1980s, and the DSR model was proposed by the United Nations based on the PSR model. The DPSIR framework is an effective tool for systematically examining the correlation between human activities and the natural environment. It is widely used to analyze environmental resource-economic society issues.

The DPSIR model can provide a reference framework for the management and regulation of human intervention in the coastal zone. Because the coastal ecosystem is a living organism, the DPSIR model should be modified and extended from the physical category to the life sciences category. The ecosystem is embodied with a trophic structure that includes producers, consumers, and decomposers, that enables it to conduct material cycling, energy transfer, and information exchange. Human beings are incorporated into the ecosystem to study the impacts of human activities as the regulator. A coastal ecosystem is unique because of the regulatory role of humans. As a living organism, it possesses active regulation functions. Therefore, the management and regulation of the coastal complex ecosystem should be changed from a human passive response to human active adaptation.

Human activities as producer, consumer, and decomposer can be divided into two categories: development (utilization) activities, with socioeconomic development as the goal, and protection (management) activities, with ecological health as the latter's goal. These two goals are the driving forces for generating different human activities that can cause different pressures and impacts on the coastal ecosystem. The coastal complex ecosystem can internalize the external impact driving force in systems engineering into goal driving, to generate different human activities to teach and guide human behaviors at the conscious level. In a narrow sense, marine environment management in a coastal zone is a conservation and management activity aiming to maintain the health of the ecosystem. In a broad sense, it is a regulatory action that can take advantage of human actions to coordinate socioeconomic development and ecosystem health.

The regulation model can then be constructed under the guidance of the coastal ecosystem based on the classic DPSIR model of systems engineering, which is called the target-pressure-state-impact-regulation model in this book. For the coastal zone, the PSI represents a series of natural physical, chemical, and biological processes in the coastal water, e.g., the coastal marine ecosystem. The changes of an ecosystem are the important feedback to the regulation, which also accords with the basic concept of ecosystem-based management. The regulation in the model mainly includes three aspects:

(1) *Pressure regulation.* This relates to reducing the interference of human activities on coastal ecosystems. Specific measures include controlling the reclamation intensity and total pollutant discharge to the ocean, changing and controlling the mode and scale of aquaculture, and changing and controlling the fishery methods and amount;
(2) *Impact regulation.* This means directly taking actions beneficial to the health of the marine ecosystem, according to the ecological processes of the coastal marine ecosystems. It is usually realized by establishing marine pastures and ecological restoration;
(3) *Target regulation.* This refers to altering human behaviors to facilitate both ecosystem health and positive socioeconomic development (Shi *et al.*, 2017).

Such regulatory measures also include educating the public on ecological environmental protection, as well as the optimization and adjustment of development modes and industrial structures. It is an important manifestation of ecological civilization. According to the current state of marine development and utilization, pressure regulation and impact regulation are most urgently needed in order to solve the severe problems faced by the ecological environment.

1.6 Contents of this Book

This book will first clarify the research objective of the study of marine ecosystem dynamics before presenting a basic research strategy. It will then clarify the discipline's position and propose basic research approaches of the study of marine ecosystem dynamics from ecology and dynamics perspectives.

It then discusses the basic theories, methods, and case studies for marine ecosystem dynamics modeling, with a focus on model stability, reliability, prediction accuracy, and parameter optimization. Key bottlenecks affecting model application, including the stability and accuracy of each model, are analyzed, and solutions are put forward.

Classification methods of the aforementioned models, which are constructed based on artificial neural networks, cellular automata, statistical dynamics, and other calculation methods, are next recommended, and the key ecological processes and theories affecting marine ecosystem dynamics modeling are sorted. These models will be classified from different angles, with the authors analyzing the research and development strategies of coupled human-natural marine ecosystem dynamics models, outlining the characteristics of the coastal ecosystem, and offering a conceptual framework for coastal ecosystem dynamics modeling.

In addition, a technical framework for improving model performance is demonstrated, and models, such as the growth model of *Ulva prolifera* at different growth stages, a marine planktonic ecosystem model for coastal seaweed farming, a coupled biophysical model, and a water quality evaluation model for marine ranches, are constructed, and their application explored.

The authors also analyze the difficulties and challenges encountered in model construction and propose a dimensionality reduction theorem for analyzing the stability of a class of self-conserving marine ecological dynamics models.

A new marine ecosystem dynamics research approach, which avoids the difficulties in parameter acquisition and optimization encountered by traditional methods and takes full advantage of existing observation data, is then put forward for the inversion modeling of marine ecosystems based on inverse problem research.

In summary, differential dynamics, artificial neural networks, and statistical-dynamics models to predict the trends and/or direction of possible future changes in marine ecosystems are established in this book. These different models complement each other, and their applications are explored using sensitivity analysis and parameter optimization. The models constructed in this book are highly accurate and applicable and are valuable in academic study and for promoting the development of research methods to advance marine ecosystem dynamics. The conclusions obtained from this empirical study can also provide a scientific foundation for areal marine ecosystem protection and planning.

References

Chai F, Liu G, Xue H, Shi L, *et al.* 2009. Seasonal and interannual variability of carbon cycle in South China Sea: A three-dimensional physical biogeochemical modeling study. *Journal of Oceanography.* **65**(5): 703–720.

Chen C. 2003. *Marine Ecosystem Dynamics and Modeling.* Beijing: Higher Education Press.

Chen C, Liu H, Beardsley R C. 2003. An unstructured grid, finite-volume, three-dimensional, primitive equations ocean model: Application to coastal ocean and estuaries. *Journal of Atmospheric and Oceanic Technology.* **20**(1): 159–186.

Chen C S, Ji R B, Zheng L Y, *et al.* 1999. Influences of physical processes on the ecosystem in Jiaozhou Bay: A coupled physical and biological model experiment. *Journal of Geophysical Research.* **104**(C12): 29925–29949.

Constanza R, Arge R, Groot R, *et al.* 1997. The value of the world's ecosystem services and natural capital. *Nature* **386**: 253–260.

Cui M, Zhu H. 2001. Coupled physical — Ecological modeling of the central part of Jiaozhou bay II: Coupled with ecological model. *Chinese Journal of Oceanology Liminology.* **19**(1): 21–28.

Ding D, Shi H, Zhang X, *et al.* 2009. *Water Quality Change Mechanism and Eco-Environmental Effects in Coastal Waters.* Beijing: China Ocean Press.

Evans G T, Parslow J S. 1985. A model of annual plankton cycles. *Biological Oceanography*. **3**: 327–347.

Feng S, Zhang J, Wei H, *et al.* 2007. *Introduction to the Environmental Dynamics of the Bohai Sea*. Beijing: Science Press.

Fennel W, Neumann T. 2004. Introduction to the modelling of marine systems. Amsterdam: Elsevier.

Franks P J S. 2002. NPZ models of plankton dynamics: Their construction, coupling to physics and application. *Journal of Oceanography*. **58**: 379–387.

Fulton E A. 2010. Approaches to end-to-end ecosystem models. *Journal of Marine Systems*. **81**(1–2): 171–183.

Gao H, Feng S, Guang Y. 2000. Progress in marine planktonic ecosystem modeling. *Oceanologia ET Limnologia Sinica*. **31**(3): 341–348.

Gao H, Zhai X. 1999. Sensitivity analysis of aquatic ecosystem dynamic model parameters. *Periodical of Ocean University of China*. **3**: 398–404.

Gibson G A, Spitz Y H. 2011. Impacts of biological parameterization, initial conditions, and environmental forcing on parameter sensitivity and uncertainty in a marine ecosystem model for the Bering Sea. *Journal of Marine Systems*. **88**(2): 214–231.

Han R, Chen Q, Wang L, *et al.* 2016. Analysis of the ecosystem structure and energy flow of the Yangtze River stuary and adjacent seas, based on the Ecopath model. *Acta Ecologia Sinica*. **36**(15): 4907–4918.

Helgason T, Gislason H. 1979. VPA — Analysis with species interaction due to predation. International Council for the Exploration of the Sea. CM 1979/G: 52.

Kaplan I C, Marshall K N. 2016. A guinea pig's tale: Learning to review end-to-end marine ecosystem models for management applications. *ICES Journal of Marine Science*. DOI: 10.1093/icesjms/fsw047.

Kishi M J, Kashiwai M, Ware D M. 2007. NEMURO — A lower trophic level model for the North Pacific marine ecosystem. *Ecological Modelling*. **202**(1–2): 12–25.

Liu G, Chai F. 2009. Seasonal and interannual variability of primary and export production in South China Sea — A three-dimensional physical biogeochemical model study. *ICES Journal of Marine Science*. **66**(2): 420–431.

Liu H, Yin B. 2006. Model study on Bohai ecosystem. Model description and primary productivity. *Acta Oceanologica Sinica*. **28**(6): 21–31.

Lu D, Guo X. 1998. Man earth areal system — the core of geographical study — on the geographical thoughts and academic contributions of Academician Wu Chuanjun. *Acta Geographia Sinica*. **53**(2): 97–105.

Ma S, Wang R. 1984. The social-economic-natural complex system. *Acta Ecologica Sinica.* **4**(1): 1–9.

Odum E P. 1997. *Ecology: A Bridge between Science and Society.* Sinauer Associate, Inc.

Pei T, Jin C, Guan D. 2003. *Ecological-Cybernetic Principle.* Beijing: Science Press.

Pei T, Jin C, Guan D. 2004. *Ecological Kinetics and Cybernetic Principle.* Beijing: Science Press.

Pei T, Yu X, Jin C, *et al.* 2001. *Ecosystem Dynamics.* Beijing: Science Press.

Plaganyi E E. 2007. Models for an ecosystem approach to fisheries. FAO (Food and Agriculture Organization of the United Nations) Fisheries Technical Paper 477.

Pope J G. 1979. A modified cohort analysis in which constant natural mortality is replaced by estimates of predation levels. International Council for the Exploration of the Sea. CM 1979/H: 16.

Qiao F, Yuan Y, Zhu M, *et al.* 2000. Study on hab dynamical model and hab limitation factors for the sea area adjacent to Chanjiang River Estuary. *Oceanologia ET Limnologia Sinica.* **31**(1): 93–100.

Riley G A, Stommel H, Bumpus D F. 1949. Quantitative ecology of the plankton of the Western North Atlantic. In: *Bulletin Bingham Oceanographic Collection.* New Haven: Peabody Museum of Natural History, **12**: 1–169.

Rombouts I, Beaugrand G, Artigas L F, *et al.* 2013. Evaluating marine ecosystem health: Case studies of indicators using direct observations and modelling methods. *Ecological Indicators.* **24**: 353–365.

Rose K A, Allen J I, Artioli Y, *et al.* 2010. End-to-end models for the analysis of marine ecosystems: Challenges, issues, and next steps. *Marine & Coastal Fisheries Dynamics Management & Ecosystem Science.* **2**(1): 115–130.

Shen G, Huang L, Guo F, *et al.* 2010. *Marine Ecology (Ed III).* Beijing: Science Press.

Shen C, Shi H, Liu Y, *et al.* 2016. Discussion of skill improvement in marine ecosystem dynamic models based on parameter optimization and skill assessment. *Chinese Journal of Oceanology and Limnology.* **34**(4): 683–696.

Shi H, Ding D, Zheng W. 2012. *Evaluation, Simulation and Regulation Techniques of Coastal Zone Models and Their Applications.* Beijing: China Ocean Press.

Shi H, Ding D, Huo Y, *et al.* 2017. *Building Strategies of China Marine Ecosystem Civilization Based on the Integrated Land-Sea Planning-Fundamental Study and Case Study.* Beijing: China Ocean Press.

Shi H, Hu L, Fang G, *et al.* 2016. Research on stability and Hopf bifurcation of marine ecosystem dynamics models. *Acta Oceanologica Sinica.* **35**(4): 124–132.

Shi H, Shen C, Li F, *et al.* 2014. Parameter sensitivity analysis of a coupled biological-physical model in Jiaozhou Bay. *Acta Ecologica Sinica.* **34**(1): 41–49.

Shi H H, Wang Z L, Fang G H. 2011. A statistical dynamics model of marine ecosystem and its application in Jiaozhou Bay. *Chinese Journal of Oceanology Liminology.* **29**(4): 905–911.

Shin Y J, Cury P. 2001. Exploring fish community dynamics through size-dependent trophic interactions using a spatialized individual-based model. *Aquatic Living Resources.* **14**, 65–80.

Shin Y J, Cury P. 2004. Using an individual-based model of fish assemblages to study the response of size spectra to changes in fishing. *Canadian Journal of Fisheries and Aquatic Sciences.* **61**: 414–431.

Shu J, Tang Q, *et al.* 2002. *Research Progress of China Marine Ecosystem Dynamics II. Bohai Sea Ecosystem Dynamics.* Beijing: Science Press.

Steele J. 1974. *The Structure of Marine Ecosystem.* Cambridge: Harvard University Press, MA: 128.

Sun R. 2001. *Animal Ecology (Ed III).* Beijing: Beijing Normal University Publishing Group.

Tang Q, Su J, Sun S, *et al.* 2005. A study of marine ecosystem dynamics in the coastal ocean of China. *Advances in Earth Science.* **20**(12): 1288–1299.

Travers M, Shin Y J, Jennings S, *et al.* 2007. Towards end-to-end models for investigating the effects of climate and fishing in marine ecosystems. *Progress in Oceanography.* **75**(4): 751–770.

Wan Z. 1999. 2-order turbulence closure plankton ecosystem dynamics model and its application in small to medium scales. Qingdao, Institute of Oceanology, Chinese Academy of Sciences.

Wang H. 1998. Several basic problems of marine ecosystem modeling. *Oceanologia ET Limnologia Sinica.* **29**(4): 341–346.

Wei H, Sun J, Moll A, *et al.* 2004. Phytoplankton dynamics in the Bohai Sea- observations and modelling. *Journal of Marine Systems.* **44**(3): 233–251.

Wroblewski J S, Sarmiento J L, Flierl G R. 1988. An ocean basin scale model of plankton dynamics in the North Atlantic. 1. Solutions for the climatological oceanographic condition in May. *Global Biogeochemical Cycles.* **2**: 199–218.

Wu C. 1991. Man-earth areal system — The core of geographical study. *Economic Geography.* **11**(3): 1–4.

Wu Z, Yu G. 1996. Research advances and basic features of the marine ecosystem dynamical models. *Advances in Earth Sciences.* **11**(1): 13–18.

Zhao L, Wei H, Xu Y, *et al.* 2005. An adjoint data assimilation approach for estimating parameters in a three-dimensional ecosystem model. *Ecological Modelling.* **186**: 234–249.

Zheng W, Shi H H, Fang G H, *et al.* 2012. Global sensitivity analysis of a marine ecosystem dynamic model of the Sanggou Bay. *Ecological Modelling.* **247**: 83–94.

Chapter 2

Primary Ecological Processes and Their Theoretical Basis of Marine Ecosystem Dynamics Models

The research of marine ecological dynamics rests on the basic theory of marine ecology. In this chapter, according to the need to construct and study marine ecological dynamics models, we organize and analyze the key ecological processes and theories of marine ecosystems, and prepare a theoretical basis for subsequent chapters on which marine ecological dynamics models are constructed and developed.

2.1 Marine Ecological Factors

2.1.1 *Ecological factors*

Ecological factors (e.g., light, temperature, oxygen, carbon dioxide, etc.) refer to environmental factors that affect the growth, development, reproduction, behavior, and distribution of organisms in the environment (Shen *et al.*, 2010; Allaby, 2010). Different ecological factors affect organisms in their own way. Some ecological factors are essential for the survival of organisms. For example, food and oxygen are indispensable for animals, while water and carbon dioxide are indispensable for plants.

2.1.2 *Classification of ecological factors*

Ecological factors can be generally divided into the two following categories:

(1) *Abiotic factors*: The abiotic factors in the marine environment are mainly light, temperature, salinity, nutrients, and ocean currents. These physical and chemical factors have a strong impact on the survival, reproduction, and distribution of marine life.
(2) *Biological factors*: Organisms living in the environment of other different species are biological factors to each other. Biological factors include intra-species (between the same species) and inter-species (between different species) relationships.

In addition, human activities are also a key factor that significantly affect the environment and should be taken into account when certain specific ecological processes are analyzed. For example, reclamation, inshore farming, and other anthropogenic activities are indispensable considerations in studying the gulf ecosystems of China.

2.1.3 *The minimum factor law and the limiting factor*

Liebig's law of the minimum states that the growth of organisms is not controlled by all factors but restricted by the necessary factors that are supplied at a minimum level. A limiting factor denotes a certain environmental factor that approaches or exceeds the tolerance limit of an organism and thus hinders the survival, growth, reproduction, or distribution of the organism.

2.1.4 *Tolerance limit*

The survival, reproduction, and distribution of organisms rely on a series of complex environmental factors. According to Liebig's law of the minimum and the limiting factor, the amount of each environmental factor exerts different effects on the organism. To any given organism, an environmental factor must be within the tolerance limits (e.g., between the ecological maximum and minimum) for the organism to survive and reproduce. The "law of tolerance" essentially states the limiting effect of the maximum and minimum of ecological factors on the survival of organisms.

2.1.5 *The practical significance of limiting factors*

It is essential to evaluate the tolerance limits and limiting factors to study environmental factors in complex environments. The primary factors that cause ecosystem changes in a particular environment are limiting factors and should be considered with priority. For example, in studying coastal water ecosystems, nutrients in the seawater are often the main limiting factor for phytoplankton growth, and ecological disasters such as harmful algal bloom tends to occur when the content of nutrients in the water body fluctuates.

2.2 Major Abiotic Marine Ecological Factors and Their Ecological Effects

2.2.1 *Light*

Solar radiation is partly reflected when it reaches the sea surface, and incident infrared ($\lambda > 780\,\text{nm}$) and ultraviolet ($\lambda < 380\,\text{nm}$) light is scattered and absorbed in the shallow water layer. Only about 50% of the visible light ($\lambda = 400 - 700\,\text{nm}$) can reach the deeper water layer, and that is called photosynthetically active radiation (PAR) because it is essential for marine life to carry out photosynthesis.

The light intensity in seawater decreases exponentially with water depth. The water body can be divided into the photic zone, disphotic zone, and aphotic zone, according to the level of light attenuation. The photic zone is the major water layer for marine plants to afford primary productivity through photosynthesis.

2.2.2 *The relationship between light and algal photosynthetic rate*

Photosynthetically active radiation is the light that plants can harvest ($\lambda = 380 - 710\,\text{nm}$). The intensity of light radiation determines the rate at which marine algae carry out photosynthesis. The rate of photosynthesis peaks at the saturation light intensity, and excessively strong radiation will inhibit photosynthesis. The saturation light intensity varies for different algae. When the light intensity becomes increasingly excessive, the photosynthetic rate decreases gradually. For example, dinoflagellates usually grow better than diatoms under strong light because the former has a higher saturation light intensity (Ryther, 1956).

2.2.3 *Temperature*

Seawater has a much smaller temperature variation than air because of its higher specific heat capacity. The diurnal temperature variation is usually less than 2°C in the shallow waters and less than 0.3°C in the deep ocean. The temperature variation of the shallow waters within a year depends on the latitude, usually less than 5°C in polar and equatorial waters but higher in the temperate and subtropical sea areas. In particular, the annual temperature variation may exceed 20°C for the inshore water affected by a continental climate.

The seawater temperature is also dependent on depth and shows a stratified pattern vertically. In low latitude areas, the temperature gradient is small in the shallow layer (<100 m), remarkable at about 200–1000 m, and essentially constant at >1000 m. The warm water at the surface and the cold water at the bottom are separated by the thermocline in between. In mid-latitude areas, the thermocline is additionally strongly affected by the season. The thermocline is more evident in the summer when the surface water becomes a lot warmer, but weakens or even disappears in the winter when the surface water becomes cooler and convection is stronger. The seawater temperature range in high latitude areas is relatively uniform vertically and varies from −1.8 to 1.8°C from the surface to the bottom.

Seawater temperature has a significant impact on the metabolism of marine organisms and regulates their activities. Most plants and ectotherms increase their metabolic rate with rising temperature (usually by 2 to 3 times for every 10°C increments) when the temperature falls in a suitable range. Different organisms and even different growth stages of the same species require different environmental temperatures, and normal development is possible only when the temperature exceeds a certain level that is considered the biological zero. Moreover, for many marine organisms, the seawater temperature must also exceed a biological zero for a certain period of time to accumulate enough heat (i.e., the total effective temperature) for its corresponding development process to complete. This is known as the law of effective temperature.

2.2.4 *Nutrients*

The content of inorganic nutrients (phosphate, nitrate, etc.) in seawater is an important factor that affects the primary productivity of marine plants, whose growth is often restricted when nutrients in the environment are insufficient. Because marine phytoplanktons are mainly distributed in the photic zone of seawater, more nutrients are consumed in the shallow water layer, and the nutrients in the shallow water can become the main limiting factor for the growth of phytoplanktons.

2.3 The Primary and Secondary Production of Marine Ecosystems

2.3.1 *Primary productivity*

The marine primary productivity refers to the rate at which organic matter is generated or accumulated by primary producers (e.g., marine phytoplankton, benthic plants, chemosynthetic bacteria, etc.) through photosynthesis or other anabolism such as chemical synthesis. The gross primary productivity is the total production of autotrophic organisms, and the net primary productivity is the gross primary productivity less the primary producer's respiratory consumption. Primary productivity can be measured by methods such as radioisotope tracing, chlorophyll assimilation index, etc. (Yan, 2000; Shen, 2010).

2.3.2 *Main factors that affect primary productivity*

Ecological factors that affect marine primary productivity include light irradiation, mineral nutrients required for plant growth, and related hydrological conditions (Sigman *et al.*, 2012).

2.3.2.1 *Light*

As the basic condition for primary production by plants, light is the most important factor that affects marine primary productivity.

Marine primary productivity is mainly generated in the photic zone. Because excessive irradiation can inhibit the photosynthesis of plants, the maximum primary productivity in natural seas appears in the subsurface layer. The primary productivity gradually decreases with water depth, and the compensation depth is reached where all organic matter produced by the photosynthesis of the plant is used for its own metabolic consumption and the net productivity becomes zero.

2.3.2.2 *Nutrients*

Inorganic salts (nitrate, phosphate, silicate, etc.) in seawater are important nutrients for primary producers such as marine phytoplankton to complete photosynthesis. In shallow seawater, nutrients in the photic zone are continuously lost due to the sedimentation of debris and other processes and thus become an important factor that affects primary productivity. In the presence of upwelling or turbulence, deep eutrophic water can rise and supplement nutrients to the surface layer. During this time, phytoplankton in the shallow seawater can multiply massively and strongly improve the primary productivity.

2.3.2.3 *Iron*

As an indispensable element for the synthesis of chlorophyll and an important cofactor for the production of nitric acid and nitrite reductase, iron has a strong impact on the primary productivity of some sea areas (Martin and Fitzwater, 1988). Iron is supplemented mainly by land sources near the shore in coastal waters but by atmospheric dustfall in the waters away from the coast. For some sea areas (e.g., the South Pacific) with insufficient atmospheric dust deposition, iron is a major limiting factor for primary productivity (Cooper *et al.*, 1996).

2.3.2.4 *Temperature*

The increase of water temperature in a certain range can increase the rate of photosynthesis by marine plants and then increase the primary productivity of the seawater (Davison, 1991). In addition, primary productivity and its spatial-temporal distribution are indirectly affected by temperature as a result of seawater stratification, since the vertical distribution of seawater temperature determines if eutrophic, deep seawater can rise to the surface. For tropical seawater, the surface layer has a high temperature and the underlying thermocline is relatively stable. As a result, deep seawater rich in nutrients cannot readily rise to the surface, where nutrients are lacking and primary productivity becomes limited. Nevertheless, the seasonal fluctuation of productivity is not obvious because the temperature and light in tropical seawaters do not have seasonal variation. For temperate seawater, both the surface layer and thermocline are highly seasonal. As a result, temporary seawater stratification appears in the summer, having a strong impact on the primary productivity, and then disappears in the winter due to the vertical mixing of seawater. That is, the rise of deep seawater to the surface supplements the surface layer with abundant nutrients in the winter to improve primary productivity. On the other hand, the lower temperature and weaker solar irradiation in the winter limit the primary productivity (Shen *et al.*, 2010).

2.3.3 *Distribution of marine primary productivity*

Geographically distinct sea areas with varying hydrological and other characteristics have different light irradiation, water temperature, and nutrients in the photic zone and thus have different primary productivity.

2.3.3.1 *Seasonal distribution of marine primary productivity with regard to latitude*

In low latitude areas, surface seawater maintains a high temperature and receives strong irradiation throughout the year. Because the stable thermocline below the surface seawater prevents the rise of deep eutrophic water, the shortage of nutrients in surface seawater inhibits the primary productivity, and there is no obvious seasonal change of primary productivity (Liu *et al.*, 2012).

In the mid-latitude area, the primary productivity is higher in spring and autumn. In spring, along with intensifying sunlight irradiation, water temperature continues to increase and the stratification of seawater gradually stabilizes. Because the surface seawater is supplemented with abundant nutrients from convective mixing in the winter, phytoplankton multiplies massively and primary productivity increases rapidly. Then in summer, with rising temperature and intensifying irradiation, the seasonal thermocline appears below the surface seawater and the nutrient supply thus falls behind; primary productivity declines as a result. In autumn, the seasonal thermocline gradually disappears as the temperature of the surface seawater drops continuously, and the vertical convective mixing escalates to replenish nutrients for the surface seawater; primary productivity rebounds as a result. In winter, the vertical convective mixing is strengthened and the deep eutrophic water is brought to the surface in large quantities. However, the photosynthesis of phytoplankton is not as active because of low water temperature and weak light intensity, and primary productivity is the lowest of the year.

In high latitude areas, because low surface water keeps a low temperature throughout the year, light becomes the main factor that affects primary productivity. Primary productivity is lower in winter when the irradiation is weak and temporarily higher in summer when the light intensity is stronger (Shen *et al.*, 2010).

2.3.3.2 *The primary productivity of inshore waters*

Inshore waters are unlike seawater in the ocean areas because plenty of land-based nutrients are imported through continental runoff, which lessens the potential restriction on primary productivity from nutrients. In addition, the depth of inshore waters is generally shallow enough not to reach the compensation depth of plants, and persistent thermocline rarely exists. Therefore, photosynthesis is actively carried out throughout the year, and there is generally no peak primary productivity in spring and autumn. Besides, its yearly average productivity is much higher than that of the ocean area at the same latitude.

2.3.4 *Secondary productivity and the energy flow in marine ecosystems*

2.3.4.1 *Secondary productivity*

Secondary productivity refers to the rate at which heterotrophic organisms (consumers such as animals, protozoa, fungi, heterotrophic bacteria, etc., that function as secondary producers) generate marine secondary production through assimilation and absorption (Allaby, 2010). Secondary productivity can be measured by the increment method, instantaneous growth method, among others (Allen, 1951).

2.3.4.2 *Food chain and food web*

The marine food chain refers to a linear combination that starts from debris and phytoplankton to connect consumers at all levels and draws predator-prey relationships. For a given ecosystem, the food chain is a basic functional unit to describe the nutritional relationship between producers and consumers and a basic representation of the energy flow. The food web, with a mesh structure, is formed by interweaving food chains.

The marine food chains include grazing and detritus food chains. Grazing food chains start from marine plants and can be divided into three basic types (Ryther, 1956) based on the number of nutritional conversions during the energy flow: ocean (6 trophic levels), coastal continental shelf (4 trophic levels), and upwelling areas (3 trophic levels). A higher number of nutritional levels generally indicates a smaller particle size of primary producers. Detritus food chains start from detritus or particulate organic matter in the ocean and describe the material-energy transfer between trophic levels starting from the uptake of detritus by consumers. Detrital food chains are essential for the structural diversity and functional stability of marine ecosystems, and also play an important role in linking the material circulation and the energy flow between surface and bottom layers, and between inshore and offshore areas. In the estuary, about 50% of total primary productivity enters the food chain via detritus (Shen *et al.*, 2010).

2.3.4.3 *Trophic level and ecological efficiency*

The trophic level corresponds to the ranks in the escalating individual steps of the food chain in the energy flow. Each trophic level includes a series of organisms. The food chain in ecosystems cannot be extended infinitely because energy is dissipated at all trophic levels due to effects such as the breathing of the organisms at the trophic level. A marine food chain typically has 4 to 5 trophic levels.

Ecological efficiency can be expressed as the ratio of energy obtained at a trophic level to the energy input received by that trophic level. High ecological efficiency reflects low energy loss as energy gets passed along trophic levels.

2.3.4.4 *Functional group*

In an ecosystem, a functional group is the combination of dominant species with similar functional status or tropic level (Reynolds *et al.*, 2002). Plenty of species exist in the marine ecosystem, food relationships are complicated between species, and all species are susceptible to the physicochemical changes of the environment. Therefore, using functional groups to study the structure of the biological community helps simplify the energy flow within communities, facilitate the examination of the marine food web and trophic dynamics, and thus assist in understanding the biodiversity and the structure and function of biological communities.

2.3.4.5 *Particle size spectrum and biomass spectrum*

In the marine ecosystem, different organisms in a biome can be segmented according to their particle size to calculate the biomass contained in each segment. The logarithm of the particle size of each segment can be plotted against the logarithm of the corresponding biomass to give the particle size spectrum, which reflects the distribution of biomass at all levels of particle size (Sheldon *et al.*, 1972). The biomass size spectra can be plotted analogously by changing the particle size in the particle size spectrum to the size of body weight (including energy).

In the biomass size spectrum, the slope reflects the ecological conversion efficiency of the ecosystem, and the intercept reflects the level of productivity. The curve thus reflects the status and dynamics of different marine ecosystems and allows the corresponding macroscopic studies. Thus, for a given ecosystem in a stable state, the biomass size spectrum typically shows a straight line with a small slope.

Peaks in the curves correspond to spatiotemporal excess or accumulation in the energy flow (for example, due to algal bloom and red tide), and discontinuities or disconnections correspond to the reverse. The difference between the spectra of different ecosystems (e.g., waters with different trophic levels) reflects differences in productivity, food web structure, energy flow characteristics, etc. (Sheldon and Parsons, 1967; Shen *et al.*, 2010).

Modeling is an important method for the applied research of particle size and biomass size spectrums. The influence of various factors on the biological particle size can be readily simulated by changing model assumptions. Modeling is popular in the research of particle size and biomass size spectrums because it is stable, efficient, convenient, and straightforward (Zhou *et al.*, 2010).

2.4 Decomposition in Marine Ecosystems and Biogeochemical Cycles

2.4.1 *Decomposition*

Decomposition refers to the gradual degradation of organic matter in the ecosystem, including excreta and dead bodies of animals and plants, into inorganic substances by decomposers, during which energy is released. It is an important function of the ecosystem. The main decomposers in the ecosystem include bacteria, fungi, microprotozoa, etc. Decomposition is in a dynamic balance that an ecosystem maintains, with the production and promotion of continuous circulation of nutrients in the ecosystem (Allaby, 2010).

2.4.2 *Settlement and decomposition of organic particulate matter*

Particulate organic matter in seawater mainly comes from terrestrial input, marine autogeneity, and resuspension of seafloor sediments. In the upper layer of seawater, particulate organic matter mainly includes the corpses, secretions, and excretions of phytoplankton, which are continuously decomposed and gradually settled at a rate related to the size of debris particles. In the oceanic region where nutrition is deficient, most particulate organic matter in the seawater completes mineralization and decomposition in the photic zone and releases abundant nutrients, and a small amount of particulate organic matter settles to the ocean floor. In contrast, in shallow waters, because of the shallowness of the water and larger size of particulate organic matter, particulate organic matter tends to settle to the ocean floor and complete the decomposition process in the sedimentary environment of the ocean floor (Shen *et al.*, 2010; Zhang *et al.*, 2006).

2.4.3 *Nutrient cycling in sediments*

Two types of decomposition occur after the particulate organic matter arrives at the ocean floor. On the surface of the sediment (including the overlying water at the interface between the sediment and seawater), continuous oxidative degradation by heterotrophic bacteria eventually generates an oxidation product (e.g., CO_2, NO_3^-). In contrast, within the hypoxic environment in the sediment, the particulate organic matter is decomposed by anaerobic bacteria to eventually give a reduction product (e.g., CH_4, H_2S, NH_3). In addition, some of the particulate organic matter deposited on the ocean floor may also be resuspended in the seawater by turbulence.

2.4.4 *Carbon cycle*

The carbon cycle is the biogeochemical cycling of carbon mainly within the atmosphere, terrestrial biosphere, hydrosphere, and lithosphere (Xu *et al.*, 2007). Marine waters are a major part of the hydrosphere of the Earth, and the marine ecosystem thus plays a decisive role in the Earth's carbon cycle (Zhou, 2003).

The marine carbon cycle consists of sea–air interface processes and internal cycle processes. Interface processes involve the CO_2 exchange between the ocean surface and the atmosphere, and the direction of the CO_2 flux is determined by the partial pressure difference of CO_2 between seawater and the atmosphere. The atmospheric CO_2 that enters the ocean from the interface process is affected by marine physical, chemical, and biological processes, which all contribute to distinct types of carbon movement referred to as the physical pump, biological pump, and carbonate pump (Volk and Hoffert, 1985). The physical pump refers to the migration and transportation of dissolved carbon in seawater under physical actions of ocean circulation such as advection and diffusion. The biological pump is realized by producers such as phytoplankton in the photic zone. The producer absorbs CO_2 from photosynthesis and fixes solar energy. The generated particulate organic carbon is then passed along the food chain to each trophic level, and large amounts of organic debris that are generated during the growth, metabolism, death, and decomposition of organisms on each trophic level generate sediments that sink to deeper water layers. In this way, some organic carbon fixed by photosynthesis leaves the photic zone and accumulates on the ocean floor. Meanwhile, the massive amount of dissolved organic carbon produced during biological metabolism has two destinations: one part is converted to inorganic carbon and re-enters the carbon cycle, while the other re-enters the food chain through the action of heterotrophic microorganisms via the micro-food web, and is eventually converted to larger particles of organic carbon and sinks to the ocean floor. The carbonate pump refers to the distinct process where some carbon, in the form of biological calcium carbonate in the bodies of marine organisms such as shells and bones and formed during the transfer of organic carbon along the food chain, leaves the carbon cycle of the ecological system after the death of marine organisms and deposits on the ocean floor.

The marine carbon cycle is an important part of the global carbon cycle and has an intimate relationship with global climate change. The CO_2 on the ocean surface can be converted to particulate organic carbon and deposited on the ocean floor through processes such as biological pumps that collect carbon into the sea, thereby reducing the CO_2 partial pressure of surface seawater and increasing the absorption of atmospheric CO_2 by the ocean. These have strong impacts on resolving the global warming problem that results from the increased discharge of greenhouse gases such as CO_2. The relationship between the marine carbon cycle and climate change has become one of the hottest topics in the research of marine ecology (Jiao *et al.*, 2016).

2.4.5 *Nitrogen cycle*

Nitrogen may be replenished in the ocean by terrestrial input (e.g., continental runoff, urban sewage, etc.), atmospheric supplementation (precipitation), and biological nitrogen fixation (e.g., nitrogen-fixing bacteria, cyanobacteria, etc.).

In seawater, soluble nitrogen exists in various forms such as inorganic nitrogen salt (NH_4^+, NO_3^-, NO_2^-, etc.), organic compounds (amino acids, urea, etc.), and molecular nitrogen (N_2). Nitrification and denitrification may occur to inorganic nitrogen in the presence of nitrifying bacteria. Inorganic nitrogen is an important nutrient for the growth of marine plants. It has a major impact on and is the main limiting factor for primary marine productivity.

2.4.6 *Phosphorus cycle*

Phosphorus exists in seawater as particulate phosphorus, dissolved organic phosphorus, and inorganic phosphate. Dissolved organic phosphorus is mainly phosphate esters. Dissolved inorganic phosphorus in seawater almost always exists in the form of orthophosphate (HPO_4^{2-} or PO_4^{3-}) and readily combines with metal ions (such as Fe^{3+}, Al^{3+}, Ca^{2+}, etc.) to form insoluble compounds. Dissolved phosphorus has a low concentration in seawater because it is easily adsorbed by amorphous particles.

Phosphorus is an important component of living organisms. Because phytoplankton in the surface layer of seawater heavily absorbs and utilizes inorganic phosphates, the content of phosphorus is low in the surface layer of the seawater layer. In addition, some phytoplankton can also reabsorb and utilize organic phosphorus via hydrolysis with its phosphatase. Inorganic phosphorus can also be generated from dead plants through hydrolysis and then become available to plants again. Phosphorus in plant cells can be passed to animals through the food chain. The phosphorus obtained by animals is partly excreted as inorganic phosphorus in metabolic activities, and dissolved organic phosphorus and particulate organic phosphorus can be broken down into inorganic phosphorus gradually by bacterial decomposition. Some particulate organic phosphorus that settles on the ocean floor may return to surface water under the influence of upwelling or turbulence. Other phosphorus in the sediment combines with metal ions to form a permanent precipitate to leave the phosphorus cycle.

Terrestrial input is the main source of phosphorus in the ocean, especially in coastal waters. Continental runoff and human activities such as sewage have an important influence on the phosphorus replenishment in coastal waters.

2.4.7 *Sulfur cycle*

The sulfur in seawater mainly comes from two sources. Firstly, under natural erosion and biological decomposition, sulfur in the sediment can enter the marine ecosystem in the form of dissolved sulfate. Secondly, sulfur in the atmosphere (e.g., SO_2, H_2S, etc.) can enter marine ecosystems through precipitation (i.e., acid rain).

The dissolved sulfur in seawater that can be absorbed and utilized by plants mainly exists in the form of SO_4^{2-} and becomes a component of some amino acids in plants before it is passed into animals through the food chain. The proteins in the remains of animals and plants after death are converted by microorganisms to SO_4^{2-} under aerobic conditions through decomposition and oxidization, and mainly to H_2S under reductive conditions, and the sulfur is then recycled and picked up by plants again.

Marine algae can absorb sulfur in the environment to synthesize dimethyl sulfur (DMS), which is then released into seawater. Part of the DMS in seawater will enter the atmosphere and absorb water to form cloud condensed nodules (CCN) to increase the cloud reflection of solar radiation. This process will reduce the temperature of the Earth's surface, help regulate the global climate, and thus have a significant impact on the study of global climate change and other issues (Li *et al.*, 2001).

2.5 Typical Marine Ecosystems

2.5.1 *Estuarine and bay ecosystems*

An estuarine ecosystem is the assembly of a water environment and associated biological communities at the estuary of a river where fresh water and seawater mix and interact. A bay ecosystem is the natural assembly of a semi-enclosed water environment near the shore surrounded by land and the associated biological communities (National Oceanic Administration, 2005). These areas are where seawater and freshwater join and mix, and are constantly affected by factors such as river runoff, tidal currents, wind, waves, etc. As a result, they exhibit complex environmental characteristics. Their salinity usually varies periodically and seasonally. During the high water season in summer, as a result of the high temperature, warm surface water that is low in salt is formed, which is in sheer contrast with the deep water that is cool and high in salt. Strong thermocline and halocline are formed and the water body becomes clearly stratified. In addition, because the estuarine and bay waters are usually shallow, the temperature of the water body is easily affected by the external environment and varies greatly seasonally.

As result of the large amounts of terrestrial input of nutrients, estuarine and bay areas form unique ecosystems with exceedingly high primary productivity. They are complex ecosystems combining the features of freshwater ecosystems, brackish water ecosystems, seawater ecosystems, tidal flat wetland ecosystems, estuary island ecosystems, etc., and offer unique habitats covering wide ranges of salinity and temperature along with hypoxia resistance for diverse estuarine and bay species. Because of the high primary productivity and richness of feed, the estuarine areas and nearby waters are excellent for the reproduction and growth of organisms. They are often the spawning and feeding areas for various marine economic species and thus have high economic value for fisheries.

Estuarine and bay areas have always been the main areas for human development and utilization. They have dense populations and frequent economic activities, and are therefore heavily disturbed by human activities. Present human activities in these areas mainly include land reclamation, port and channel transportation, fishery and aquaculture production, and coastal industries. With accelerating urbanization and industrialization in coastal areas, the impact of human activities on estuarine and bay ecosystems is intensifying, which calls for more efforts on the research and protection of estuarine and bay ecosystems (Chen, 2011).

2.5.2 *Salt marsh wetland ecosystem*

The formation and development of salt marsh wetlands may proceed through two pathways. Some salt marsh wetlands are formed mainly under the influence of marine dynamics and distributed in sand dams and sandbars. Others are formed under the influence of land runoff (sand transported by runoff) and occur as various large estuary deltas, examples of which in China include the Liaohe Estuary, the Yellow River Estuary, and the Yangtze River Estuary (Tong, 2004). The dominant vegetation in the salt marsh wetland is usually *spartina* or *salicornia*. The competition between species is an important biological factor affecting the structure and productivity of salt marsh ecosystems (Hacker and Bertness, 1999). Competitive substitution is a major community succession process (Sun *et al.*, 2003).

The main biological communities in salt marsh wetlands include birds, fishes, some crustaceans, various benthic animals, and microorganisms including bacteria and fungi. Salt marsh plants are mostly converted to debris and then taken up in the debris food chain by organisms such as benthic animals, and only a small amount of plants is ingested by terrestrial animals. Unused debris builds up on wetland sediments. In addition, tidal water, sea wind, and other forces transfer some wetland debris to adjacent sea areas.

In terms of ecosystem service functions, salt marsh wetlands are places where many birds, fishes, and amphibians breed, inhabit and migrate, and have rich biological resources and tourism value. Besides, saltmarsh wetlands serve as a sink for various nutrients entering the sea and exert significant influence on the seawater quality by purifying terrestrial runoff (Mitsch and Gosselink, 2000).

2.5.3 *Island ecosystem*

The island ecosystem is a relatively independent ecosystem composed of four small habitats, namely island land, island base, island beach, and shallow sea around the island, and their respective biological communities (Yang, 2000). An island ecosystem generally has three subsystems, i.e., the island terrestrial ecosystem, intertidal ecosystem, and surrounding seawater ecosystem. It both has characteristics of a terrestrial ecosystem and is affected by marine climate and hydrology, and thus has unique features that reflect a hybrid of both terrestrial and marine ecosystems (Wang, 2006; Wang *et al.*, 2014; Chi *et al.*, 2017). The main features of an island ecosystem are:

(1) **Duality**. Part of the island is above the high tide line and almost never exposed to seawater. This part of the land belongs to the island terrestrial ecosystem because it has a biological community and an environment resembling that of a continental ecosystem. Islands relatively unaffected by human activities usually have habitat patches joined by plant communities, along with a certain number of terrestrial organisms. These animals and plants, along with the island environment, together form a relatively complete island terrestrial ecosystem. On the contrary, the climate in the offshore area of the islands is completely controlled by the sea, and the ecological factors are greatly influenced by

marine climate, hydrology, etc. The intertidal zone and offshore area are dense habitats with high biomass, having abundant crustaceans, mollusks, fishes, etc. These areas are the most productive area in the island ecosystem. The living environment of these organisms and the material flow in these areas fully resemble the characteristics of a marine ecosystem.

(2) **Self-sufficiency**. Because islands are surrounded by water and far from continental land, species in the island ecosystem have limited interactions with the world outside. Besides, islands are typically small with simple territorial structures and scarce freshwater. Consequently, biological diversity is generally poor in an island ecosystem, where there are only a few types of species. Nevertheless, the simple ecological structure of island ecosystems does not affect the normal workings of ecological services, which remain structurally and functionally independent and sufficient. The self-sufficiency of island ecosystems is primarily exemplified by the diversity of both habitat types and biological resources. Habitats in island ecosystems include terrestrial land, wetlands, and waters that correspond to landforms of land area, sea-land transition zone, and sea area, respectively. Island ecosystems are often a miniature of many types of habitats worldwide. The diversity of biological resources results from the varicose landform units and the characteristics of zonal distribution, leading to the diversity in the distribution of species population.

(3) **Uniqueness**. Islands are surrounded by water and have limited exchanges with the continent in both material and information. This special geography of islands fosters a relatively independent regional unit, and often helps preserve various unique and rare species with high research value within independent habitats. In addition, islands have distinct landforms, geological structures, and mineral resources compared with the continent because the genesis varies greatly from one island to another.

(4) **Vulnerability**. Island ecosystems are generally unstable and have fragile ecological features because of geographical isolation and the relative infertility of resources. They are thus highly vulnerable to damage and susceptible to serious ecological and environmental hazards (Zhu and Li, 2004). The major aspects that threaten the wellbeing of island ecosystems are twofold. In terms of natural factors, islands are generally not robust when faced

with natural disasters such as typhoons, tsunamis, droughts, etc. On the other hand, intensifying human activities such as land development seriously threaten the fragile island ecosystem. Because island ecosystems are particularly fragile, they are often difficult to remedy and restore once damage occurs (Chi *et al.*, 2015). China has a large number of widely distributed islands (e.g., the Changshan Islands, Zhoushan Islands, Miaodao Islands, etc.) that are rich in resources. In China, islands play a pivotal role in ecosystem protection, socio-economic development, and defending national interests. However, with rising levels of development, the natural resources of islands show clear signs of damage. Research still needs to be strengthened with regard to the protection and rational development of island ecosystems (Shi *et al.*, 2009; 2016).

2.5.4 *Seagrass bed ecosystem*

Temperate seagrass is mainly composed of *Zostera* that typically grows on soft substrates between the intertidal and subtidal zones. The resulting seagrass bed is a typical marine ecosystem with high productivity and biodiversity, as well as a valuable function in ecosystem service (Adams, 1976).

The seagrass bed has high primary productivity and provides food for a variety of fishes, marine mammals, etc. It also has a high level of secondary productivity and provides a nursery for a variety of economic fishes, shellfish larvae, etc. Dead seagrass mostly enters the energy flow of the detrital food chain and provides rich nutrients for the nearby sea area. The seagrass bed is also an important attachment site for various epiphytic algae. Such attachment further increases the primary productivity of the seagrass bed area, making it a habitat for a variety of animals and plants with high species diversity.

The seagrass bed not only provides food and habitat for a variety of organisms but also helps stabilize the substrate and mitigate the impact of wind and waves. The densely clustered underground stems and roots of the seagrass readily help fix the substrate. The floating leaves not only slow down the impact of waves but also filter the seawater and accelerate the sedimentation of suspended matter.

Therefore, the seagrass bed plays an important role in stabilizing the coastal ecosystem (Barbier *et al.*, 2011).

2.5.5 *Mangrove ecosystem*

The mangrove ecosystem is a woody biome consisting of perennial bushes and trees, mainly mangrove plants, in tidal flat wetlands in the upper part of the intertidal zone of tropical and subtropical coasts flooded periodically by tides (Zhang and Sui, 2001). Mangrove ecosystems have very high net primary productivity.

Mangrove plants mostly grow on fine alluvial soil in inner bays of estuaries where seawater salinity changes greatly, sediments contain rich organic debris, and tidal effects are strong. As the mangrove plants adapt to their special and variable habitat, they form unique characteristics such as highly developed root systems (surface roots, respiratory roots, aerial roots, etc.), viviparity, xerophytic leaf structures, salt tolerance, etc. Additional adaptive features in the physiology and ecology of mangrove vegetation include high osmotic pressure on leaves, rich tannins in bark, strong photosynthesis, special salt metabolism mechanisms, etc. (Tomlinson, 1986).

The subtidal zone of the mangrove ecosystem is rich in organic matter and has high primary and secondary productivity since the root system of mangrove plants provides good habitats and food sources for attached organisms. Overall, the community structure of the mangrove ecosystem is relatively simple, but the species and biodiversity are rich.

Aside from having high primary productivity, the mangrove ecosystem is also an ecologically important coastal area that helps resist the impact of waves and storms and thus stabilize the coast. Besides, it can also filter organic matter and pollutants from terrestrial runoff and land source output to help purify seawater and reduce marine pollution.

2.5.6 *Coral reef ecosystem*

The coral reef ecosystem is a natural unification of coral reef habitats and biological communities in tropical seas (water temperature of 20–28°C) (National Oceanic Administration, 2005). Coral reefs are usually formed by reef-building corals as a result of the interaction between coral insects, algae (e.g., zooxanthellae), and some mollusks (e.g., giant clam). The types of coral reef ecosystems that can be segmented according to the relationship between the reef body and the shoreline include shore reefs, atolls, and barrier reefs. Alternatively, a coral reef ecosystem can also be classified based on its form as a table reef, point reef, reef pinnacle, reef flat, etc. The formation of coral reefs generally has strict requirements on natural environmental conditions such as water temperature, light, salinity, water depth, wave, etc. Large-scale coral reefs usually appear at sea areas with clean water, suitable salinity, sufficient irradiation, good circulation, and an average water temperature of 25°C. In China, coral reefs are mainly distributed from the southern Taiwan Strait to the South China Sea.

Coral reef ecosystems have a variety of primary producers (phytoplankton, benthic algae, symbiotic zooxanthellae, etc.) and have a relatively higher level of primary productivity among ecosystems. Besides, the richness of species in the biome of coral reefs provides a good habitat for marine life, and the biodiversity is high. Coral reefs contain many representative species of marine life (Shen *et al.*, 2010).

Thanks to their special structure and high biodiversity, coral reef ecosystems provide unique ecological service functions. Firstly, coral reefs provide numerous fishery products, medicines, and building materials. Secondly, coral reefs act as a hydraulic fence in the coastal zone, which effectively reduces the energy of the wave and water flow to protect the coastline by ameliorating or preventing coastal erosion. In addition, coral reefs provide not only the above ecological benefits but also important tourism resources in tropical seas. Their unique structure and rich biodiversity create considerable socioeconomic benefits by attracting tourism.

2.5.7 *Deep ocean ecosystem*

The deep ocean ecosystem involves the water body and seafloor outside the continental shelf. Compared with the onshore water body, the deep ocean ecosystem is relatively stable and has sufficient light in the photic zone for the photosynthesis of phytoplankton. Often between the surface water and deep water in the ocean area is a thermocline where temperature decreases rapidly with increasing depth, and the low temperature of the water underneath the thermocline is less variable. The pressure in the ocean area increases with depth and can attain $2 - 6 \times 10^4$ kPa in the deep ocean area.

In ocean areas, the primary producers are mainly picophytoplankton, and the productivity level is relatively low. The food chain is relatively extended and can reach six trophic levels on average. The upper layer of ocean waters usually has a large number of economic fishes, e.g., *Scomberomorus niphonius*, tuna, pomfret, etc. The middle layer of ocean waters is inhabited by a large number of planktonic crustaceans such as the Antarctic krill. These planktonic crustaceans are often an important link in the marine food chain. The bottom layer of ocean waters has high species diversity with a variety of abyssal fishes, crustaceans, polychaete, echinoderms, etc. Many deep-sea animals have developed unique mechanisms to adapt to extreme conditions such as a dark seafloor, scarce food, low temperature, high pressure, etc. For example, many abyssal fishes and cephalopods have specialized light emitters. Because the deep ocean has a low temperature, high pressure, and scarce light, there is almost no primary productivity. Benthic animals mostly obtain food from the remains of plants and animals as well as organic debris deposited from the upper waters and that has not yet completely decomposed. Some marine animals that migrate vertically also bring some food for the benthic animals.

The deep ocean area is the largest ecological unit in the world, but the composition and ecological dynamics of the deep ocean ecosystem remain to be further understood. An important future direction for marine ecology research is the exploration of deep-ocean ecosystems.

References

Adams S M. 1976. The ecology of eelgrass, Zostera marina (L.), fish communities. I. Structural analysis. *Journal of Experimental Marine Biology & Ecology*. **22**: 269–291.

Allaby M. 2010. A dictionary of ecology. Oxford University Press.

Allen K R. 1951. *The Horokiwi Stream: A Study of a Trout Population*. New Zealand Marine Department.

Barbier E B, Hacker S D, Kennedy C, *et al.* 2011. The value of estuarine and coastal ecosystem services. *Ecological Monographs*. **81**:169–193.

Chen X Y. 2011. *Ecosystem Health Assessment for Estuary and Bay: Methodologies and Application* (in Chinese). Ocean University of China, Qingdao.

Chi Y, Shi H H, Guo Z. 2015. Connotation, features and causes of island ecological vulnerability (in Chinese). *Acta Oceanologica Sinica.* **37**(12): 93–105.

Chi Y, Shi H H, Wang Y Y. 2017. Evaluation on spatial heterogeneity of island ecosystem carrying capacity — A case study of southern Miaodao Archipelago (in Chinese). *China Environmental Science.* **37**(3): 1188–1200.

Cooper D J, Watson A J, Nightingale P D, 1996. Large decrease in ocean-surface C02 fugacity in response to in situ iron fertilization. *Nature.* **383**: 511.

Davison I R. 1991. Environmental effects on algal photosynthesis temperature. *Journal of Phycology.* **27**: 2–8.

Hacker S D, Bertness M D. 1999. Experimental evidence for factors maintaining plant species diversity in a New England salt marsh. *Ecology.* **80**: 2064–2073.

Jiao N Z, Li C, Wang X X. 2016. Response and feedback of marine carbon sink to climate change (in Chinese). *Advances in Earth Science.* **31**(7): 668–681.

Li H Y, Wang D Z, Lin Y M. 2001. Progress in Study of marine dimethylsulfide (in Chinese). *Xiamen University Natural Science.* **40**(3): 725–733.

Liu H X, Song X Y, Huang H H. 2012. A comparison study on primary production in typical low-latitude seas (South China Sea and Bay of Bengal) (in Chinese). *Acta Ecologica Sinica.* **32**(18): 5900–5906.

Martin J H, Fitzwater S E. 1988. Iron deficiency limits phytoplankton growth in the north-east Pacific subarctic. *Nature.* **331**: 341–343.

Mitsch W J, Gosselink G J. 2000. *Wetlands.* Canada: John Wiley & Son, 259–305.

National Oceanic Administration. 2005. The guidance for the assessment of coastal marine ecosystem health (in Chinese).

Reynolds C S, Huszar V, Kruk C, *et al.* 2002. Towards a functional classification of the freshwater phytoplankton. *Journal of Plankton Research.* **24**: 417–428.

Ryther J H. 1956. Photosynthesis in the ocean as a function of light intensity. *Limnology & Oceanography.* **1**: 61–70.

Sheldon R W, Parsons T R. 1967. A continuous size spectrum for particulate matter in the sea. *Journal of the Fisheries Research Board of Canada.* **24**: 909–915.

Sheldon R W, Prakash A, Sutcliffe W H. 1972. The size distribution of particles in the ocean. *Limnology & Oceanography.* **1**: 327–340.

Shen G Y, Huang L F, Guo F. 2010. *Marine Ecology*, 3rd Edition (in Chinese). Beijing: Science Press.

Shi H H, Chi Y, Zhen W. 2016. Basic idea of balance sheet design of island natural resources. *Marine Economy in China*. **2**: 138–145.

Shi H H, Zhen W, Ding D W. 2009. Valuation and ecosystem services of typical island (in Chinese). Marine Environmental Science. **28**(6): 743–748.

Sigman D M, Hain M P. 2012. The biological productivity of the ocean. *Nature Education Knowledge*. **3**(10): 21.

Sun S, Cai Y, Tian X. 2003. Salt marsh vegetation change after a short — Term tidal restriction in the Changjiang Estuary. *Wetlands*. **23**: 257–266.

Tomlinson P B. 1986. The botany of mangroves. Cambridge tropical biology series.

Tong C F. 2004. *Structure, Function and Services of Estuarine Wetland Ecosystem: A case study in Yangtze Estuary* (in Chinese). Shanghai: East China Normal University.

Volk T, Hoffert M I. 1985. Ocean carbon pumps: Analysis of relative strengths and efficiencies in ocean-driven atmospheric CQ changes. In: *The Carbon Cycle and Atmospheric CO: Natural Variations Archean to Present*, 99–110.

Wang X L. 2006. *Research on Risk Assessment Methods and Application of Island Ecosystem*. Qingdao: Institute of Oceanology, Chinese Academy of Sciences.

Wang X L, Wang Y, Shi H H. 2014. Discussion of carbon sequestration estimates in the island terrestrial ecosystems (in Chinese). *Acta Ecologica Sinica*. **34**(1): 88–96.

Xu Y F, Zhao L, Li Y C. 2007. Ocean carbon cycle and marine ecosystem dynamics (in Chinese). *Marine Environmental Science*. **26**(5): 495–500.

Yan X Z. 2000. The different methods for determining primary production (in Chinese). *Chinese Journal of Fisheries*. **13**(1): 81–86.

Yang H W. 2000. *Chinese Island* (in Chinese). Beijing: China Ocean Press.

Zhang N X, Song J M, He Z P. 2006. Biogeochemical mechanism of particulate organic carbon (POC) variations in seawaters (in Chinese). *Acta Ecologica Sinica*. **26**(7): 2328–2339.

Zhang Q M, Sui S Z. 2001. The mangrove wetland resources and their conservation in China (in Chinese). *Journal of Natural Resources*, **16**(1): 28–36.

Zhou G S. 2003. *Global Change* (in Chinese). Beijing: China Meteorological Press.

Zhou L B, Tan Y H, Huang L M. 2010. The advances in the aquatic particle/biomass size spectra study (in Chinese). *Acta Ecologica Sinica.* **30**(12): 3319–3333.

Zhu X D, Li Y F. 2004. Research on the proposition of micro-island and its sustainable development. *Marine Sciences.* **2**(8): 1–4.

Chapter 3

Stability Analysis of Marine Ecosystem Dynamics Models

The service and its value of marine ecosystems are indispensable for the survival and development of humans (Costanza *et al.*, 1997). However, human activities nowadays severely affect the structure and function of marine ecosystems (Jackson *et al.*, 2001; Myers and Worm, 2003; MA, 2005; Lotze *et al.*, 2006). The latest research shows that human activity has impacted entire marine ecosystems, and up to 41% of marine areas worldwide are affected by multiple stress factors (Halpern *et al.*, 2008). With regard to coastal zones, the harmful algal blooms caused by eutrophication and other factors have become a major marine disaster that endangers both the marine ecosystem and human health. Marine algal blooms have become a topical issue in the research of marine biology and marine ecology (Bricker *et al.*, 2003; Fujii *et al.*, 2005; Granéli *et al.*, 2008; Gao *et al.*, 2009). Frequent harmful algal blooms not only seriously impede the economic development of coastal areas but also threaten the health and living environment of humans.

Research of marine ecosystem dynamics is playing an increasingly important role to reveal the change mechanism and to simulate and predict future changes of marine ecosystems. Mathematical models are important tools for studying the dynamics of marine ecosystems (Franks and Chen, 1996; Gao *et al.*, 1998; Chen *et al.*, 1999; Liu *et al.*, 2007; Vichi *et al.*, 2007). Predictability is the key to restricting the effectiveness of marine ecosystem dynamics models, and system stability is required to ensure predictability. Therefore, a key area in the research of marine ecosystem dynamics focuses on the stability analysis of system dynamics models, a classic concern in ecosystem dynamics models as well (Holling, 1973; McCann and Hastings, 1997; Edwards and Bees, 2001). However, most marine ecosystem models are highly complicated, which makes it difficult to directly analyze the stability of such systems. Numerical simulation has become a common alternative, but it suffers from inherent uncertainties because of its high sensitivity to factors such as time step, experimental cycle, to name a few (Heimsund and Berntsen, 2004; Cossarini and Solidoro, 2008).

This chapter firstly introduces a class of self-conserving marine ecosystem dynamical models. The characteristics of these models are then analyzed to verify how well-posed such systems are. The limitations of linear approximation and the Lyapunov–Krasovskii method are also discussed. The dimensionality reduction theorem is then introduced to analyze model stability, and the nutrient-phytoplankton-zooplankton (NPZ) and nutrient-phytoplankton-zooplankton-detritus (NPZD) models are analyzed as examples. The results from the theoretical analysis are then verified by numerical simulation. Finally, the Hopf bifurcation conditions of the discussed models are analyzed.

3.1 A Class of Self-Conserving Marine Ecosystem Dynamical Models

The current chapter examines the NPZ and NPZD models, in which N, P, Z, and D represent nutrients, phytoplankton, zooplankton, and organic debris, respectively. The two models are expressed as follows:

$$\begin{cases} \dfrac{dP}{dt} = U_m \cdot \dfrac{N}{k_N + N} \cdot f(I) \cdot g(T) \cdot P - G_m \cdot h(T) \cdot Z \cdot (1 - e^{-\lambda P}) \\ \qquad - M_P \cdot P \\ \dfrac{dZ}{dt} = \gamma \cdot G_m \cdot h(T) \cdot Z \cdot (1 - e^{-\lambda P}) - M_Z \cdot Z \\ \dfrac{dN}{dt} = -U_m \cdot \dfrac{N}{k_N + N} \cdot f(I) \cdot g(T) \cdot P + (1-\gamma) \cdot G_m \cdot h(T) \cdot Z \\ \qquad \cdot (1 - e^{-\lambda P}) + M_Z \cdot Z + M_P \cdot P \end{cases}, \tag{3.1}$$

$$
\begin{cases}
\dfrac{dP}{dt} = U_m \cdot \dfrac{N}{k_N + N} \cdot f(I) \cdot g(T) \cdot P - G_m \cdot h(T) \cdot Z \cdot (1 - e^{-\lambda P}) \\
\qquad - M_p \cdot P \\
\dfrac{dZ}{dt} = \gamma \cdot G_m \cdot h(T) \cdot Z \cdot (1 - e^{-\lambda P}) - M_Z \cdot Z \\
\dfrac{dN}{dt} = -U_m \dfrac{N}{k_N + N} \cdot f(I) \cdot g(T) \cdot P + \theta \cdot G_m \cdot h(T) \cdot Z \\
\qquad \cdot (1 - e^{-\lambda P}) + e \cdot D \\
\dfrac{dD}{dt} = (1 - \gamma - \theta) G_m \cdot h(T) \cdot Z(1 - e^{-\lambda P}) + M_p \cdot P \\
\qquad + M_Z \cdot Z - e \cdot D
\end{cases},
\tag{3.2}
$$

where $g(T)$ and $h(T)$ are the temperature-dependent growth function and feeding function, respectively, and can be expressed as follows based on the "Q10" rule (Fransz and Verhagen, 1985):

$$g(T) = Qg_{10}^{\left(\frac{T-10}{T}\right)}, \quad h(T) = Qh_{10}^{\left(\frac{T-10}{T}\right)}. \tag{3.3}$$

The light limit function $f(I)$ can be expressed as:

$$f(I) = \frac{1}{4.53} \cdot \frac{I_s}{I_0}(1 - e^{-4.53}) \cdot e^{\left[1 - \frac{1}{4.53} \cdot \frac{I_s}{I_0}(1 - e^{-4.53})\right]}. \tag{3.4}$$

In fact, Steele (1962) gave the photosynthetic response function as follows:

$$f(I) = \frac{I}{I_0} e^{\left[1 - \frac{I}{I_0}\right]}, \tag{3.5}$$

where I is the effective light radiation (PAR), I_0 is the optimal light intensity, k_{ext} is the light attenuation coefficient of seawater, I_s is the light radiation intensity on the sea surface, and z is the water depth.

In general, k_{ext} is considered a function of water transparency (S) in offshore waters (Fei, 1984):

$$k_{\text{ext}} = 1.51/S. \tag{3.6}$$

Because photosynthesis generally occurs in the photic zone, the effective optical radiation I (PAR) can be calculated as follows, with the depth of the photic zone denoted as H:

$$
\begin{aligned}
I &= \frac{1}{H}\int_0^H I_s \cdot \epsilon^{-k_{\text{ext}}\cdot z} dz = \frac{I_s}{Hk_{\text{ext}}}(1 - e^{-k_{\text{ext}}\cdot H}) \\
&= \frac{I_s \times S}{H \times 1.51} \cdot (1 - e^{-\frac{1.51 \times H}{S}}),
\end{aligned} \tag{3.7}
$$

Fei (1984) estimated that H and S are linearly correlated. Similarly, we assume $H = 3S$. Therefore, we can derive the full expression of $f(I)$.

Table 3.1 lists other parameters along with their default values.

Table 3.1. Parameters in the NPZ and NPZD models.

Parameter	Default value	Definition	Reference
U_m	2.0	Maximum phytoplankton nutrient intake	(Franks *et al.*, 1986)
k_N	1.0	Half-saturation constant of DIN uptake by the phytoplankton	(Franks *et al.*, 1986)
G_m	1.5	Zooplankton maximum feeding rate	(Gao H W, 1998)
λ	0.2	Ivlev uptake constant	(Franks *et al.*, 1986)
γ	0.2	Zooplankton growth factor	(Gao H W, 1998)
θ	0.4	Zooplankton excretion coefficient	(McCreary *et al.*, 1996)
I_s	82.0	Average sea surface light intensity	—
I_0	150.0	Optimal light intensity	(Feng *et al.*, 2007)
T	13.1	Sea surface temperature	—
Qg_{10}	2.08	Growth temperature dependence coefficient	(Fransz and Verhagen, 1985)
Qh_{10}	3.1	Feeding temperature dependence coefficient	(Fransz and Verhagen, 1985)
M_P	0.1	Phytoplankton mortality	(McCreary *et al.*, 1996)
M_Z	0.2	Zooplankton mortality	(Franks *et al.*, 1986)
e	0.05	Debris remineralization rate	(McCreary *et al.*, 1996)

3.2 The Existence of a Unique Solution

3.2.1 *Model simplification*

Assume that

$$\vec{F_1}(X) = \begin{pmatrix} F_{11} \\ F_{12} \\ F_{13} \end{pmatrix} = \begin{pmatrix} U_m \cdot \dfrac{N}{k_N + N} \cdot f(I) \cdot g(T) \cdot P - G_m \cdot h(T) \\ \cdot Z \cdot (1 - e^{-\lambda P}) - M_P \cdot P \\ \gamma \cdot G_m \cdot h(T) \cdot Z \cdot (1 - e^{-\lambda P}) - M_Z \cdot Z \\ -U_m \cdot \dfrac{N}{k + N} \cdot f(I) \cdot g(T) \cdot P + (1 - \gamma) \cdot G_m \\ \cdot h(T) \cdot Z \cdot (1 - e^{-\lambda P}) + M_Z \cdot Z + M_P \cdot P \end{pmatrix} \tag{3.8}$$

$$\vec{F_2}(Y) = \begin{pmatrix} F_{21} \\ F_{22} \\ F_{23} \\ F_{24} \end{pmatrix} = \begin{pmatrix} U_m \cdot \dfrac{N}{k_N + N} \cdot f(I) \cdot g(T) \cdot P - G_m \cdot h(T) \cdot Z \cdot (1 - e^{-\lambda P}) \\ - M_p \cdot P \\ \gamma \cdot G_m \cdot h(T) \cdot Z \cdot (1 - e^{-\lambda P}) - M_Z \cdot Z \\ -U_m \dfrac{N}{k_N + N} \cdot f(I) \cdot g(T) \cdot P + \theta \cdot G_m \cdot h(T) \\ \cdot Z \cdot (1 - e^{-\lambda P}) + e \cdot D \\ (1 - \gamma - \theta) G_m \cdot h(T) \cdot Z(1 - e^{-\lambda P}) \\ + M_p \cdot P + M_Z \cdot Z - e \cdot D \end{pmatrix} \tag{3.9}$$

which may be simplified as follows when initial conditions are given:

$$\begin{cases} \dfrac{d\vec{X}}{dt} = \vec{F_1}(X) \\ \vec{X}(t_0) = \vec{X_0} \end{cases}, \tag{3.10}$$

$$\begin{cases} \dfrac{d\vec{Y}}{dt} = \vec{F_2}(Y) \\ \vec{Y}(t_0) = \vec{Y_0} \end{cases}. \tag{3.11}$$

3.2.2 *Existence and uniqueness of model solutions*

Theorem 3.1. *With given initial values and parameters, the NPZ system* (3.10) *has a solution that is unique.*

Proof. Because $\vec{F_1}$ is continuous on $R \times R^3$ and the partial derivatives of P, Z, and N exist and are continuous, according to the mean value theorem, for $\forall\ X_1,\ X_2 \in R^3,\ \vec{F_1}(X_2) - \vec{F_1}(X_1) = (\nabla \vec{F_1}(\xi))^T \cdot (X_2 - X_1)$ where $\xi = X_1 + \theta \cdot (X_2 - X_1)$, $0 < \theta < 1$.

Thus, $\exists G_0 = \left\{ (t, \vec{X}) | |t - t_0| \le a, \left\| \vec{X} - \vec{X_0} \right\| \le b \right\}$, where a and b are arbitrary finite positive numbers such that $\forall (t, \vec{X_1}), (t, \vec{X_2}) \in G_0$.

By the consistency of the matrix norm and vector norm:

$$\begin{aligned} &\left\| \vec{F_1}(X_2) - \vec{F_1}(X_1) \right\| \\ &= \left\| (\nabla \vec{F_1}(\xi))^T \cdot (X_2 - X_1) \right\| \\ &\le \left\| (\nabla \vec{F_1}(\xi))^T \right\| \cdot \|X_2 - X_1\| \le L \cdot \|X_2 - X_1\|, \end{aligned}$$

in which L is the maximum value of $\left\| \nabla \vec{F}_1^T(\cdot) \right\|$ on G_0.

Therefore, $\vec{F_1}$ satisfies the Lipschitz condition on G_0, and the initial value problem of (3.10) becomes a Cauchy problem. Hence, according to the Cauchy existence and uniqueness theorem, a unique solution must exist near the initial point. □

Theorem 3.2. *The initial value problem* (3.11) *has a unique solution.*

3.3 Stability of the Model's Nonsingular Equilibrium Solution

A stable solution is not disturbed by a small variation of the initial value. Asymptotic stability further reduces the impact of the initial value on the solution by deriving a perturbation solution from the initial solution with $t \to \infty$. Stability analysis is the cornerstone in modeling the dynamics of marine ecosystems and ensures the accuracy and predictability of equations.

3.3.1 *Problem description*

Consider the general equation:

$$\frac{d\vec{x}}{dt} = \vec{f}(t, \vec{x}). \tag{3.12}$$

In order to simplify, only the stability of the zero solution of Eq. (3.12) will be discussed (i.e., assuming $f(t,\vec{0}) \equiv 0$). In fact, the general non-zero solution can be transformed into the zero solution by the transformation $\vec{y} = \vec{x} - \vec{\varphi}(t)$.

Equation (3.12) can be regarded as the sum of the linear part $A(t)\vec{x}$ and the nonlinear part $\vec{N}(t,\vec{x})$ (i.e., the higher order terms of $\vec{x}$),

$$\frac{d\vec{x}}{dt} = A(t)\vec{x} + \vec{N}(t,\vec{x}), \tag{3.13}$$

where A(t) is a matrix function of order n that is continuous for all $t \geq t_0$, and $\vec{N}(t,\vec{x})$ is continuous for $\{G|t \geq t_0, \|\vec{x}\| \leq M)$ and satisfies the Lipschitz condition for all $\vec{x}$. In addition,

$$\vec{N}(t,\vec{0}) \equiv \vec{0}(t \geq t_0), \lim_{\|\vec{x}\|\to 0} \frac{\|\vec{N}(t,\vec{x})\|}{\|\vec{x}\|} = 0 \quad \text{(valid for all } t \geq t_0).$$

Under certain conditions, the stability of the original Eq. (3.13) is consistent with the stability of the linearized Eq. (3.14):

$$\frac{d\vec{x}}{dt} = A(t)\vec{x}. \tag{3.14}$$

3.3.2 *Limitations of linear approximation in analyzing model stability*

In what follows, we will see that linear approximation cannot directly evaluate the stability of the nonsingular stationary solutions of the NPZ (3.1) and NPZD (3.2) models. Instead, we will discuss in a more general way the stability of the zero solution of self-conserving equations, which contains the aforementioned ecosystems. For the self-conserving equations $\vec{f}(t,\vec{x}) = \vec{f}(\vec{x})$ with $\vec{f}(\vec{x}) = (f_1(\vec{x}), f_2(\vec{x}), f_3(\vec{x}), \ldots, f_n(\vec{x}))$ and

$$f_1(\vec{x}) + f_2(\vec{x}) + f_3(\vec{x}) + \cdots + f_n(\vec{x}) = 0. \tag{3.15}$$

Without loss of generality, one can assume that $\vec{f}(\vec{0}) = \vec{0}$. Indeed, by the transformation $y = x - x^*$, the above general problem can always be transformed into the special case. Therefore, $\frac{d}{dt}(x_1 + x_2 + x_3 + \cdots + x_n) = 0$, and consequently,

$$\begin{aligned} &x_1(t) + x_2(t) + x_3(t) + \cdots + x_n(t) \\ &\quad = x_{01} + x_{02} + x_{03} + \cdots + x_{0n} = c, \end{aligned} \tag{3.16}$$

in which c is a constant, and $X_0 = (x_{01}, x_{02}, x_{03}, \ldots, x_{0n})$ are the initial values. Hence, the equations of the NPZ and NPZD models are clearly self-conserved.

The following proves that adopting linear approximation directly cannot determine the stability of the zero solution of the self-conserving Eq. (3.12).

Theorem 3.3. *If the system of Eqs.* (3.12) *is self-conserved, then its linearized system of Eqs.* (3.14) *has a coefficient matrix that has zero eigenvalue.*

Proof. Because (3.12) is self-conserved, the Taylor expansion then gives

$$A(t) = A = (\nabla \vec{f}(\vec{0}))^T = \begin{pmatrix} \frac{\partial f_1}{\partial x_1}, \frac{\partial f_1}{\partial x_2}, \frac{\partial f_1}{\partial x_3}, \ldots, \frac{\partial f_1}{\partial x_n} \\ \frac{\partial f_2}{\partial x_1}, \frac{\partial f_2}{\partial x_2}, \frac{\partial f_2}{\partial x_3}, \ldots, \frac{\partial f_2}{\partial x_n} \\ \vdots \qquad\qquad \vdots \qquad\qquad \vdots \\ \frac{\partial f_n}{\partial x_1}, \frac{\partial f_n}{\partial x_2}, \frac{\partial f_n}{\partial x_3}, \ldots, \frac{\partial f_n}{\partial x_n} \end{pmatrix}_{\vec{x}=\vec{0}}, \tag{3.17}$$

and

$$f_1(\vec{x}) f_2(\vec{x}) + f_3(\vec{x}) + \cdots + f_n(\vec{x}) = 0 \ \forall \ (\vec{x}) \in R^n. \tag{3.18}$$

Hence,

$$\begin{aligned} &\frac{\partial(f_1(\vec{x}) + f_2(\vec{x}) + f_3(\vec{x}) + \cdots + f_n(\vec{x}))}{\partial x_i} = 0, \\ &i = 1, 2, \ldots, n. \quad \forall \vec{x} \in R^n. \end{aligned}$$

The rank of matrix A is thus $r(A) \leq n - 1 < n$, and the determinant of A has $\det(A) = |A| \equiv 0$. The characteristic equation of matrix A

then has

$$\begin{aligned}\det(\eta E - A) &= \eta^n + a_1\eta^{n-1} + a_2\eta^{n-2} + \cdots + a_{n-1}\eta + (-1)^n|A| \\ &= \eta^n + a_1\eta^{n-1} + a_2\eta^{n-2} + \cdots + a_{n-1}\eta \\ &= \eta(\eta^{n-1} + a_1\eta^{n-2} + a_2\eta^{n-3} + \cdots + a_{n-1}) = 0. \quad (3.19)\end{aligned}$$

Therefore, the coefficient matrix of the linearized system of Eqs. (3.14) for the self-conserving Eqs. (3.12) must have zero eigenvalue.

Thus, by following Wang *et al.* (1983), linear approximation cannot determine the stability of the zero solution of the self-conserving Eq. (3.12). □

3.3.3 *Limitations of the Lyapunov–Krasovskii method*

Theorem 3.4. *The Lyapunov–Krasovskii method is limited in analyzing the nonsingular equilibrium solutions of self-conserving equations.*

Proof. Assume $\frac{d\vec{x}}{dt} = \vec{f}(\vec{x})$ is self-conserved. Let $\vec{Q}(\vec{x}) = (\frac{\partial \vec{f}(\vec{x})}{\partial \vec{x}^T})^T + \frac{\partial \vec{f}(\vec{x})}{\partial \vec{x}^T}$. Therefore, according to the Lyapunov–Krasovskii theorem (Zhong, 2004), the zero solution of the original Eqs. (3.12) is asymptotically stable if $\vec{Q}(\vec{x})$ is negatively definite for any $\vec{x} \in D$, where D is the domain of $\vec{f}(\vec{x})$. That is, $\forall \vec{x} \in D, \vec{x} \neq \vec{0}$

$$\begin{aligned}\vec{x}^T \cdot \vec{Q}(\vec{x}) \cdot \vec{x} &= \vec{x}^T \cdot \left[\left(\frac{\partial \vec{f}(\vec{x})}{\partial \vec{x}^T}\right)^T + \frac{\partial \vec{f}(\vec{x})}{\partial \vec{x}^T}\right] \cdot \vec{x} \\ &= \vec{x}^T \cdot \left(\frac{\partial \vec{f}(\vec{x})}{\partial \vec{x}^T}\right)^T \cdot \vec{x} + \vec{x}^T \cdot \frac{\partial \vec{f}(\vec{x})}{\partial \vec{x}^T} \cdot \vec{x} \\ &= \left\{\vec{x}^T \cdot \frac{\partial \vec{f}(\vec{x})}{\partial \vec{x}^T} \cdot \vec{x}\right\}^T + \vec{x}^T \cdot \frac{\partial \vec{f}(\vec{x})}{\partial \vec{x}^T} \cdot \vec{x} \\ &= 2 \cdot \vec{x}^T \cdot \frac{\partial \vec{f}(\vec{x})}{\partial \vec{x}^T} \cdot \vec{x} < 0, \qquad (3.20)\end{aligned}$$

which yields $\frac{\partial \vec{f}(\vec{x})}{\partial \vec{x}^T}$ defined by:

$$\frac{\partial \vec{f}(\vec{x})}{\partial \vec{x}^T} = \begin{pmatrix} \frac{\partial f_1}{\partial x_1}, \frac{\partial f_1}{\partial x_2}, \frac{\partial f_1}{\partial x_3}, \ldots, \frac{\partial f_1}{\partial x_n} \\ \frac{\partial f_2}{\partial x_1}, \frac{\partial f_2}{\partial x_2}, \frac{\partial f_2}{\partial x_3}, \ldots, \frac{\partial f_2}{\partial x_n} \\ \vdots \qquad \vdots \qquad \vdots \\ \frac{\partial f_n}{\partial x_1}, \frac{\partial f_n}{\partial x_2}, \frac{\partial f_n}{\partial x_3}, \ldots, \frac{\partial f_n}{\partial x_n} \end{pmatrix} \tag{3.21}$$

is also negative definite.

However, for the self-conserving equation

$$f_1(\vec{x}) + f_2(\vec{x}) + f_3(\vec{x}) + \cdots + f_n(\vec{x}) = 0 \; \forall x \in R^n, \tag{3.22}$$

which implies $\frac{\partial (f_1(\vec{x}) + f_2(\vec{x}) + f_3(\vec{x}) + \cdots + f_n(\vec{x}))}{\partial x_i} = 0, i = 1, 2, \ldots, n.\ \forall x \in D$.

Hence, $\frac{\partial \vec{f}(\vec{x})}{\partial \vec{x}^T}$ is not a negative definite matrix since $rank(\frac{\partial \vec{f}(\vec{x})}{\partial \vec{x}^T}) \leq n - 1 < n$. That is, $\vec{Q}(\vec{x})$ cannot be negative definite regardless of the selected initial values and parameters. Hence, the Lyapunov–Krasovskii method is limited in evaluating the asymptotic stability of self-conserving equations.

The above demonstrates that both linear approximation and the Lyapunov–Krasovskii method are only a sufficient condition for stability; the stability of the equation remains indeterminate if neither condition is met. Other methods are thus needed for evaluating a self-conserving equation. □

3.3.4 *Applicability of dimensionality reduction*

According to the definition of stability, the dimensionality reduction theorem for self-conserving equations can be derived as follows.

Theorem 3.5. *The dimensionality reduction theorem.*

Consider the following n-dimensional self-conserving equations

$$\begin{cases} \dfrac{dx_1}{dt} = f_1(x_1, x_2, x_3, \ldots, x_{n-1}, x_n) \\ \dfrac{dx_2}{dt} = f_2(x_1, x_2, x_3, \ldots, x_{n-1}, x_n) \\ \dfrac{dx_3}{dt} = f_3(x_1, x_2, x_3, \ldots, x_{n-1}, x_n) \\ \vdots \qquad\qquad \vdots \\ \dfrac{dx_n}{dt} = f_n(x_1, x_2, x_3, \ldots, x_{n-1}, x_n) \end{cases}, \tag{3.23}$$

and

$$\begin{cases} \dfrac{dx_1}{dt} = f_1(x_1, x_2, x_3, \ldots, x_{n-1}, -x_1 - x_2 - x_3 - \cdots - x_{n-1}) \\ \dfrac{dx_2}{dt} = f_2(x_1, x_2, x_3, \ldots, x_{n-1}, -x_1 - x_2 - x_3 - \cdots - x_{n-1}) \\ \dfrac{dx_3}{dt} = f_3(x_1, x_2, x_3, \ldots, x_{n-1}, -x_1 - x_2 - x_3 - \cdots - x_{n-1}) \\ \vdots \qquad\qquad \vdots \\ \dfrac{dx_{n-1}}{dt} = f_{n-1}(x_1, x_2, x_3, \cdots, x_{n-1}, -x_1 - x_2 - x_3 - \cdots - x_{n-1}) \end{cases}. \tag{3.24}$$

The zero solution to (3.24) being stable is equivalent to the zero solution to (3.23) being stable on the manifold $x_1(t) + x_2(t) + x_3(t) + \cdots + x_n(t) = 0$.

Proof. (sufficient condition) Since the zero solution to (3.24) is stable, then from the definition of the Lyapunov stability:

$\forall \varepsilon > 0, \exists \delta_1 > 0$, such that

$$\|x_0\| \le \delta_1, \forall x_0 = (x_{01}, x_{02}, x_{03}, \ldots, x_{0n-1}). \tag{3.25}$$

For all solutions $\vec{x}(t)$ to (3.24) with the initial condition $\vec{x}(t_0) = x_0$,

$$\|\vec{x}(t)\| < \varepsilon, \ \ \forall t \ge t_0 \tag{3.26}$$

Hence, for $\varepsilon > 0$, $\exists\delta = \delta_1 > 0$, such that $\forall X_0 = (x_{01}, x_{02}, x_{03}, \ldots, x_{0n-1}, x_{0n})$

$$\|X_0\|^2 = x_{01}^2 + x_{02}^2 + x_{03}^2 + \cdots + x_{0n-1}^2 + x_{0n}^2 < \delta^2. \tag{3.27}$$

Evidently, $\sqrt{x_{01}^2 + x_{02}^2 + x_{03}^3 + \cdots + x_{0n-1}^2} < \delta = \delta_1$.

Thus, for all solutions $\vec{X}(t)$ to (3.23) with the initial condition $\vec{x}(t_0) = X_0$,

$$\begin{aligned}\left\|\vec{X}(t)\right\|^2 &= x_1^2 + x_2^2 + x_3^2 + \cdots + x_{n-1}^2 + x_n^2\\ &= x_1^2 + x_2^2 + x_3^2 + \cdots + x_{n-1}^2 + (-x_1 - x_2 - x_3 - \cdots - x_{n-1})^2\\ &\le x_1^2 + x_2^2 + x_3^2 + \cdots + x_{n-1}^2 + (n-1)\cdot(x_1^2 + x_2^2 + x_3^2 + \cdots + x_{n-1}^2)\\ &\le n\cdot(x_1^2 + x_2^2 + x_3^2 + \cdots + x_{n-1}^2) < n\cdot\varepsilon^2.\end{aligned} \tag{3.28}$$

That is, the zero solution to (3.23) is stable on the manifold $x_1(t) + x_2(t) + x_3(t) + \cdots + x_n(t) = 0$.

(**necessary condition**) Because the zero solution to (3.23) is stable on the manifold

$x_1(t) + x_2(t) + x_3(t) + \cdots + x_n(t) = 0$,

$\forall\varepsilon > 0, \exists\delta > 0$, such that $\forall X_0 = (x_{01}, x_{02}, x_{03}, \ldots, x_{0n-1}, x_{0n})$ satisfying

$$\|x_0\|^2 = x_{01}^2 + x_{02}^2 + x_{03}^2 + \cdots + x_{0n-1}^2 + x_{0n}^2 < \delta^2, \tag{3.29}$$

then for all solutions $\vec{X}(t)$ to (3.23) with the initial condition $\vec{x}(t_0) = X_0$,

$$\|\vec{X}(t)\| < \varepsilon \quad (\forall t \ge t_0).$$

$$\forall\varepsilon > 0, \exists\delta_1 = \delta > 0, \forall x_0 = (x_{01}, x_{02}, x_{03}, \ldots, x_{0n-1})$$

$$\|x_0\| \le \|X_0\| < \delta_1. \tag{3.30}$$

Because on the manifold $x_1(t)+x_2(t)+x_3(t)+\cdots+x_n(t) = 0(\forall t \ge t_0)$, the first $n-1$ dimensions of the solutions $\vec{X}(t)$ of Eqs. (3.23) with initial conditions $\vec{X}(t_0) = X_0$ are the solutions to (3.24) with initial conditions $\vec{x}(t_0) = x_0$, $\vec{X}(t) = (\vec{x}(t), x_n(t))$. Hence,

$$\|\vec{x}(t)\| \leq \left\|\vec{X}(t)\right\| < \varepsilon, \quad \forall t \geq t_0.$$

Therefore, the zero solution to Eqs. (3.24) is stable. □

3.4 Model Stability and Verification Based on Dimensionality Reduction

3.4.1 *The NPZ model*

(1) *Stability analysis*

According to Theorem 3.5, we can reduce the dimension of the self-conservation problem by one and then use linear approximation to evaluate the stability. Accordingly, let $\vec{F}_1(X) = \vec{0}$, and the nonsingular equilibrium point of the dynamic equation be $(P^*Z^*N^*)$. The following only considers $P^* > 0, Z^* > 0, N^* > 0$.

Because $\frac{dP}{dt} + \frac{dZ}{dt} + \frac{dN}{dt} = 0$, which means $\frac{d}{dt}(P + Z + N) = 0$, we have $P + Z + N = c$, where c is a constant. Hence:

$$\begin{cases} P^* = -\dfrac{\ln(1 - \frac{Mz}{r\cdot Gm\cdot h(T)})}{\lambda} \\ Z^* = c - P^* - N^* \\ N^* = \dfrac{k_N \cdot \dfrac{Gm \cdot h(T) \cdot Z^* \cdot (1 - e^{-\lambda P}) + Mp \cdot P^*}{Um \cdot f(I) \cdot g(T) \cdot P^*}}{1 - \dfrac{Gm \cdot h(T) \cdot Z^* \cdot (1 - e^{-\lambda P^*}) + Mp \cdot P^*}{Um \cdot f(I) \cdot g(T) \cdot P^*}} \end{cases}. \tag{3.31}$$

Let $X^* = (P^*, Z^*, N^*)$. From the Taylor expansion of the right hand side function $\vec{F}_1(X)$ in (3.8) at $X = X^*$, we have:

$$\begin{aligned} \vec{F}_1(X) &= \vec{F}_1(X^*) + (\nabla \vec{F}_1(X^*))^T \cdot (X - X^*) + o(\|X - X^*\|) \\ &= (\nabla \vec{F}_1(X^*))^T \cdot (X - X^*) + o(\|X - X^*\|). \end{aligned} \tag{3.32}$$

Assume $x = X - X^* = (x_1, x_2, x_3) = (P - P^*, Z - Z^*, N - N^*)$.

$$\tag{3.33}$$

The zero solution of the NPZ system can be obtained by substituting (3.31) and (3.33) into (3.10):

$$\begin{cases} \dfrac{dx}{dt} = \vec{F}_1(x + X^*) = (\nabla \vec{F}_1(X^*))^T \cdot x + o(\|x\|) \\ \vec{x}(t_0) = \vec{0} \end{cases}. \tag{3.34}$$

The Jacobi matrix of $\vec{F}_1$ about (P, Z, N), $(\nabla \vec{F}_1(X^*))^T$, is as follows:

$$(\nabla \vec{F}_1(X^*))^T = \begin{bmatrix} a_{11} & a_{12} & a_{13} \\ a_{21} & a_{22} & a_{23} \\ a_{31} & a_{32} & a_{33} \end{bmatrix} = [F_1^* \; F_2^* \; F_3^*], \tag{3.35}$$

where

$$F_1^* = \begin{bmatrix} U_m \cdot \dfrac{N^*}{k_N + N^*} \cdot f(I) \cdot g(T) - G_m \cdot h(T) \cdot Z^* \cdot \lambda \cdot e^{-\lambda P^*} - M_P \\ \gamma \cdot G_m \cdot h(T) \cdot Z^* \cdot \lambda \cdot e^{-\lambda P^*} \\ -U_m \cdot \dfrac{N^*}{k_N + N^*} \cdot f(I) \cdot g(T) + (1-\gamma) \cdot G_m \cdot h(T) \cdot Z^* \cdot \lambda \cdot e^{-\lambda P^*} + M_P \end{bmatrix}, \tag{3.36}$$

$$F_2^* = \begin{bmatrix} -G_m \cdot h(T) \cdot (1 - e^{-\lambda P^*}) \\ \gamma \cdot G_m \cdot h(T) \cdot (1 - e^{-\lambda P^*}) - M_Z \\ (1-\gamma) \cdot G_m \cdot h(T) \cdot (1 - e^{-\lambda P^*}) + M_Z \end{bmatrix}, \tag{3.37}$$

$$F_3^* = \begin{bmatrix} U_m \cdot \dfrac{k_N}{(k_N + N^*)^2} \cdot f(I) \cdot g(T) \cdot P^* \\ 0 \\ -U_m \cdot \dfrac{k_N}{(k_N + N^*)^2} \cdot f(I) \cdot g(T) \cdot P^* \end{bmatrix}. \tag{3.38}$$

The stability problem of (3.34) can be linearized based on the dimensionality reduction theorem, therefore the NPZ stability problem becomes the stability of the zero solution of the following:

$$\frac{d}{dt}\begin{pmatrix}x_1\\x_2\end{pmatrix}=\begin{bmatrix}a_{11} & a_{12} & a_{13}\\a_{21} & a_{22} & a_{23}\end{bmatrix}\cdot\begin{bmatrix}x_1\\x_2\\-x_1-x_2\end{bmatrix},$$

$$\text{i.e., }\frac{d}{dt}\begin{pmatrix}x_1\\x_2\end{pmatrix}=\begin{bmatrix}a_{11}-a_{13}, & a_{12}-a_{13}\\a_{21}-a_{23}, & a_{22}-a_{23}\end{bmatrix}\cdot\begin{bmatrix}x_1\\x_2\end{bmatrix}\triangleq A\cdot X. \tag{3.39}$$

Since the above is a system of linear equations with constant coefficients, according to the stability theory, when all eigenvalues of A have negative real parts, the zero solution of (3.39) is asymptotically stable.

The eigenvalue equation of the matrix has $\det(\eta E-A)=0$, that is, $\eta^2+a_1\cdot\eta+a_2=0$,

$$\text{where }\begin{cases}a_1=-(a_{11}+a_{22}-a_{13}-a_{23})\\a_2=\det(A)\end{cases}.$$

Hence, according to the Hurwitz criterion (Wang and Feng, 2006), the steady-state solution of the NPZ system is stable as long as

$$\begin{cases}H_1=a_1>0\\H_2=a_1\cdot a_2>0\end{cases},\quad\text{i.e.,}\begin{cases}a_1>0\\a_2>0\end{cases}. \tag{3.40}$$

(2) *Verification by numerical simulation*

With the default parameters in Table 3.1, we get the nonsingular equilibrium solution $(P^*, Z^*, N^*) = (3.5675 \quad 1.3619 \quad 2.3706)$ and additionally $\begin{cases}a_1=0.0950>0\\a_2=0.0636>0\end{cases}$. Hence, the steady-state solution of the NPZ system is stable when default parameters are taken. The phase diagram and time series diagram of the (3.1) system with initial values of $(P_0, Z_0, N_0) = (3.5, 2, 1.8)$ are given in Figs. 3.1 and 3.2.

3.4.2 The NPZD model

(1) *Stability analysis*

Since the NPZD equation is self-conserving, the dimensionality reduction theorem can be applied. Let $\vec{F_2}(Y) = \vec{0}$, and the

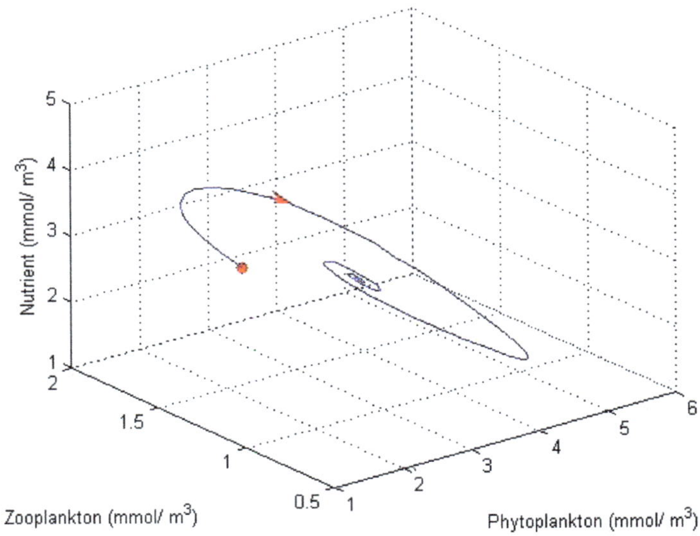

Fig. 3.1. The phase diagram of the stability of the NPZ model.

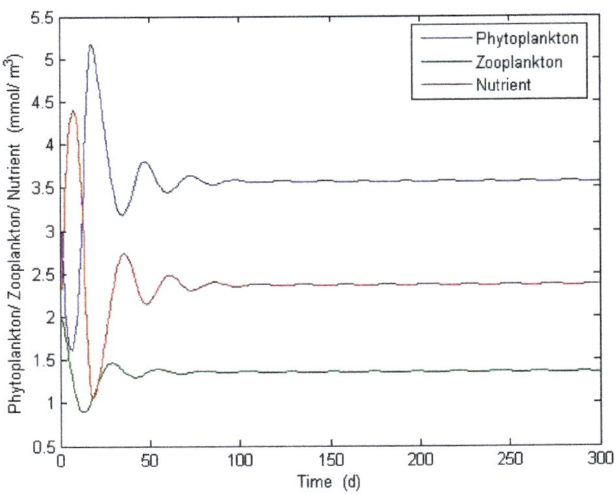

Fig. 3.2. The time series of state variables in the NPZ model.

equilibrium of the kinetic equation be (P^*, Z^*, N^*, D^*). The following only considers $P^* > 0, Z^* > 0, N^* > 0, D^* > 0$.

Let $P + Z + N + D = c$, where c is a constant. Therefore,

$$\begin{cases} P^* = -\dfrac{\ln\left(1 - \frac{Mz}{r \cdot Gm \cdot h(T)}\right)}{\lambda} \\ Z^* = \dfrac{U_m \cdot \dfrac{N^*}{k_N + N^*} \cdot f(I) \cdot g(T) \cdot P^* - M_P \cdot P^*}{G_m \cdot h(T) \cdot (1 - e^{-\lambda P})} \\ N^* = \dfrac{k_N \cdot \dfrac{\theta \cdot G_m \cdot h(T) \cdot Z^* \cdot (1 - e^{-\lambda P^*}) + e \cdot D^*}{U_m \cdot f(I) \cdot g(T) \cdot P^*}}{1 - \dfrac{\theta \cdot G_m \cdot h(T) \cdot Z^* \cdot (1 - e^{-\lambda P^*}) + e \cdot D^*}{U_m \cdot f(I) \cdot g(T) \cdot P^*}} \\ D^* = c - P^* - Z^* - N^* \end{cases} . \tag{3.41}$$

Let $Y^* = (P^*, Z^*, N^*, D^*)$. From the Taylor expansion of the right hand side function $\vec{F_2}(Y)$ in (3.11) at $Y = Y^*$, and assume

$$y = Y - Y^* = (y_1, y_2, y_3, y_4) = (P - P^*, Z - Z^*, N - N^*, D - D^*). \tag{3.42}$$

The relative zero solution to the initial value problem of the linearized equation of the NPZD system can be obtained by substituting the above into (3.11).

$$\begin{cases} \dfrac{dy}{dt} = \vec{F_2}(y + Y^*) = (\nabla \vec{F_1}(Y^*))^T \cdot y \\ \vec{y}(t_0) = \vec{0} \end{cases} . \tag{3.43}$$

The Jacobi matrix of $\vec{F_3}$ about (P, Z, N, D), $(\nabla \vec{F_3}(Y^*))^T$, can be expressed as follows:

$$(\nabla \vec{F_3}(Y^*))^T = \begin{bmatrix} b_{11} & b_{12} & b_{13} & b_{14} \\ b_{21} & b_{22} & b_{23} & b_{24} \\ b_{31} & b_{32} & b_{33} & b_{34} \\ b_{41} & b_{42} & b_{43} & b_{44} \end{bmatrix} = [F_1^* \; F_2^* \; F_3^* \; F_4^*], \tag{3.44}$$

where

$$F_1^* = \begin{bmatrix} U_m \cdot \dfrac{N^*}{k_N + N^*} \cdot f(I) \cdot g(T) - G_m \cdot h(T) \cdot Z^* \cdot \lambda \cdot e^{-\lambda P^*} - M_P \\ \gamma \cdot G_m \cdot h(T) \cdot Z^* \cdot \lambda \cdot e^{-\lambda P^*} \\ -U_m \cdot \dfrac{N^*}{k_N + N^*} \cdot f(I) \cdot g(T) + \theta \cdot G_m \cdot h(T) \cdot Z^* \cdot \lambda \cdot e^{-\lambda P^*} \\ (1-\gamma-\theta) \cdot G_m \cdot h(T) \cdot Z^* \cdot \lambda \cdot e^{-\lambda P^*} + M_P \end{bmatrix},$$

$$F_2^* = \begin{bmatrix} -G_m \cdot h(T) \cdot (1 - e^{-\lambda P^*}) \\ \gamma \cdot G_m \cdot h(T) \cdot (1 - e^{-\lambda P^*}) - M_Z \\ \theta \cdot G_m \cdot h(T) \cdot (1 - e^{-\lambda P^*}) \\ (1-\gamma-\theta) \cdot G_m \cdot h(T) \cdot (1 - e^{-\lambda P^*}) + M_Z \end{bmatrix},$$

$$F_3^* = \begin{bmatrix} U_m \cdot \dfrac{k_N}{k_N + N^*} \cdot f(I) \cdot g(T) \cdot P \\ 0 \\ -U_m \cdot \dfrac{k_N}{k_N + N^*} \cdot f(I) \cdot g(T) \cdot P \\ 0 \end{bmatrix}, \quad F_4^* = \begin{bmatrix} 0 \\ 0 \\ e \\ -e \end{bmatrix}.$$

According to the dimensionality reduction theorem, regarding the stability of (3.43), the stability of the zero solution to the following reveals the N-P-Z-D stability problem:

$$\frac{d}{dt}\begin{pmatrix} y_1 \\ y_2 \\ y_3 \end{pmatrix} = \begin{bmatrix} b_{11}\ b_{12}\ b_{13}\ b_{14} \\ b_{21}\ b_{22}\ b_{23}\ b_{24} \\ b_{31}\ b_{32}\ b_{33}\ b_{34} \end{bmatrix} \cdot \begin{bmatrix} y_1 \\ y_2 \\ y_3 \\ -y_1 - y_2 - y_3 \end{bmatrix},$$

$$\text{that is, } \frac{d}{dt}\begin{pmatrix} y_1 \\ y_2 \\ y_3 \end{pmatrix} = \begin{bmatrix} b_{11} - b_{14},\ b_{12} - b_{14},\ b_{13}\ - b_{14} \\ b_{21} - b_{24},\ b_{22} - b_{24},\ b_{23}\ - b_{24} \\ b_{31} - b_{34},\ b_{32} - b_{34},\ b_{33}\ - b_{34} \end{bmatrix} \cdot \begin{bmatrix} y_1 \\ y_2 \\ y_3 \end{bmatrix} \triangleq B \cdot Y. \quad (3.45)$$

According to the stability theory, when all eigenvalues of B have negative real parts, the zero solution of (3.45) is (asymptotically) stable.

The eigenvalue equation of the matrix has $\det(\eta E - B) = 0$, i.e., $\eta^3 + b_1 \cdot \eta^2 + b_2 \cdot \eta + b_3 = 0$, where,

$$\begin{aligned}
b_1 &= -(b_{11} + b_{22} + b_{33} - b_{14} - b_{24} - b_{34}) \\
b_2 &= -b_{11}b_{24} - b_{11}b_{34} + b_{11}b_{22} - b_{22}b_{34} - b_{14}b_{33} - b_{23}b_{32} \\
&\quad - b_{31}b_{13} + b_{11}b_{33} + b_{12}b_{24} + b_{21}b_{14} - b_{14}b_{22} + b_{23}b_{34} \\
&\quad + b_{24}b_{32} + b_{31}b_{14} + b_{22}b_{33} + b_{13}b_{34} - b_{21}b_{12} - b_{33}b_{24} \\
b_3 &= -\det(B).
\end{aligned}$$

Thus, based on the Hurwitz criterion and the dimensionality reduction theorem (Wang and Feng, 2006), the steady-state solution of the N-P-Z-D system is stable as long as

$$\begin{cases} H_1 = b_1 > 0 \\ H_2 = b_1 \cdot b_2 - b_3 > 0 \\ H_3 = b_3(b_1 \cdot b_2 - b_3) > 0 \end{cases}, \quad \text{i.e.,} \begin{cases} b_1 > 0 \\ b_1 \cdot b_2 - b_3 > 0 \\ b_3 > 0 \end{cases}.$$

(2) *Verification by numerical simulation*
With the default parameters (Table 3.1) and the initial values $(P_0, Z_0, N_0, D_0) = (3.6\,3.4\,4\,0.2)$, the nonsingular equilibrium solution is derived as $(P^*, Z^*, N^*, D^*) = (3.5233\quad 0.0337\quad 0.1905\quad 7.4525)$, and additionally $\begin{cases} b_1 = 1.7892 > 0 \\ b_1 \cdot b_2 - b_3 = 0.4763 > 0 \\ b_3 = 0.0011 > 0 \end{cases}$. That is, the steady-state solution of the N-P-Z-D system is stable with default parameters. The phase diagram and time series diagram obtained from numerical simulation with the initial values are shown in Figs. 3.3 and 3.4, respectively.

3.5 Conditions That Models Satisfy Hopf Bifurcation Under Specific Parameters

3.5.1 *Preliminaries*

Theorem 3.6. *The Hopf bifurcation theorem*

Assume that $f(X, \mu), X \in R^n, \mu \in R^1$ satisfies

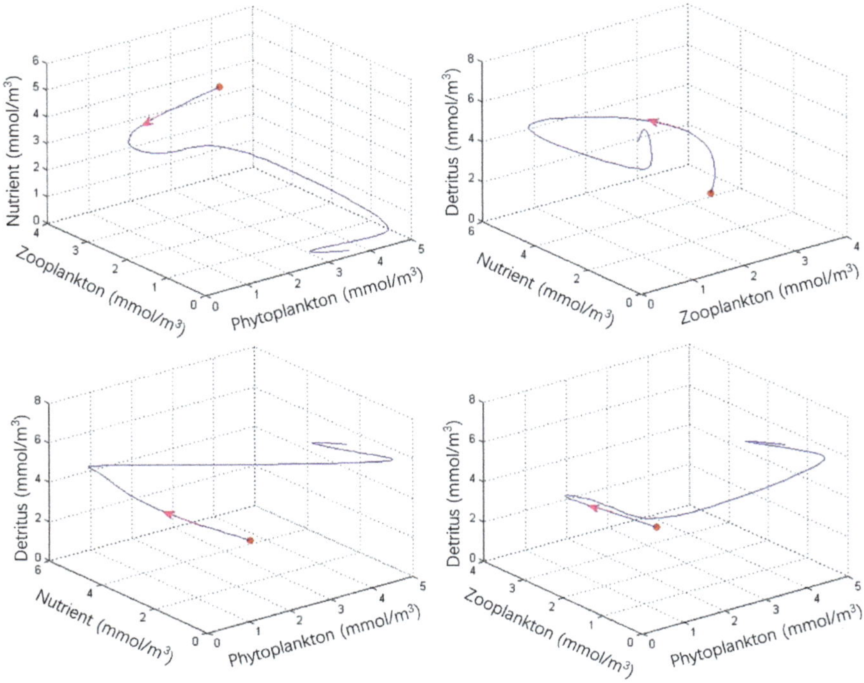

Fig. 3.3. The phase diagram of the stability of the NPZD model.

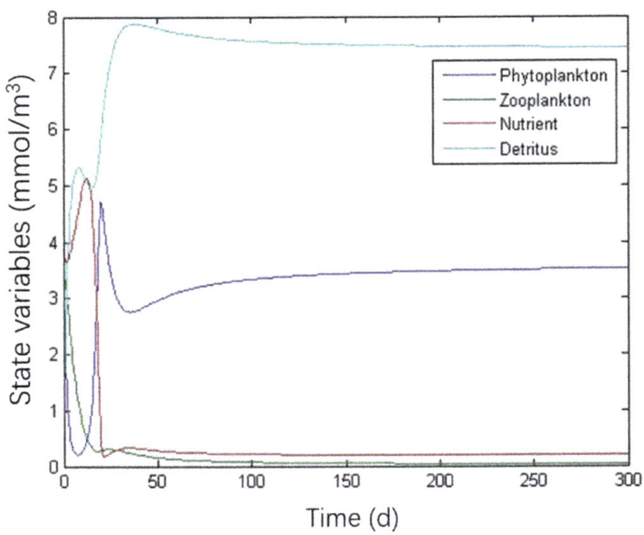

Fig. 3.4. The time series of state variables in the NPZD model.

(1) for $X = X^*(\mu), f(X^*, \mu) \equiv 0$,
(2) f can be analyzed with regard to X, μ in U, i.e., a given neighboring domain $(X, \mu) = (X^*(\mu_c), \mu_c)$,
(3) $D_X f(X^*(\mu_c), \mu_c) = \begin{pmatrix} A & \\ & B \end{pmatrix}$,

where all eigenvalues of $A : R^2 \to R^2$ have zero real part, all eigenvalues of $B : R^{n-2} \to R^{n-2}$ have negative real part, and the projection of the system's trajectory on the $x_1 - x_2$ plane approaches the origin (when $t \to +\infty$ or $t \to -\infty$). For the system

$$\frac{dX}{dt} = f(X, \mu), \tag{3.46}$$

if $X = X^*(\mu)$ is asymptotically stable (unstable) when $\mu = \mu_c$ and unstable (asymptotically stable) when $\mu > \mu_c (< \mu_c)$, then system (3.46) has asymptotically stable (unstable) closed orbit in the neighboring domain of $X = X^*(\mu)$ when $\mu > \mu_c (< \mu_c)$ and $|\mu - \mu_c|$ is sufficiently small.

In other words, the Hopf bifurcation refers to the bifurcation phenomenon of the system produces limit cycle from the equilibrium point when the bifurcation parameter changes.

For a system of marine ecological dynamics, the transition from steady-state to bifurcation indicates a change in the topology of the system, and is indicative for explaining sudden outbreaks of various marine disasters such as a red tide.

3.5.2 *NPZ model*

To examine the Hopf bifurcation of the NPZ model, according to the dimensionality reduction theorem (Theorem 3.5), only the stability of the dimensionality reduction equation is needed. According to Theorem 3.6, a pair of eigenvalues, all with zero real parts, of the linearized Eq. (3.36) needs to be found.

Therefore, consider the characteristic equation $\det(\eta E - A) = 0$, that is, $\eta^2 + a_1 \cdot \eta - a_2 = 0$, in which $\begin{aligned} a_1 &= -(a_{11} + a_{22} - a_{13} - a_{23}) \\ a_2 &= \det(A) \end{aligned}$

(the definitions of A, a_1, a_2 are the same as in 3.4.1).

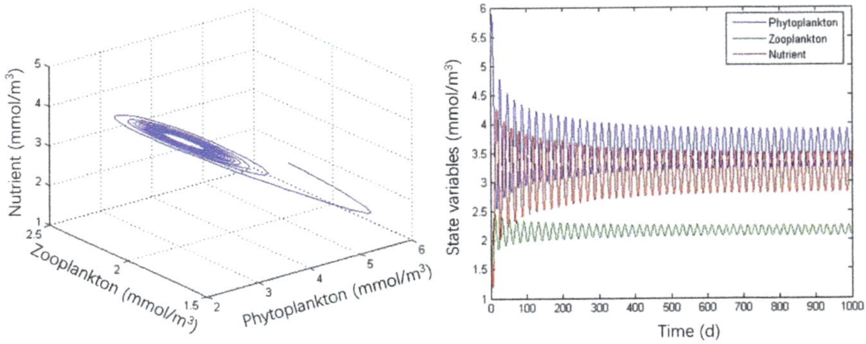

Fig. 3.5. The phase diagram of the NPZ model and time series of state variables when U_m meets the critical condition of the Hopf bifurcation.

Hence, $\eta = \frac{-a_1 \pm \sqrt{a_1^2 - 4a_1a_2}}{2}$. The system satisfies the bifurcation condition as long as $a_1 = 0$.

When U_m is used as the bifurcation parameter for analysis, the condition for the Hopf bifurcation is as follows when U_m has initial values of $(P_0, Z_0, N_0) = (3.6, 1.5, 3.8)$ and other parameters take the default values:

$$\begin{cases} \vec{F_1} = 0 \\ a_1 \ = 0 \end{cases}.$$

The equilibrium point is then $(P^*\,Z^*\,N^*) = (3.5675\,2.1643\,3.1682)$, and the critical value has $U_m = 2.7147$. The phase diagram now shows a limit cycle (Fig. 3.5).

Figures 3.6 and 3.7 show the observed bifurcation when U_m vacillates near the critical value, firstly $U_m = 2.9147$ in Fig. 3.6 and then $U_m = 2.5147$ in Fig. 3.7.

Other parameters can be calculated analogously as the above.

(1) For k (the half-saturation constant of nitrogen intake), the equilibrium point is $(P^*\,Z^*\,N^*) = (3.5675\,1.4123\,3.9203)$ when the initial value is $X_0 = [P_0\,Z_0\,N_0\,k] = [3.6\,1.5\,3.8\,1]$ and the corresponding critical value is $k = 1.4949$. The phase diagram at $k = 1.4949$ is a limit cycle (Fig. 3.8).

Figures 3.9 and 3.10 show the observed bifurcation when k vacillates near the critical value, firstly $k = 1.8949$ in Fig. 3.9 and then $k = 1.3949$ in Fig. 3.10.

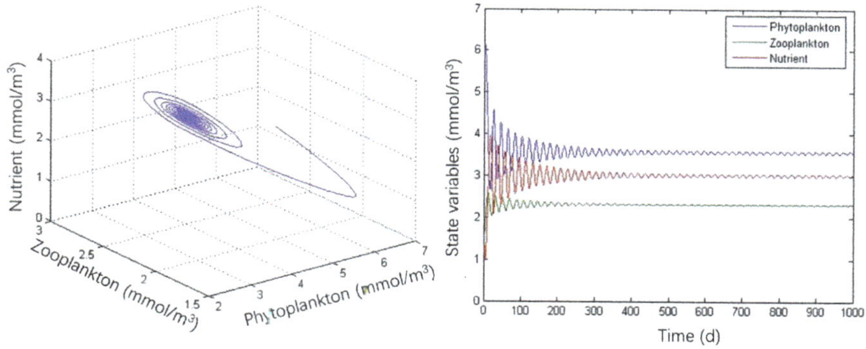

Fig. 3.6. The phase diagram of the NPZ model and time series of state variables when $U_m = 2.9147$.

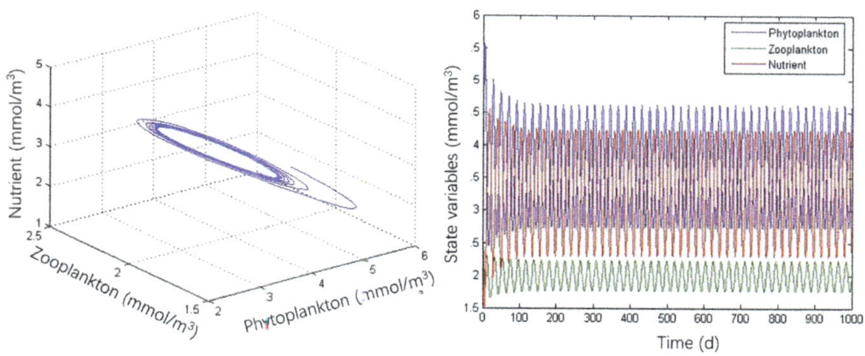

Fig. 3.7. The phase diagram of the NPZ model and time series of state variables when $U_m = 2.5147$.

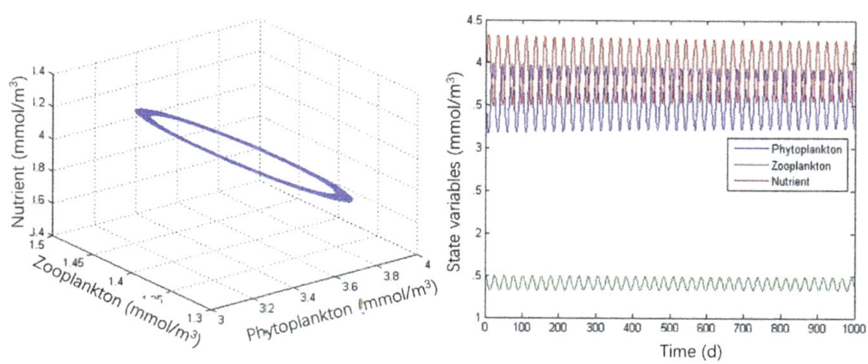

Fig. 3.8. The phase diagram of the NPZ model and time series of state variables when k meets the critical condition of the Hopf bifurcation.

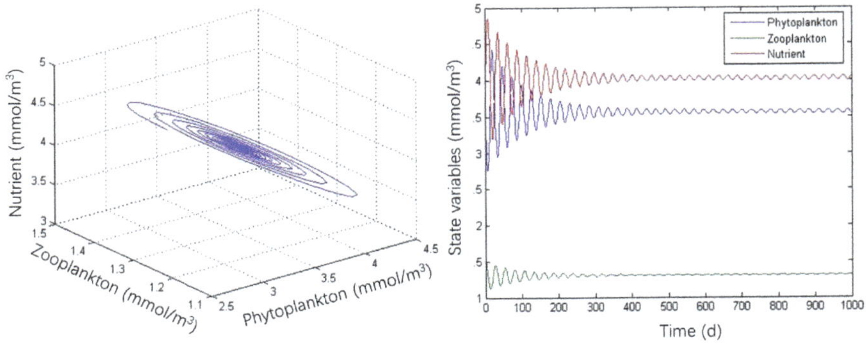

Fig. 3.9. The phase diagram of the NPZ model and time series of state variables when $k = 1.8949$.

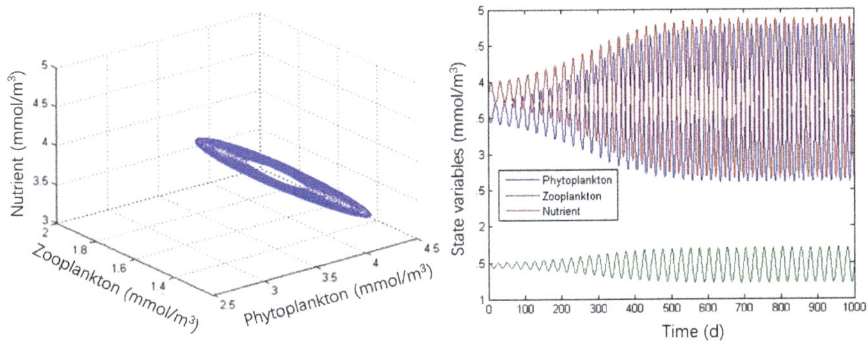

Fig. 3.10. The phase diagram of the NPZ model and time series of state variables when $k = 1.3949$.

(2) For Mp (the Phytoplankton mortality), the equilibrium point is $(P^* \, Z^* \, N^*) = (3.5675 \, 0.9596 \, 4.3729)$ when the initial value is $X_0 = [P_0 \, Z_0 \, N_0 \, Mp_0] = [3.6 \, 1.5 \, 3.8 \, 0.1]$ and the critical value is $Mp = 0.2885$. The phase diagram at $Mp = 0.2885$ is a limit cycle (Fig. 3.11).

Figures 3.12 and 3.13 show the observed bifurcation when Mp vacillates near the critical value, firstly $Mp = 0.3185$ in Fig. 3.9 and then $Mp = 0.2585$ in Fig. 3.10.

(3) For Gm (maximum feeding rate), the equilibrium point is $(P^* \, Z^* \, N^*) = (3.9282 \, 1.6721 \, 3.2998)$ when $X_0 = [P_0 \, Z_0 \, N_0] = [3.6 \;\, 1.5 \;\, 3.8]$, and the critical value is $Gm = 1.4060$. The phase diagram at $Gm = 1.4060$ is a limit cycle (Fig. 3.14).

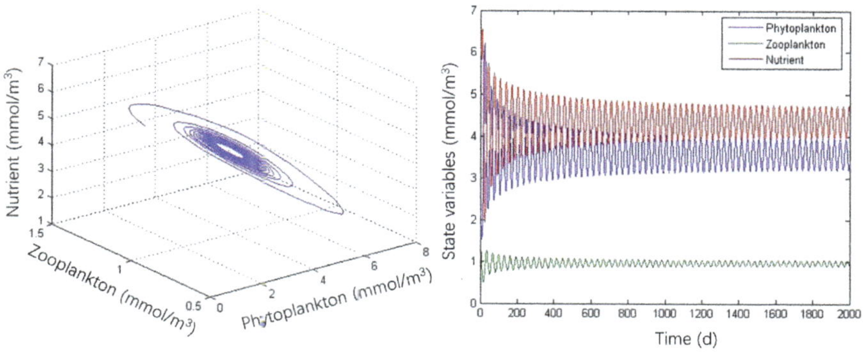

Fig. 3.11. The phase diagram of the NPZ model and time series of state variables when Mp meets the critical condition of the Hopf bifurcation.

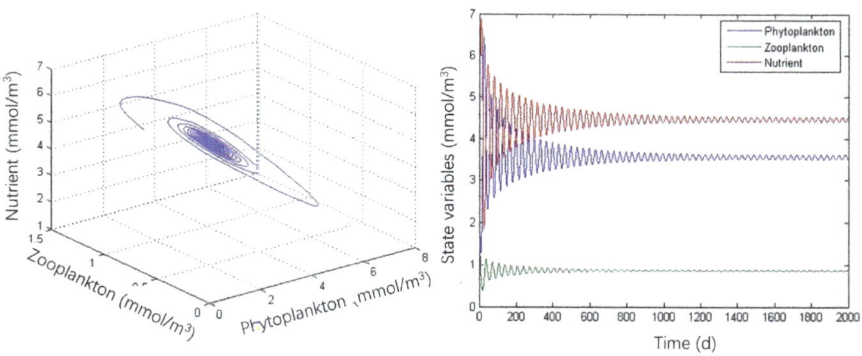

Fig. 3.12. The phase diagram of the NPZ model and time series of state variables when $Mp = 0.3185$.

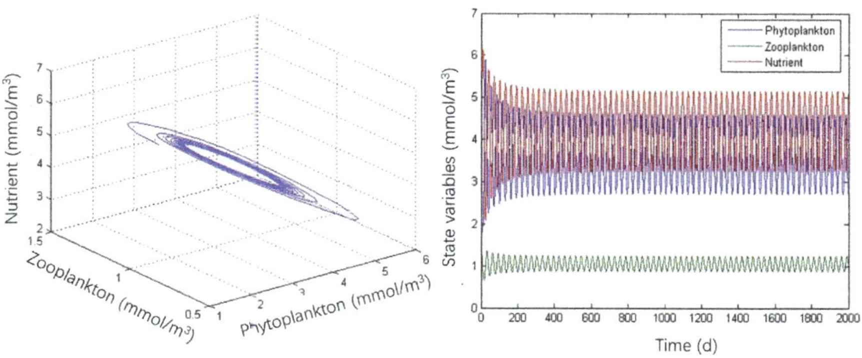

Fig. 3.13. The phase diagram of the NPZ model and time series of state variables when $Mp = 0.2585$.

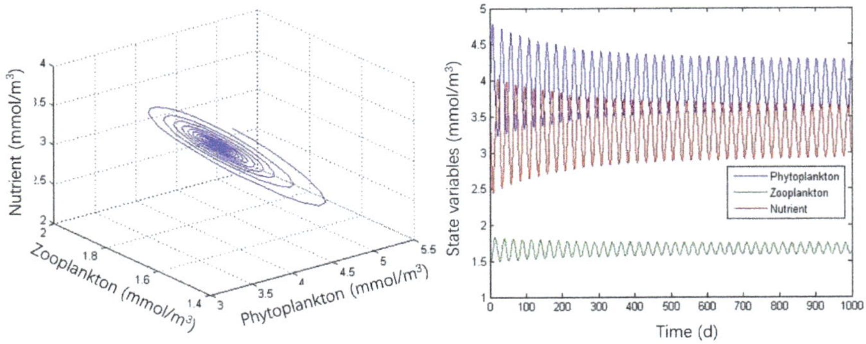

Fig. 3.14. The phase diagram of the NPZ model and time series of state variables when Gm meets the critical condition of the Hopf bifurcation.

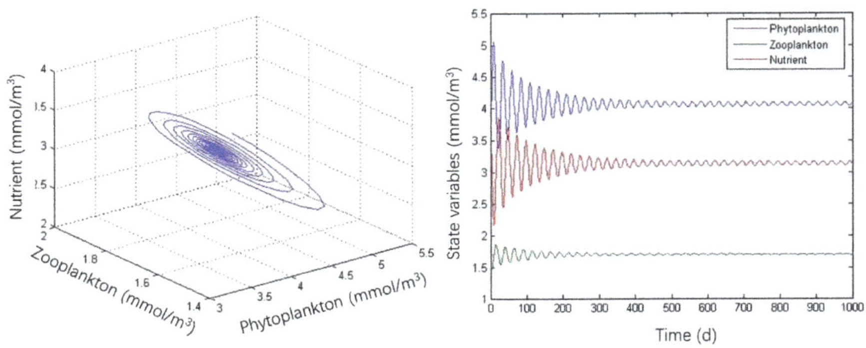

Fig. 3.15. The phase diagram of the NPZ model and time series of state variables when $Gm = 1.3760$.

Figures 3.15 and 3.16 show the observed bifurcation when Gm vacillates near the critical value, firstly $Gm = 1.3760$ in Fig. 3.15 and then $Gm = 1.4360$ in Fig. 3.16.

3.5.3 *NPZK model*

Consider the characteristic equation:

$$\det(\eta E - B) = 0, \tag{3.47}$$

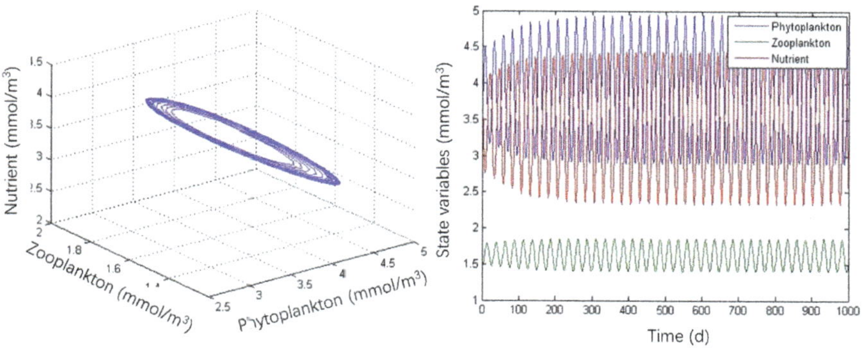

Fig. 3.16. The phase diagram of the NPZ model and time series of state variables when $Gm = 1.4360$.

that is, $\eta^3 + b_1 \cdot \eta^2 + b_2 \cdot \eta + b_3 = 0$, where:

$$\begin{aligned}
b_1 &= -(b_{11} + b_{22} - b_{33} - b_{14} - b_{24} - b_{34}) \\
b_2 &= -b_{11}b_{24} - b_{11}b_{34} + b_{11}b_{22} - b_{22}b_{34} - b_{14}b_{33} - b_{23}b_{32} \\
&\quad - b_{31}b_{13} + b_{11}b_{33} + b_{12}b_{24} + b_{21}b_{14} - b_{14}b_{22} + b_{23}b_{34} + b_{24}b_{32} \\
&\quad + b_{31}b_{14} + b_{22}b_{33} + b_{13}b_{34} - b_{21}b_{12} - b_{33}b_{24} \\
b_3 &= -\det(B),
\end{aligned}$$

and all parameters are as defined by the same way as above.

According to the Hurwitz criterion, the Hopf bifurcation occurs in the system as long as $H_2 = b_1 \cdot b_2 - b_3 = 0$, at which time two conjugate complex roots of (3.47) have zero real part (i.e., a pair of purely imaginary roots exists.) The roots of the characteristic equation are $\eta_{1,2} = \pm i w_0$, $\eta_3 = -b_1$, where $w_0^2 = b_2$.

When U_m is used as the bifurcation parameter for analysis, the condition for Hopf bifurcation becomes as follows when U_m is computed under the initial data $(P_0, Z_0, K_0, N_0) = (3.6, 1.4, 60, 0.2)$ and other parameters take the default value

$$\begin{cases} \vec{F}_2 = 0 \\ H_2 = 0 \end{cases}.$$

The critical value is $U_m = 5.5178$, at which point the phase diagram is a limit cycle (Fig. 3.17) and the time series is as shown in Fig. 3.18.

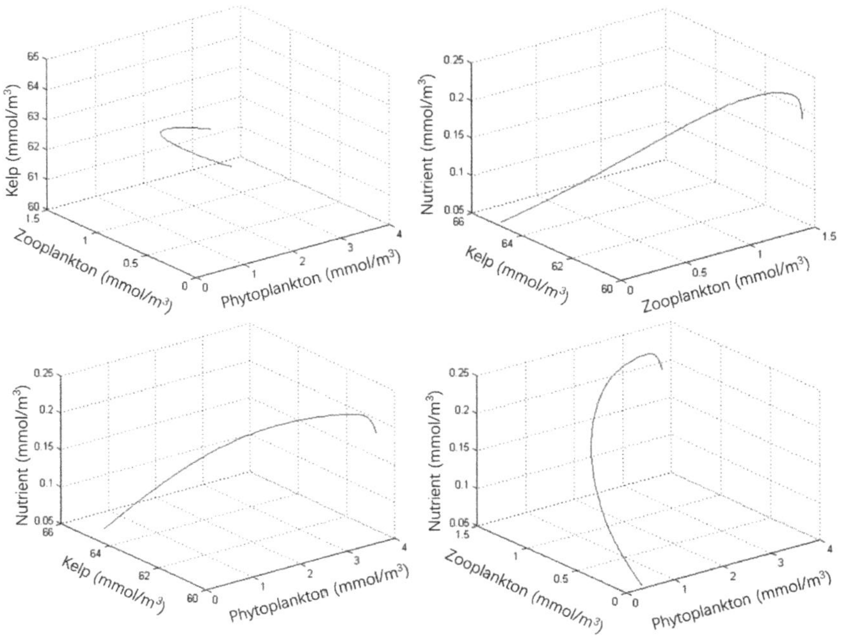

Fig. 3.17. The phase diagram of the NPZK model when U_m satisfies the critical condition of the Hopf bifurcation.

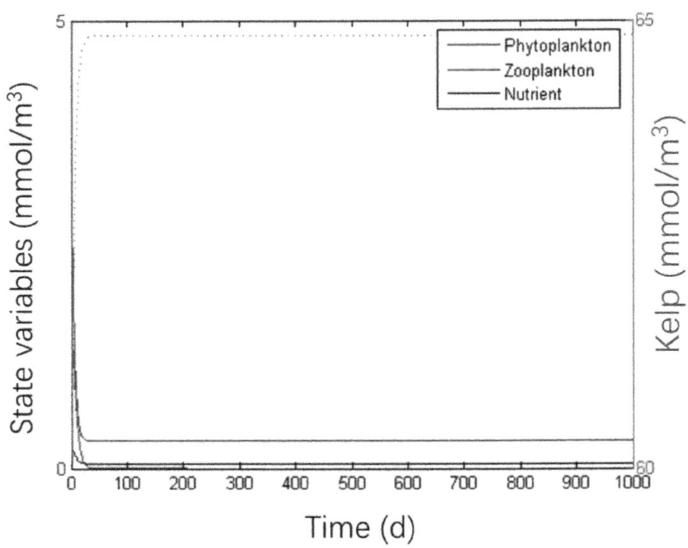

Fig. 3.18. The time series of state variables of the NPZK model when U_m satisfies the critical condition of the Hopf bifurcation.

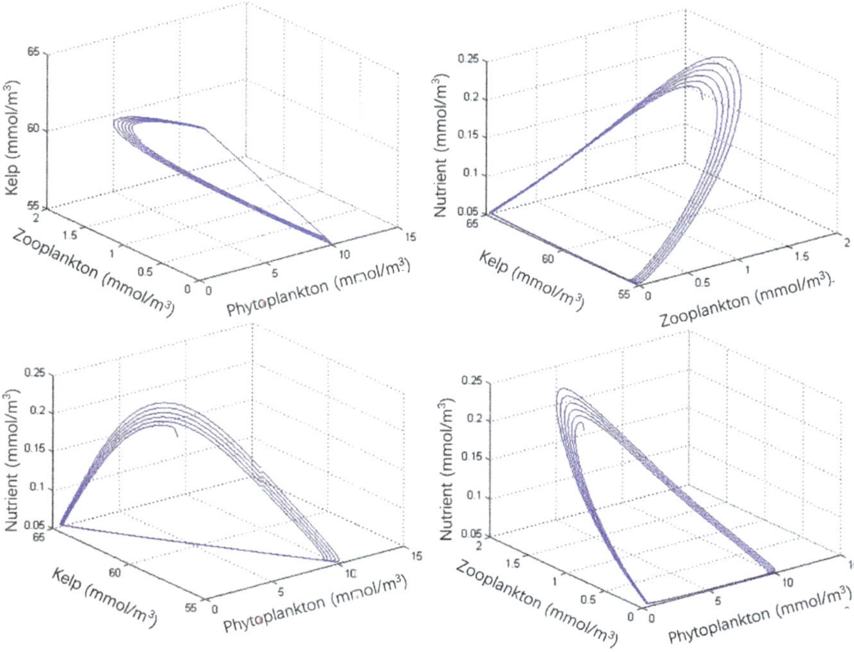

Fig. 3.19. The phase diagram of the NPZK model when $U_m = 6.0178$.

The phase diagrams and time series can be also plotted for the cases when U_m is near the critical value. Figures 3.19 and 3.20 give the phase diagram and time series, respectively, for $U_m = 6.0178$.

Figures 3.21 and 3.22 give the phase diagram and time series, respectively, for $U_m = 5.0178$.

Other parameters can be calculated analogously as the above.

(1) For K_k (the half-saturation constant of kelp nitrogen intake), the critical value is $K_k = 6.333$ when the initial value is $X_0 = [P_0\, Z_0\, K_0\, N_0] = [3.6\, 1.4\, 60\, 0.2\, 1]$. The phase diagram and time series at $K_k = 6.333$ are given in Figs. 3.23 and 3.24, respectively.

Figures 3.25 to 3.28 show the phase diagram and time series, respectively, when K_k vacillates near the critical value, firstly $K_k = 6.833$ in Figs. 3.25 and 3.26.

Figures 3.27 and 3.28 show the phase diagram and time series when $K_k = 5.833$.

(2) For Mp (the phytoplankton mortality), the critical value is $Mp = 0.0362$ when the initial value $X_0 = [P_0\, Z_0\, K_0\, N_0] =$

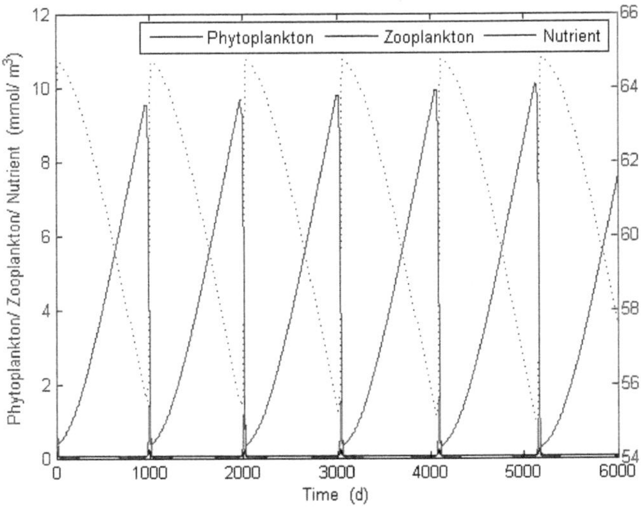

Fig. 3.20. The time series of state variables of the NPZK model when $U_m = 6.0178$.

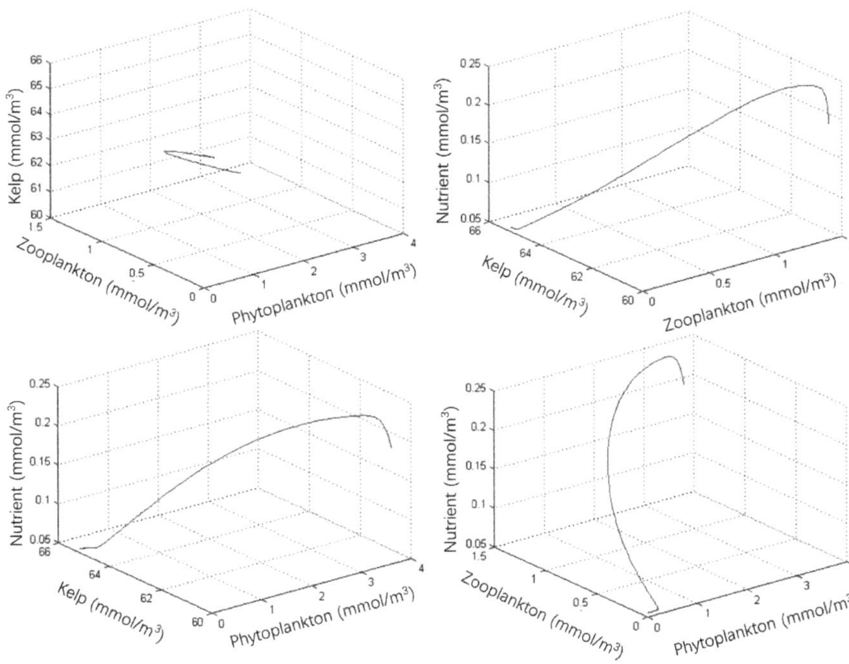

Fig. 3.21. The phase diagram of the NPZK model when $U_m = 5.0178$.

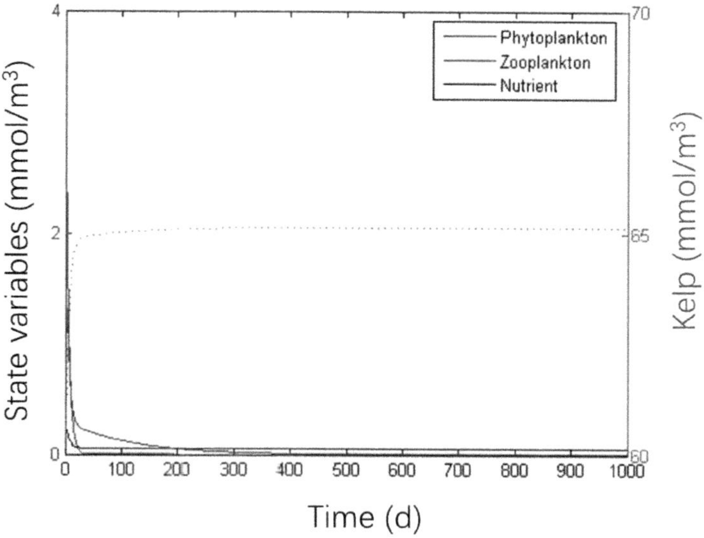

Fig. 3.22. The time series of state variables of the NPZK model when $U_m = 5.0178$.

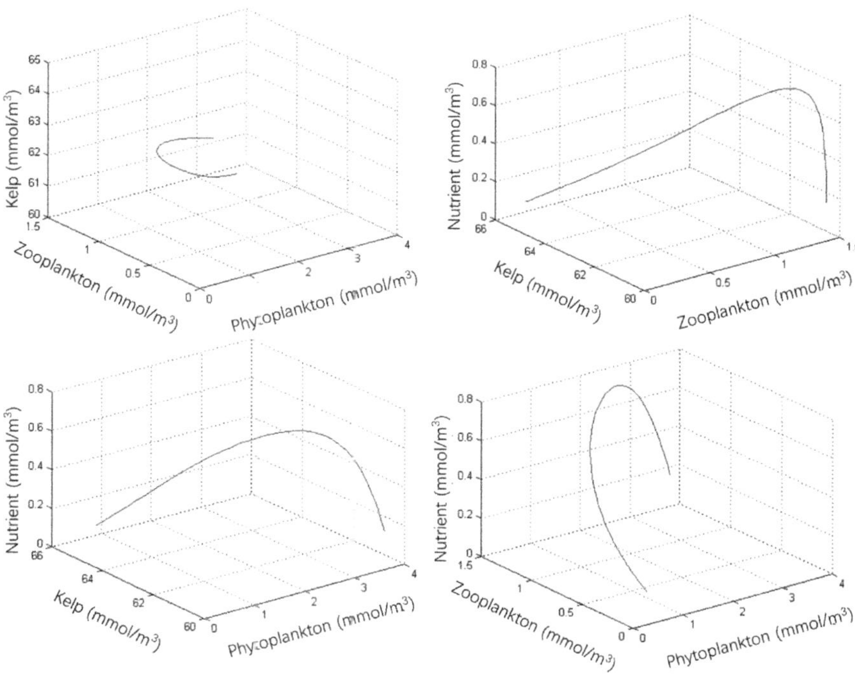

Fig. 3.23. The phase diagram at $K_k = 6.333$.

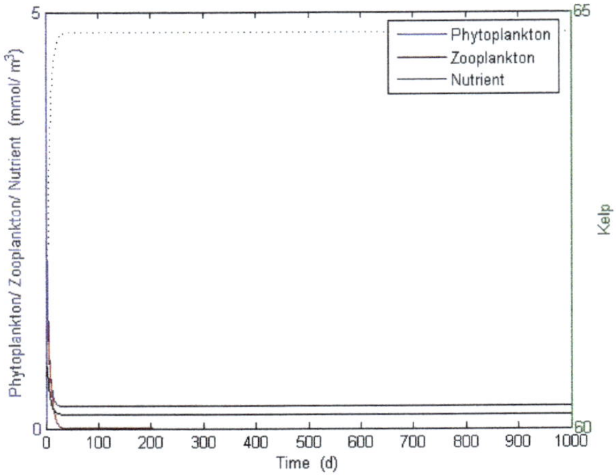

Fig. 3.24. The time series of state variables with $K_k = 6.333$.

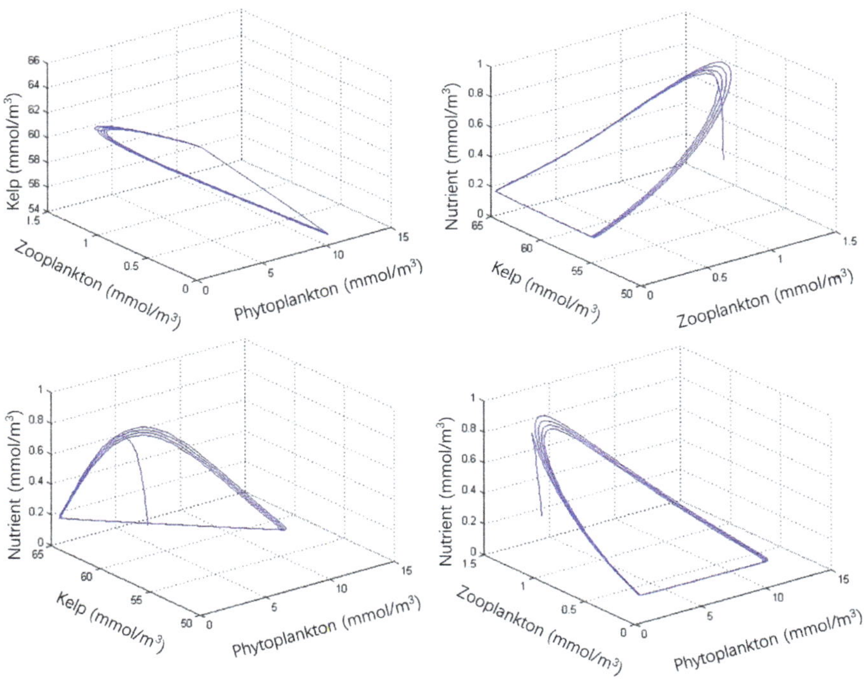

Fig. 3.25. Phase diagram with $K_k = 6.833$.

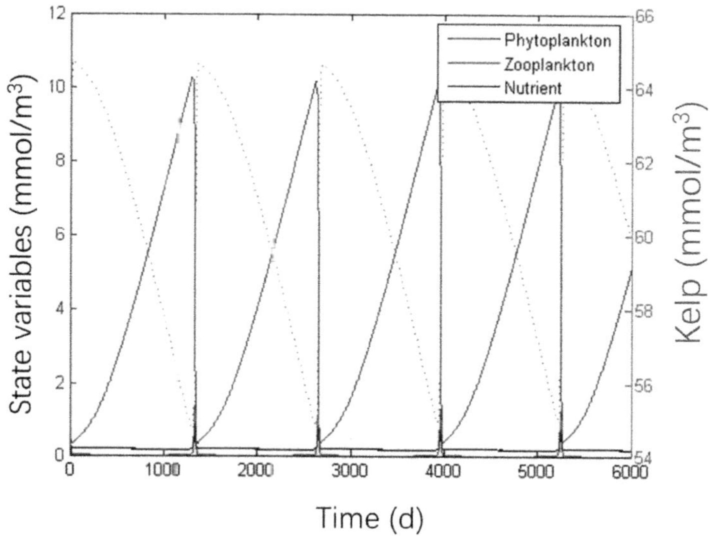

Fig. 3.26. Time series of the state variables with $K_k = 6.833$.

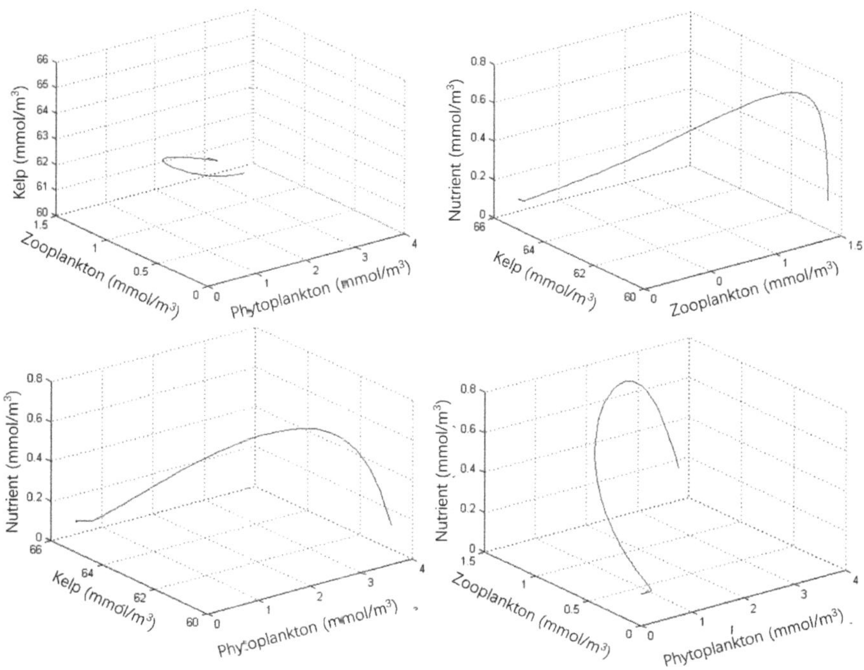

Fig. 3.27. Phase diagram with $K_k = 5.833$.

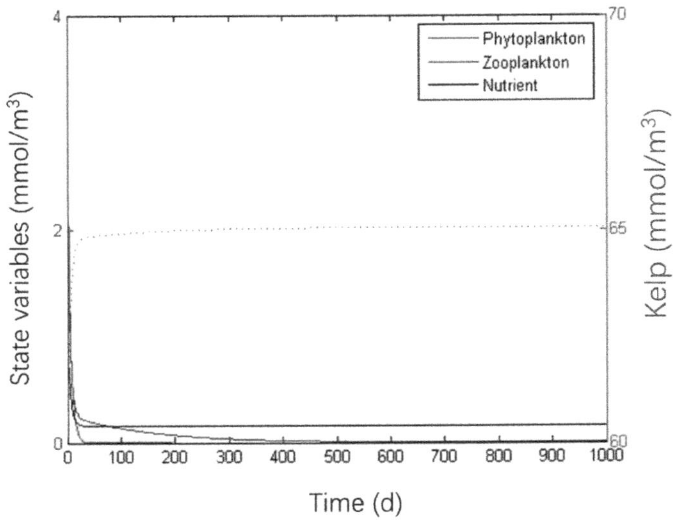

Fig. 3.28. Time series of the state variables with $K_k = 5.833$.

[3.6 1.4 60 0.2]. The phase diagram and time series with $Mp = 0.0362$ are given in Figs. 3.29 and 3.30, respectively.

Figures 3.31 to 3.34 show the phase diagram and time series when M_p vacillates near the critical value, firstly $M_p = 0.0302$ in Figs. 3.31 and 3.32.

Figures 3.33 and 3.34 show the phase diagram and time series, respectively, when $M_p = 0.0422$.

3.5.4 *The NPZD model*

Consider the characteristic equation:

$$\det(\eta E - C) = 0. \tag{3.48}$$

That is, $\eta^3 + c_1 \cdot \eta^2 + c_2 \cdot \eta + c_3 = 0$, where:

$$\begin{aligned} c_1 &= -(c_{11} + c_{22} + c_{33} - c_{14} - c_{24} - c_{34}) \\ c_2 &= -c_{11}c_{24} - c_{11}c_{34} + c_{11}c_{22} - c_{22}c_{34} - c_{14}c_{33} - c_{23}c_{32} \\ &\quad - c_{31}c_{13} + c_{11}c_{33} + c_{12}c_{24} + c_{21}c_{14} - c_{14}c_{22} + c_{23}c_{34} + c_{24}c_{32} \\ &\quad + c_{31}c_{14} + c_{22}c_{33} + c_{13}c_{34} - c_{21}c_{12} - c_{33}c_{24} \\ c_3 &= -\det(C), \end{aligned}$$

and all parameters are defined by the same way as above.

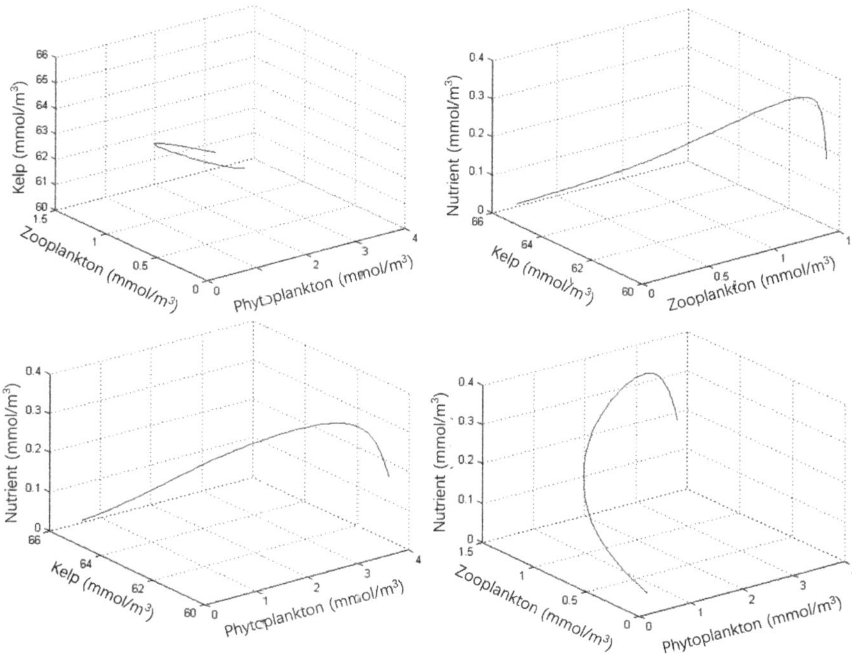

Fig. 3.29. Phase diagram when M_p is at critical value.

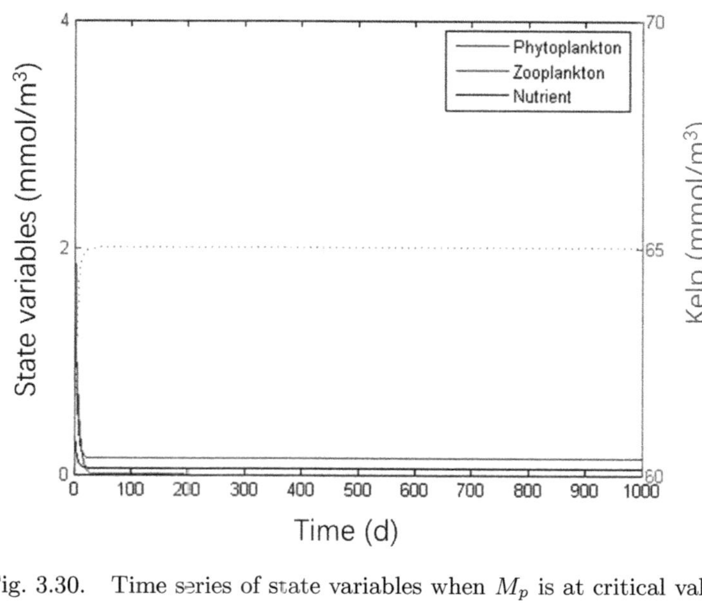

Fig. 3.30. Time series of state variables when M_p is at critical value.

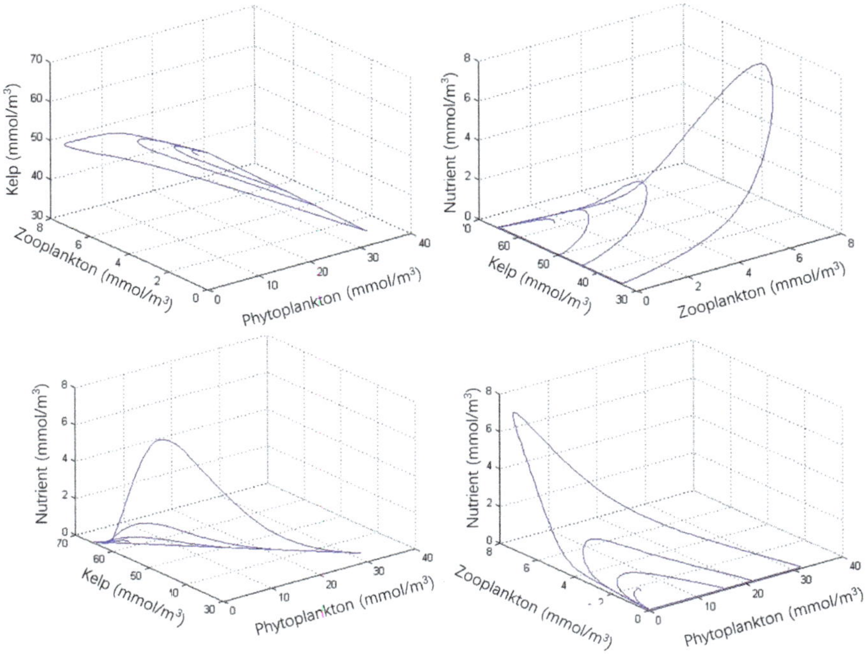

Fig. 3.31. Phase diagram with $M_p = 0.0302$.

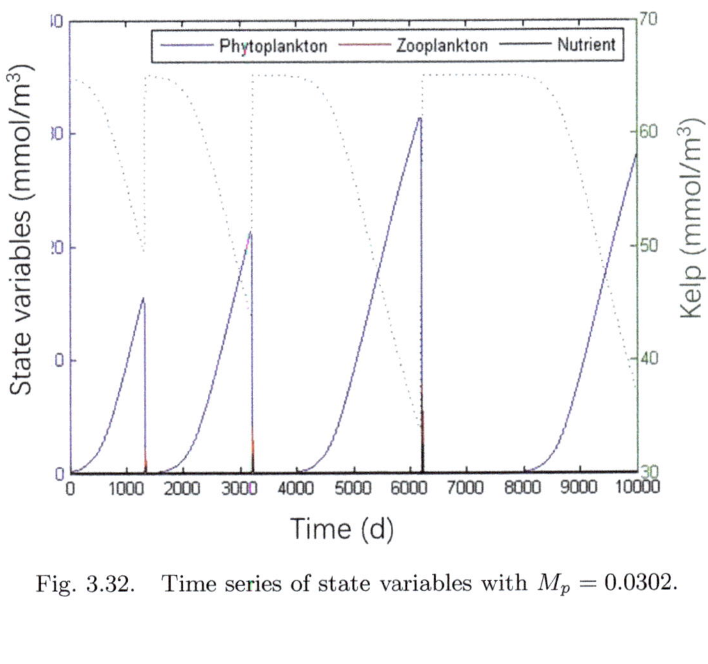

Fig. 3.32. Time series of state variables with $M_p = 0.0302$.

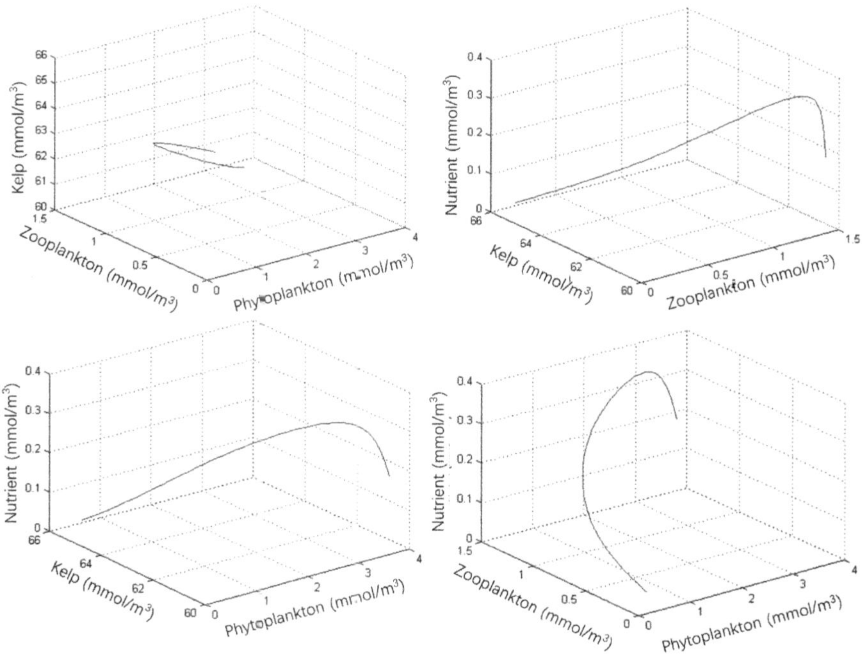

Fig. 3.33. Phase diagram with $M_p = 0.0422$.

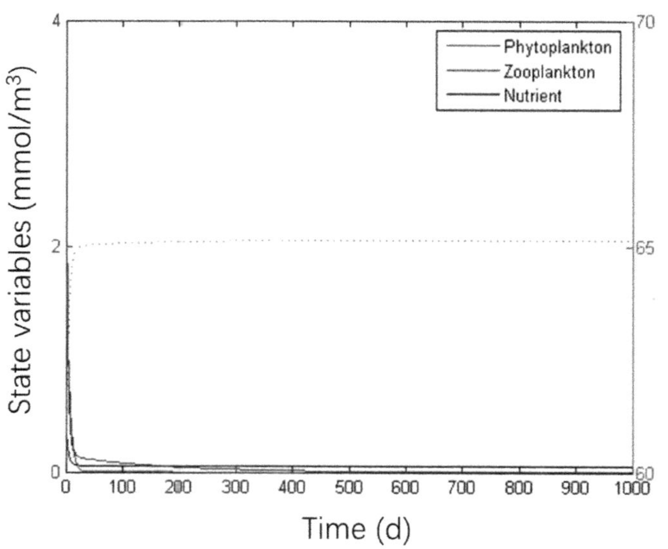

Fig. 3.34. Time series of the state variables with $M_p = 0.0422$.

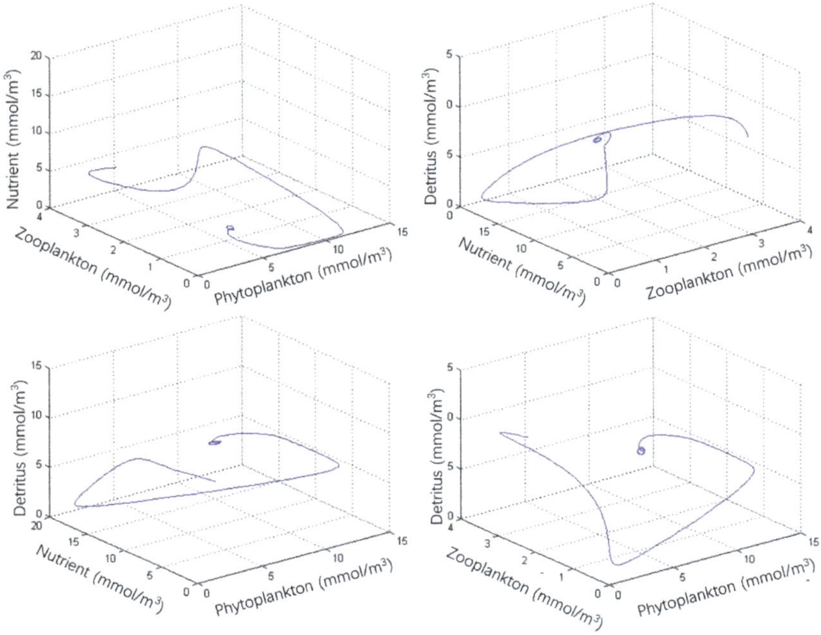

Fig. 3.35. The phase diagram when U_m satisfies the critical condition of the Hopf bifurcation.

Set U_m as the bifurcation parameter analogously as in the analysis of the N-P-Z-K model. With the initial condition of $(P_0, Z_0, N_0, D_0) = (4, 3.6, 4.5, 7.5)$ and all other parameters taking default value, the Hopf bifurcation condition for U_m is

$$\begin{cases} \vec{F}_3 = 0 \\ H_2 = c_1 \cdot c_2 - c_3 = 0 \end{cases}.$$

The critical condition is thus $U_m = 0.7529$, at which point the phase diagram is a limit cycle as shown in Fig. 3.35 and the time series is shown in Fig. 3.36.

The phase diagrams and time series can be also plotted for the cases when U_m is near the critical value. Figures 3.37 and 3.38 give the phase diagram and time series for $U_m = 0.6529$.

Figures 3.39 and 3.40 give the phase diagram and time series, respectively, with $U_m = 0.8529$.

For the parameter G_m (maximum intake rate), the critical value is $G_m = 2.2553$ when the initial value is $X_0 = [P_0\, Z_0\, N_0\, D_0] =$

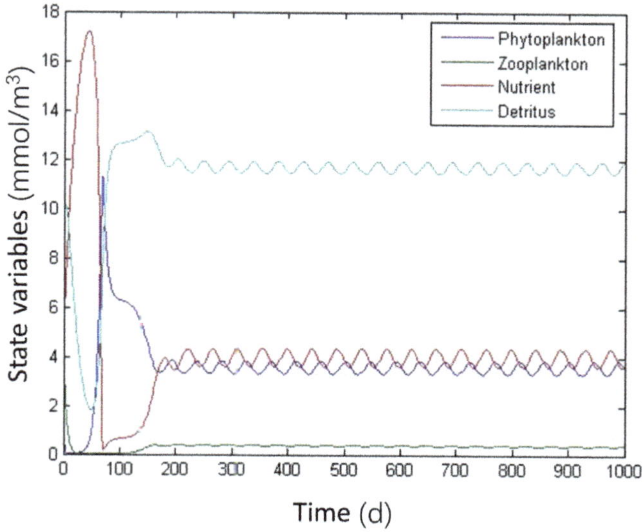

Fig. 3.36. The time series of state variables when U_m meets the critical condition of the Hopf bifurcation.

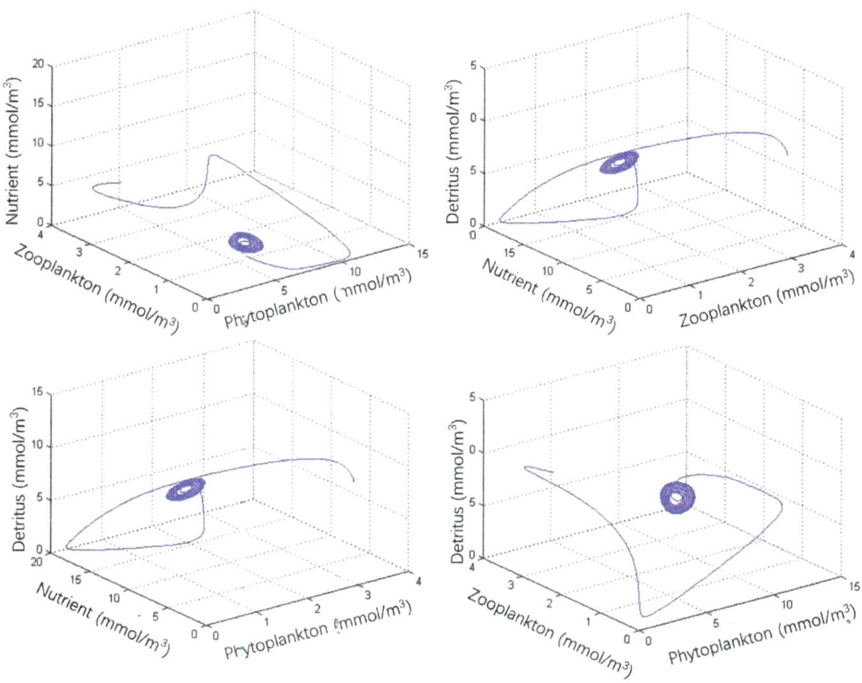

Fig. 3.37. The phase diagram when $U_m = 0.6529$.

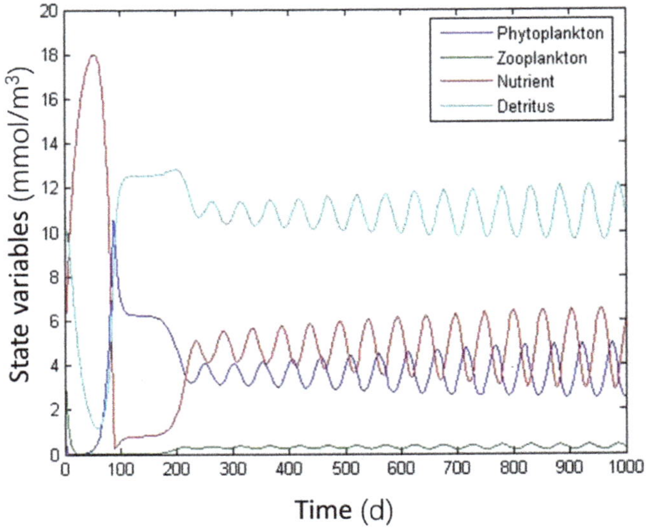

Fig. 3.38. The time series of state variables with $U_m = 0.6529$.

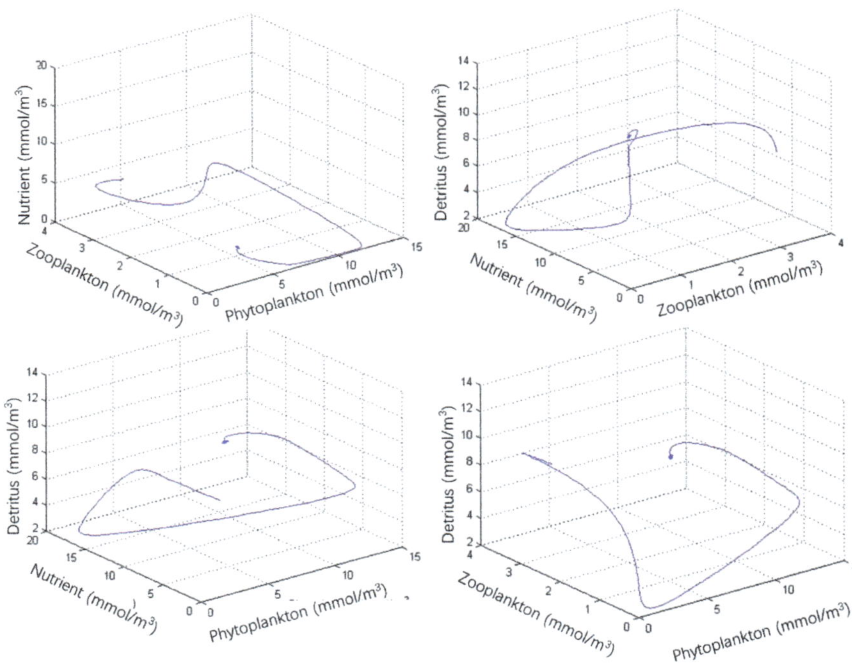

Fig. 3.39. The phase diagram with $U_m = 0.8529$.

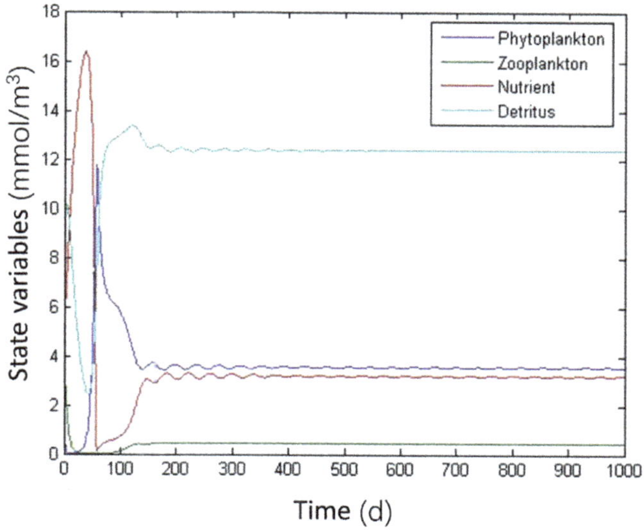

Fig. 3.40. The time series of state variables with $U_m = 0.8529$.

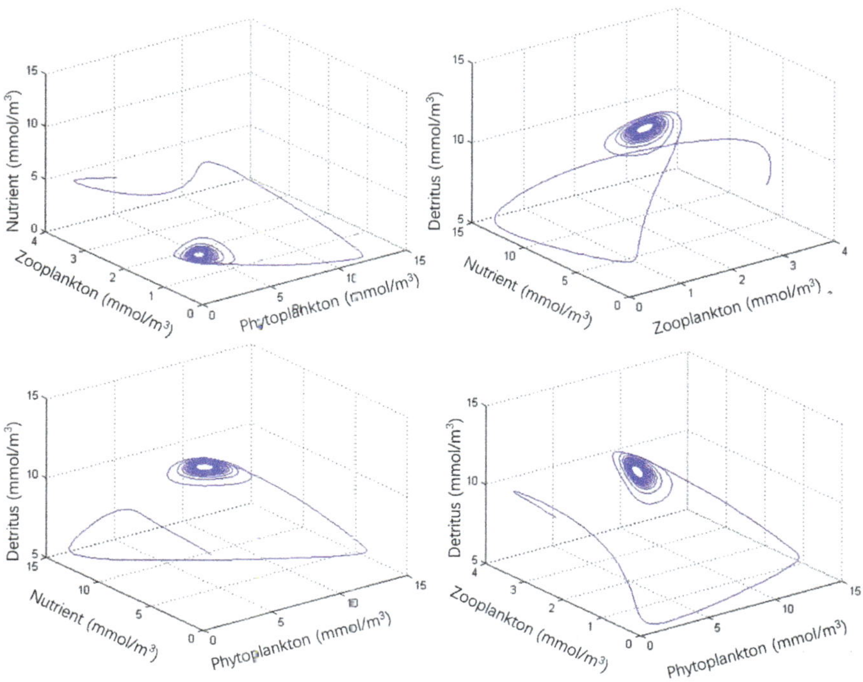

Fig. 3.41. The phase diagram when G_m satisfies the critical condition.

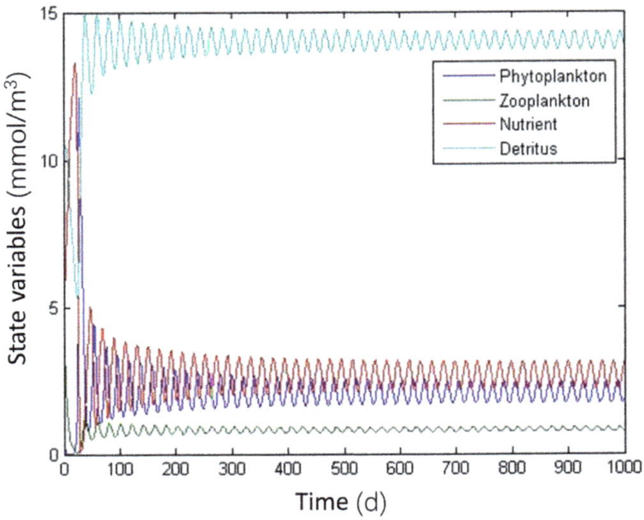

Fig. 3.42. The time series of state variables when G_m meets the critical condition.

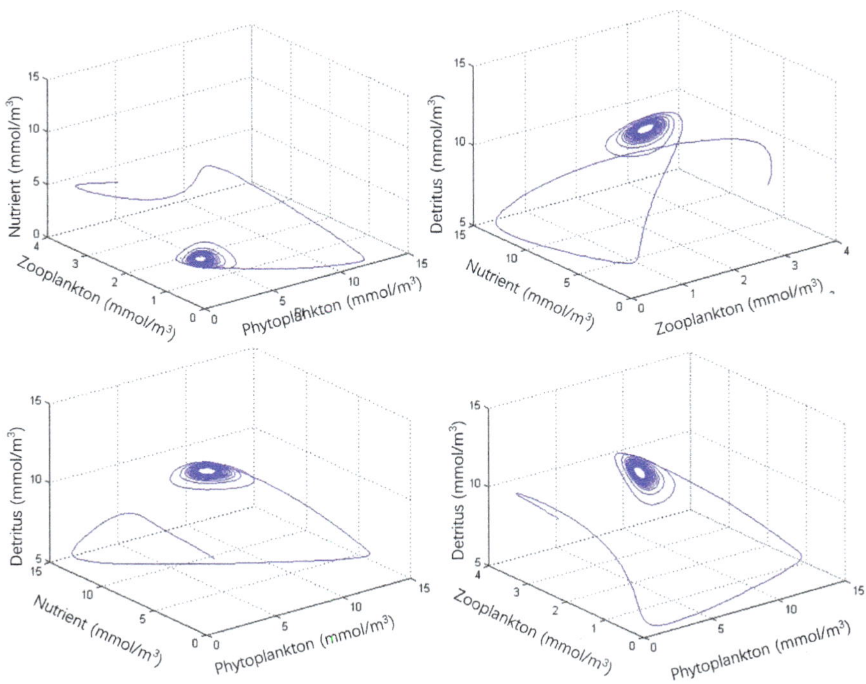

Fig. 3.43. Phase diagram with $G_m = 2.1553$.

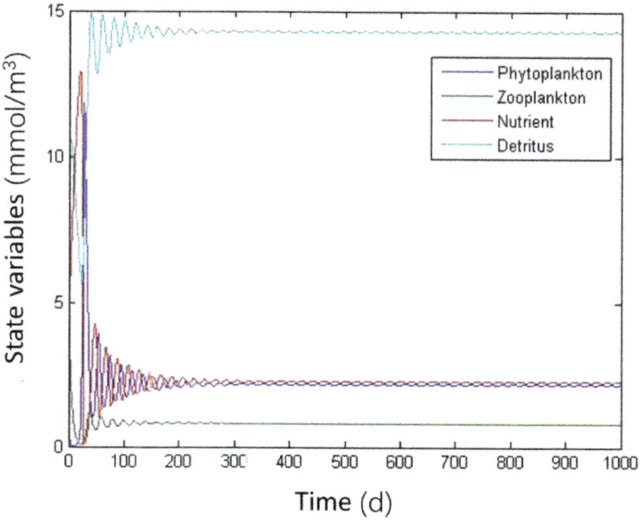

Fig. 3.44. Time series of state variables with $G_m = 2.1553$.

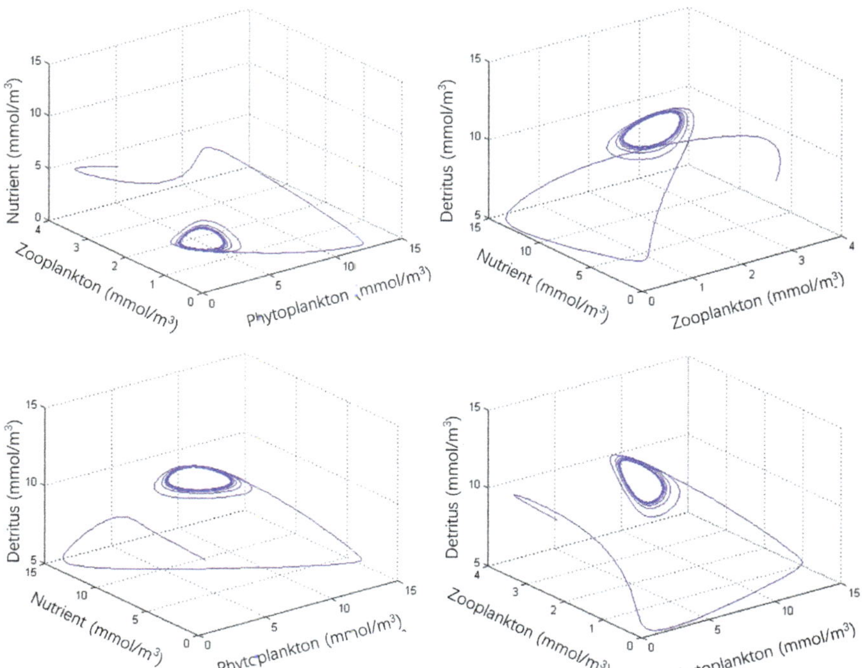

Fig. 3.45. Phase diagrams with $G_m = 2.3553$.

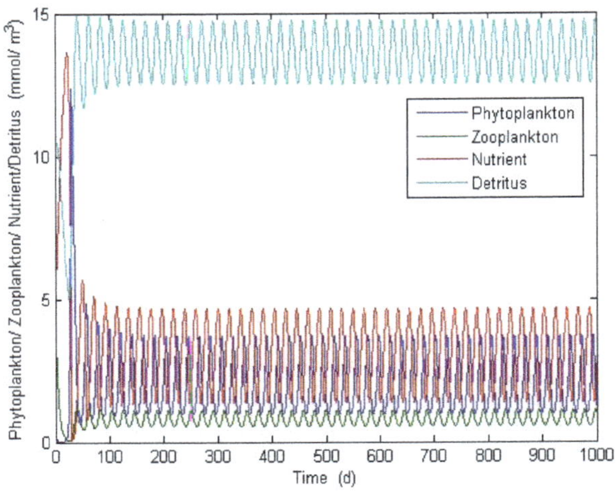

Fig. 3.46. Time series of state variables with $G_m = 2.3553$.

[4 3.6 4.5 7.5]. The corresponding phase diagrams and time series are given in Figs. 3.41 and 3.42, respectively.

The following results can be obtained when G_m vacillates near the critical value, Figures 3.43 and 3.44 show the phase diagrams and time series, respectively, for $G_m = 2.1553$.

Figures 3.45 and 3.46 show the phase diagram and time series, respectively, when $G_m = 2.3553$.

3.6 Conclusions

This chapter analyzes a class of self-conserving marine ecosystem dynamics models and proves the existence and uniqueness of their solutions, then demonstrates the limitations of linear approximation and the Lyapunov–Krasovskii method in analyzing the stability of these models. The dimensionality reduction theorem is proposed and proved, and then combined with linear approximation to analyze the stability of the solutions to the NPZ and NPZD models. Finally, with ecologically significant parameters and initial conditions, the two ecosystem models are numerically simulated, and the results from simulation and theoretical analysis are verified (Shi *et al.*, 2016).

The main conclusions of this chapter are as follows:

(1) The two ecological models studied in this chapter have unique solutions under a set of ecologically significant parameters and initial conditions.
(2) Conventional stability analysis methods for dynamic systems, such as linear approximation and the Lyapunov–Krasovskii method, have some limitations in analyzing the stability of self-conserving systems.
(3) The dimensionality reduction theorem is a necessary and sufficient condition that serves as an important basis in analyzing the stability of solutions of self-conserving systems. With the dimensionality reduction theorem, the stability problem of the original high-dimensional manifold becomes equivalent to the stability of the system of equations with one less dimension in the entire low-dimensional space.
(4) The numerical simulation results are consistent with the theoretical analysis results, which validates the proposed dimensionality reduction theorem.

The research of marine ecosystem dynamics is interesting and complex. Mathematical models based on differential equations are important tools for studying marine ecosystem dynamics (Gao *et al.*, 2009). The stability of marine ecosystems is one of the most important research topics in studying marine models (Chen, 2003). There is currently limited research on the stability of marine ecosystem dynamical models, although an increasing number of scientists now recognize the importance of predictability and stability in these models. The results in this chapter show that conventional methods for evaluating system stability are limited in some aspects. Consequently, the dimensionality reduction theorem is introduced and proved to clarify the stability of the zero solution of a particular type of self-conserving equations. The stability of the zero solution of non-self-conserving equations still awaits further study.

Acknowledgment

The reference (Shi *et al.*, 2016) was reproduced with permission from Springer Nature.

References

Bricker S B, Ferreira J G, Simas T. 2003. An integrated methodology for assessment of estuarine trophic status. *Ecological Modelling*. **169**: 39–60.

Chen C S. 2003. *Marine Ecosystem Dynamics and Modeling* (in Chinese). Beijing: Higher Education Press, Beijing.

Chen C S, Ji R B, Zheng L Y, *et al.* 1999. Influences of physical processes on the ecosystem in Jiaozhou Bay: A coupled physical and biological model experiment. *Journal of Geophysical Research*. **104**: 29925–29949.

Cossarini G, Solidoro C. 2008. Global sensitivity analysis of a trophodynamic model of the Gulf of Trieste. *Ecological Modelling*. **212**: 16–27.

Costanza R, Agre R, de Groot R, *et al.* 1997. The value of the world's ecosystem and natural capital. *Nature*. **387**: 253–260.

Edwards A M, Bees M A. 2001. Generic dynamics of a simple plankton population model with a non-integer exponent of closure. *Chaos Solitons Fractals*. **12**: 289–300.

Fei Z L. 1984. An estimation of the diffuse attenuation coefficient in offshore waters (in Chinese). *Yellow Sea and Bohai Sea*. **2**(1): 26–29.

Feng S Z, Zhang J, Wei H. 2007. *Introduction to Environmental Dynamics of Bohai Sea* (in Chinese). Beijing: Science Press.

Franks P J S, Chen C. 1996. Plankton production in tidal fronts: A model of Georges Bank in summer. *Journal of Marine Research*. **54**: 631–651.

Franks P J S, Wroblewski J S, Flierr G R. 1986. Behavior of a simple plankton model with food-level acclimation by herbivores. *Marine Biology*. **91**: 121–129.

Fransz H G, Verhagen J H G. 1985. Modelling research on the production cycle of phytoplanton in the southern Bight of North Sea in relation to river-borne nutrient loads. *Netherlands Journal of Sea Research*. **19**: 241–250.

Fujii M, Yoshie N, Yamanaka Y, *et al.* 2005. Simulated biogeochemical responses to iron enrichments in three high nutrient, low chlorophyll (HNLC) regions. *Progress in Oceanography*. **64**: 307–324

Gao H, Feng S, Guan Y. 1998. Modelling annual cycles of primary production in different regions of the Bohai Sea. *Fisheries Oceanography* (3/4): 258–264.

Gao M, Shi H, Li Z. 2009. Chaos in a seasonally and periodically forced phytoplankton-zooplankton system. *Nonlinear Analysis Real World Applications*. **10**: 1643–1650.

Graneli E, Weberg M, Salomon P S. 2008. Harmful algal blooms of allelopathicmicroalgal species: The role of eutrophication. *Harmful Algae.* **8**: 94–102.

Halpern B S, Walbridge S, Selkoe K A, *et al.* 2008. A global map of human impact on marine ecosystems. *Science.* **319**: 948–952.

Heimsund B O, Berntsen J. 2004. On a class of model instabilities that may steps, implicit methods, and low viscosities. *Ocean Modelling.* **7**: 135–144.

Holling C S. 1973. Resilience and Stability of Ecological Systems. *Annual Review of Ecology, Evolution, and Systematics.* **4**: 1–23.

Jackson J B C, Kirby M X, Berger W H, *et al.* 2001. Historical overfishing and the recent collapse of coastal ecosysocean occur when applying small time terns. *Science.* **293**: 629–638

Liu Z, Wei H, Bai J, *et al.* 2007. Nutrients seasonal variation and budget in Jiaozhou Bay, China: A 3-Dimensional physical-biological coupled model study. *Water, Air, & Soil Pollution.* **7**: 607–623.

Lotze H K, Lenihan H S, Bourque B J, 2006. Depletion, degradation, and recovery potential of estuaries and coastal seas. *Science.* **312**: 1806–1809.

McCann K, Hastings A. 1997. Re-evaluating the omnivory-stability relationship in food webs. *Proceedings of the Royal Society of London. Series B.* **264**: 1249–1254.

McCreary J P, Kohler K E, Hood R R, 1996. A four-component ecosystem model of biological in Arabian Sea. *Progress in Oceanography.* **37**: 193–240.

Millennium ecosystem assessment (MA). 2005. *Ecosystems and Human Well-Being: Synthesis.* Washington DC: Island Press.

Myers R A, Worm B. 2003. Rapid worldwide depletion of predatory fish communities. *Nature.* **423**: 280–283.

Shi H, Hu L, Fang G, *et al.* 2016. Research on stability and Hopf bifurcation of marine ecosystem dynamics models. *Acta Oceanologica Sinica.* **35**(4): 124–132. DOI: 10.1007/sl3131-016-0809-x.

Steele J H. 1962. Environmental control of photosynthesis in the sea. *Limnology and Oceanography.* **7**: 137–150.

Vichi M, Masina S, Navarra A. 2007. A generalized model of pelagic biogeochemistry for the global ocean ecosystem. Part II: Numerical simulations. *Journal of Marine Systems.* **64**: 110–134.

Wang G X, Zhou Z M, Zhu S M, Wang S S. 1983. *Ordinary Differential Equations* (in Chinese). Beijing: Higher Eduation Press.

Wang H L, Feng J F. 2006. *Ecological Dynamics and Prediction of Red Tide* (in Chinese). Tianjin: Tianjin University Press.

Zhong Q H. 2004. *Modern Control Theory* (in Chinese). Beijing: Higher Eduation Press.

Chapter 4

Research on Skill Improvement Technology of Marine Ecosystem Dynamics Models

High demands are imposed on marine ecosystem dynamics models according to the needs of marine management of the ecosystem. The application of models, especially prediction applications serving management and decision making, require models to perform well (Stow *et al.*, 2009). Therefore, learning how to improve and evaluate a model's skill should be an important research topic in marine ecosystem dynamics models and their applications. However, current research on marine ecosystem dynamics models mainly focuses on model construction and application, which results in insufficient research attention or a lack of scientific guidance on model skill improvement.

The *Journal of Marine Systems* devoted two issues in 2009 to discuss skill assessment for coupled biological/physical models of marine systems (Stow *et al.*, 2009). It is pointed out that model skill and skill evaluation, refer to the fidelity of model behavior to natural reality and human judgment of model skill, respectively. Since natural reality is generally obtained by the approximation of observation, model skill can be considered as how well the model fits the observational data. In general, model skill is a general description of model behavior, which is the basis for judging the reliability of model application results. Therefore, it is necessary to optimize model skill to improve the accuracy of a model's results.

The evaluation of model skill through quantitative analysis can guide model optimization and application. Understanding the optimized model skill is a prerequisite for applying models.

4.1 Construction of Technology Framework for Improving Skill of Marine Ecosystem Dynamics Model

4.1.1 *Analysis of factors affecting skill of marine ecosystem dynamics model*

A model's skill generally depends on the model's structure, parameters, solution, and inputs such as the model's initial conditions, boundary conditions, and external physical forcing forces (Gregg *et al.*, 2009). Fundamentally, the greatest difference between models is the model structure. The structure of a marine ecosystem dynamics model is mainly represented by biological processes and their coupling mechanisms with physical and chemical processes, model space dimensions, model complexity, and so on. Because marine ecosystem dynamics processes are often carried out in the context of marine hydrodynamics, the coupling mechanism and spatial dimension of the model are usually determined by whether the ecosystem dynamic processes of concern are largely influenced by hydrodynamic conditions such as horizontal convection and vertical mixing. Model complexity is generally characterized by the number of state variables, which mainly depends on the type of ecosystem being simulated (Arhonditsis and Brett, 2004).

Currently, in research on physical processes, it is generally believed that a more comprehensive physical model is used in practice to couple biological models with relatively low complexity. From this perspective, one might theorize that increasing the complexity of the model structure, such as coupling physics-chemistry-biological processes, using three-dimensional space models, and adding state variables, could improve the model's potential to reproduce complex natural systems accurately. However, various studies have shown that a model's skill did not depend on the type of ecosystem being simulated, and that increasing model structure complexity did not necessarily improve model skill systematically (Matear, 1995; Hurtt and Armstrong, 1996; 1999; Arhonditsis and Brett, 2004;

Flynn, 2005; Friedrichs *et al.*, 2007; 2009). This phenomenon occurs mainly because the current understanding of complex natural systems, especially marine ecosystems and their observational data, is insufficient to support relatively complex models. In addition, model solving depends on numerical algorithms while model input generally depends on observations. The model parameters vary with the model and study area without a universal pattern. These model parameters are also difficult to measure, which makes it difficult to determine their specific values; and this affects a model's skill (Friedrichs, 2002). These results show that a model's structure, solution, and inputs depend on the current level of development in theoretical research and observation technology, and that the means of determining model and the selection of parameter values has some flexibility; Therefore, suggesting that the skill of marine ecosystem dynamics models is generally improved by optimizing model parameters.

4.1.2 *Strategies for improving model skill based on parameter optimization*

Model parameters can be optimized using mathematical methods, primarily because the model is a conceptual representation of natural processes. It is unscientific to simply choose parameters based on observations and experiments. For example, marine ecosystem dynamics models generally classify many different species into large functional groups (such as phytoplankton and zooplankton) and use empirical functions to describe the evolution and interaction of different state variables of the model (Hood *et al.*, 2006; Peng *et al.*, 2012). The parameterization of these functions is often based on experimental laboratory results of a single species under controlled conditions and does not represent the overall biological characteristics of the marine ecosystem (Pahlow and Oschlies, 2009). In addition, parameters generally vary both temporally and spatially, resulting in a lack of experimental data and large differences in parameter values given in different literature works. Researchers can only obtain a reasonable value range of parameters rather than a determined value. Previously, the estimation of model parameters mainly relied on manual adjustments based on experience accumulation and subjective judgment, which generally requires a large workload, and obtaining optimal results is challenging (Gregg *et al.*, 2009; Ward *et al.*, 2010).

Therefore, automatic optimization methods that aim at improving the goodness-of-fit of simulations to observations are often used to adjust model parameters to improve model skill (Gregg *et al.*, 2009; Ward *et al.*, 2010).

Since the marine ecosystem dynamics model is a functional group consisting of the multi-parameter and multi-state variable with various characteristics such as highly nonlinear and structural complexity, the parameter optimization with the marine ecosystem dynamics model as constraints are nonlinear, multi-objective, and multi-extremum, which has many challenges in the optimization process. First of all, the model involves a large number of parameters, which makes this optimization problem have a huge solution space. It thus has high requirements for the amount of observation data and searchability of the optimization algorithm. Secondly, the marine ecosystem dynamics model has multiple state variables, which belong to the multi-objective optimization problem. Thirdly, with the increase of complexity of the model structure, the analytic properties of the objective function required by common optimization methods are difficult to obtain. For example, although the gradient method has strong searchability, it requires the gradient information of the objective function, so it is not suitable for the complex model. Fourthly, the nonlinearity and complexity of the marine ecosystem dynamics model also cause the corresponding objective functions to belong to multimodal functions so that most optimization problems belong to multi-extremum problems. Methods similar to the Simplex method do not have strong randomness, and their search effects depend heavily on the starting point of the search. Frequently, it tends to fall into local extrema and converge prematurely. This disadvantage is more pronounced when the solution space of the optimization problem is large (Gregg *et al.*, 2009; Shi, 2010).

Current studies have classified the optimization process of ecosystem model parameters into two key issues:

(1) Determining which parameters have a greater impact on the model output error.
(2) How to optimize these parameters to improve the goodness-of-fit between the model simulation results and observation results.

The former can be determined by the method of parameter sensitivity analysis, while the latter needs to select an effective parameter optimization algorithm. Therefore, parameter optimization research based on sensitivity analysis is a basic step of the ecosystem model. This process of improving model skill is commonly referred to as model calibration. Since there are many challenges in the parameter optimization of marine ecosystem dynamics models, using this optimization strategy as a key step has become a hot and challenging topic in the field of marine ecosystem modeling (Fennel *et al.*, 2001; Solidoro *et al.*, 2003). As parameter optimization may cause various errors in model parameters such as physical models and unknown processes (Gregg *et al.*, 2009), validation (verification, confirmation, or corroboration) is a key step for model parameter optimization. By examining the skill of the optimized model, the validation can not only determine whether the model parameter optimization is in line with reality but also evaluate the performance of both the parameter optimization algorithm and optimized model, thereby further guiding the improvement of model skill.

4.1.3 *Technical framework for improving the skill of marine ecosystem dynamics models*

In the current field of marine ecosystem dynamics modeling, there are relatively few and uneven studies on model skill improvement and evaluation. Most of the studies on the related aspects of model skill are based on the personal interests and abilities of the modeler, but there is no standard technical framework. This section introduces the methods for improving the skill of marine ecosystem dynamics models based on parameter optimization and their research progress from aspects of model calibration and verification. It focuses on the study of sensitivity analysis and parameter optimization methods and their global treatment methods for the optimization of marine ecosystem dynamics models and explores the application of different model skill evaluation strategies in model verification and model skill improvement. Furthermore, the establishment of a logical framework to inspire research ideas on how to improve the skill of marine ecosystem dynamics models is attempted, with a view to forming a

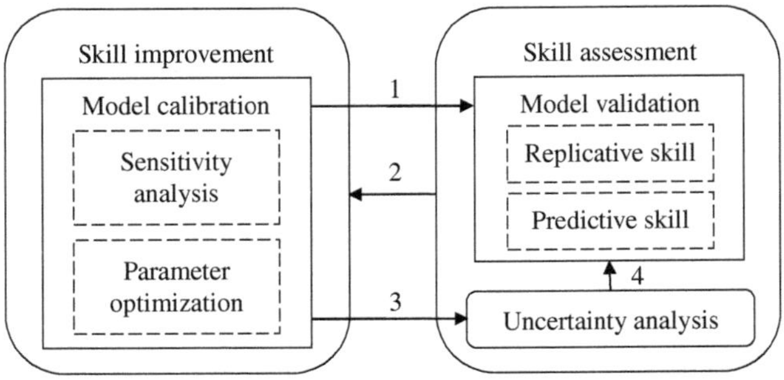

Fig. 4.1. Technical framework for skill improvement and assessment of marine ecosystem dynamics models.
Notes: Arrow 1 indicates that model calibration is the basic step to improving model skill; arrow 2 indicates that skill evaluation provides a reference for skill improvement to ensure the validity of the model after calibration; arrow 3 indicates that the algorithm used for model calibration can be used to analyze model uncertainty; arrow 4 indicates that uncertainty analysis can be used to assist model validation.

scientific and technical framework for improving the skill of marine ecosystem dynamics models (Fig. 4.1).

4.2 Design Strategy of Parameter Optimization for Model Calibration

4.2.1 *Using parameter sensitivity analysis to determine parameters to be optimized*

Many processes involving biology, physics, and chemistry are considered in the marine ecosystem dynamics model with many parameters, which make the solution space of the parameter optimization problem huge. It not only requires sufficient observation data but also demands strong searchability of the optimization algorithm. However, since ocean observation is time-consuming and labor-intensive, the amount of general observation data cannot meet the needs of the solution, and the existing search algorithms have not been developed to meet the search of solution space with high dimensionality. It can be seen that reducing the difficulty of the optimization problem by reducing the dimension of the solution space is an effective solution. Reducing the dimension of the solution space to the greatest extent without affecting the optimization effect has become the primary issue for model parameter optimization. Although the marine ecosystem dynamics model has many parameters, the study demonstrated that there were fewer parameters that significantly impacted the simulation results. Therefore, through the analysis of the influence of the parameters on the simulation results, several parameters with greater influence can be selected as parameters to be optimized. These parameters to be optimized are generally referred to as control parameters, which can reduce the solution space dimension. Parameter sensitivity analysis is an effective method for analyzing the effect of parameter changes on the results, which has been considered an important basis and key step for parameter optimization and model calibration of marine ecosystem dynamics models (Fennel *et al.*, 2001; Solidoro *et al.*, 2003; Xu *et al.*, 2004; Shi *et al.*, 2014). According to whether the interaction between parameters is considered, the sensitivity analysis is divided into two types: local sensitivity analysis and global sensitivity analysis.

(1) *Local sensitivity analysis*

The local sensitivity analysis only examines the influence of the change of a single parameter at a fixed value on the model results, and only the center values of other parameters are selected (Solidoro *et al.*, 2003; Xu *et al.*, 2004; Shi *et al.*, 2014). This method has a simple idea and strong operability that is widely used in the field of marine ecosystem dynamics models (Solidoro *et al.*, 2003; Shi *et al.*, 2014). However, this method suffers from certain limitations:

(a) The initial value of the model parameters, which directly affects the sensitivity result.
(b) The amount of parameter disturbance.
(c) The artificially selected and fixed amount of parameter disturbance may cause some parameters of the model to exceed the value range, and fail to meet the actual meaning of the parameters.
(d) Although there may be a nonlinear relationship between the parameters of the model, only the effect of a single parameter change on the model results is considered in the local sensitivity while the impact of parameter interaction on results is not tested (Xu *et al.*, 2004; Shi *et al.*, 2014).

The parameters of marine ecosystem dynamics models generally have a nonlinear relationship in which two parameters with strong correlations cannot generally be effectively optimized simultaneously (Friedrichs *et al.*, 2007), thus affecting the optimization efficiency. Local sensitivity analysis fails to identify highly relevant parameters, thus cannot be used as a scientific basis for the selection of model optimization parameters. Some studies have combined the results of local sensitivity with statistical analysis methods in an attempt to screen out appropriate control parameter sets based on this protocol (Solidom *et al.*, 2003; Shi *et al.*, 2014). Although the effectiveness of this method is still to be validated, it also avoids the correlation between the parameters to a certain extent, which is a feasible method in the case of limited calculation costs.

(2) *Global sensitivity analysis*
Global sensitivity analysis tests the effect of multiple parameter changes within the corresponding value range on the model results simultaneously. Analyzing the parameter sensitivity as a whole has certain advantages:

(a) The parameter changes in the entire value range are generally independent of the selected initial value and do not exceed the value range.
(b) The change of the model result is caused by the combined action of all parameter changes. Thus, the resulting parameter sensitivity is an overall outcome. Global sensitivity analysis not only filters the parameters that have a greater impact on the model results but also can identify the parameters that have a strong correlation, thereby further reducing the number of control parameters. This method has been gradually applied in marine ecosystem dynamics models (Zheng *et al.*, 2012; Chu *et al.*, 2007; Cossarini and Solidoro, 2008; Gibson and Spitz, 2011; Morris *et al.*, 2014), but has not been commonly used because its application in marine ecosystem dynamics models is computationally expensive.

Global sensitivity analysis methods include qualitative and quantitative methods. Qualitative global sensitivity analysis qualitatively analyzes the relative magnitude of the uncertainty of the parameters of the model-on-model results. Since this method can obtain the ranking of the sensitivity of each parameter of the model at a relatively low computational cost, it is also known as factor screening sensitivity analysis (Xu *et al.*, 2004). Qualitative global sensitivity analysis methods include the Fourier amplitude sensitivity test (FAST) (Cukier *et al.*, 1973), multiple regression methods (McKay *et al.*, 1979), Morris method (Morris, 1991), and so on. Among them, the Morris method is a discrete search based on the parameter space, in which the sampling design is ingenious and the calculation cost is low. Therefore, this protocol is widely used, also in the application of marine ecosystem dynamics models (Zheng *et al.*, 2012; Cossarini and Solidoro, 2008; Morris *et al.*, 2014).

Quantitative global sensitivity analysis uses the variance of the model result to reflect the uncertainty of the model result. According to the decomposition of the model variance, the contribution rate of the uncertainty of each parameter to the uncertainty of the model result is calculated quantitatively, and the uncertainty of the model parameter can be analyzed accordingly (Xu *et al.*, 2004). Quantitative global sensitivity analysis includes the Sobol method (Sobol, 1993) and the Fourier amplitude sensitivity test extended method (Saltelli *et al.*, 1999), which combines the Sobol method and the Fourier amplitude sensitivity test method. Among them, the Sobol method decomposes the model into a single parameter and a function of the combination of parameters, whereas the Monte-Carlo method is used to calculate the sensitivity of each order of the model parameters (Sobol, 1993; Xu *et al.*, 2004) but is seldom used in marine ecosystem dynamics models (Chu *et al.*, 2007; Gibson and Spitz, 2011; Morris *et al.*, 2014). Since the quantitative global sensitivity analysis has a large number of calculations, for complex models such as marine ecosystem dynamics models that have a large number of calculations, qualitative methods are often used to exclude parameters that hardly affect the model results, and then quantitative methods are used to analyze the contribution rate of these parameters on model uncertainty. (Chu-Agor *et al.*, 2011; Morris *et al.*, 2014).

4.2.2 *Selection of model parameter optimization methods*

Fundamentally, the optimization of model parameters is to obtain the model parameters by using inversion of existing observation data, which is classified as data assimilation in the field of meteorology and oceans. The data assimilation method is often used in ocean numerical simulation to optimize the initial conditions, boundary conditions, and model parameters of the model. It was initially applied to the assimilation and prediction of data such as the ocean tidal field and sea temperature. Since Ishizaka (1990) pioneered the assimilation of a three-dimensional model using chlorophyll observation data, marine ecosystem data assimilation has gradually emerged (Wang *et al.*, 2007; Gregg *et al.*, 2009). According to the different application orders of observation data information, data assimilation methods can be divided into two types: inverse method and sequential method (Bouttier *et al.*, 1999; Gregg *et al.*, 2009). The inverse method directly minimizes the cost function to obtain the parameter set that minimizes errors, including the steepest descent method, conjugate gradient method, variable adjoint, and metaheuristic algorithms such as simulated annealing and genetic algorithm. The sequential method re-initializes the model by assimilating and correcting the output results when the observation data is available, including the Nudging method, optimal interpolation, and the Kalman filter and its variants. After using summarized data assimilation methods in the field of marine biology, Gregg *et al.* (2009) put forward the notion that the reverse method has a long history and wider scope that is mostly used for parameter optimization, whereas the sequential method has gradually increased since 2000 and is mostly used for state variable estimation. Thus, parameter optimization is actually a means of data assimilation.

Comparing the pros and cons of optimization algorithms is to examine the skill of optimization algorithms, which mainly include optimization model accuracy and optimization algorithm efficiency that depend on the convergence and convergence speed of the algorithm, respectively. In the study of parameter optimization, it was found that currently used optimization algorithms tended to converge to the local optimum during the search. For example, the widely used variational adjoint and Kalman filter are part of the local search in algorithm design, which cannot effectively solve the problem of multi-extremum optimization in marine ecosystem dynamics, thus leading to low accuracy of the optimization model. This phenomenon has led people to seek global optimization from the global optimization capabilities of algorithms. Optimization algorithms such as simulated annealing and genetic algorithms with a certain degree of global search capabilities have gradually attracted research attention.

(1) *Common methods for the local optimization of parameter*
Currently, the optimization of marine ecosystem dynamics model parameters is mainly based on local optimization. Firstly, this is because research on the optimization of marine ecosystem dynamics model parameters was conducted relatively late, whereas the application of local optimization algorithms is relatively mature. Some models with a low nonlinear degree can still use local optimization algorithms to improve the accuracy of the model to a certain extent. Secondly, the research on global optimization algorithms is not yet mature, and algorithms with global convergence tend to converge slowly, resulting in higher calculation costs. Local optimization methods include two types of protocols: gradient-based numerical and direct search, depending on the analytic properties of the optimization problem (Tashkova *et al.*, 2012). Reverse methods such as the steepest descent method, conjugate gradient, and a variational adjoint other than sub-heuristic algorithms generally belong to gradient-based numerical methods. These algorithms have high optimization efficiency and are the most widely used types of optimization algorithms. However, these methods cannot be used for optimization problems with discontinuous, non-smooth, polymorphic, or morbid objective functions. The sequential methods such as the optimal interpolation method and Kalman filtering method generally belong to direct search methods, which do not require high analytic properties of the objective function but are inefficient when dealing with high-dimensional optimization problems. The local optimization algorithm has been widely used as an algorithm with earlier development time and mature development.

Since the optimization efficiency of most global optimization algorithms depends on local search capabilities, research on local optimization methods also contributes to the development of global optimization methods. At present, optimization algorithms applied in the field of marine ecosystem dynamics are mainly variational adjoint and Kalman filtering (Wang *et al.*, 2007; Gregg *et al.*, 2009). The variational method was first used by Sasaki (1970). Variational adjoint organically combines observational data with marine ecosystem dynamics models by employing the variational method and adjoint equations and adjusts the control parameters based on the gradient information of the objective function on the control parameters by using optimization algorithms. Bennett and Mcintosh (1982) first used the variational method to study the inverse problem of tidal models in open seas. Lawson *et al.* (1995) first used it to invert the parameters and initial field of a simple predation-prey marine ecological model, which was then widely used in marine ecosystem dynamics models (Gregg *et al.*, 2009; Ward *et al.*, 2010). Variational adjoint is heavily dependent on the initial set of parameters and the temporal and spatial scale of observation data, which cannot be optimized when parameter gradient information is not available, thus limiting the application in the field of marine ecosystem dynamics models.

Kalman filtering improves the model prediction ability by continuously updating the model prediction field with new observation data as the initial field at the next moment, which can be regarded as an extension of the optimal interpolation method. This method is different from other data assimilation methods in that it can simultaneously predict and analyze model state variables and errors. Kalman filtering was proposed by Kalman in 1960 (then known as standard Kalman filtering). Jones (1965) first introduced the method to meteorology. Since then, it has been widely used in meteorology and oceans. With the improvement of model nonlinearity, extended, simplified, and ensemble Kalman filters are gradually put forward to improve the convergence and stability of Kalman filters for nonlinear systems (Wang *et al.*, 2007). In particular, ensemble Kalman filtering was proposed by Epsteines (1969) based on the theory of stochastic dynamics prediction, which was applied to data assimilation by Evensen (1994). This method uses ensemble statistics to estimate the covariance of error, which can be used for the optimization of a strong nonlinear model and has been widely used in the field of marine ecosystem dynamics models (Wang *et al.*, 2007; Gregg *et al.*, 2009).

(2) *Global parameter optimization method in exploration*

In order to effectively deal with multi-extremum optimization problems such as marine ecosystem dynamics models and improve the accuracy of model parameter optimization, the application of global parameter optimization methods in marine ecosystem dynamics model optimization has been gradually investigated. According to whether the state variables of the optimization problem are considered random, the current global optimization methods are divided into two categories: deterministic and stochastic (Li, 2012). This method of processing randomness of state changes requires different degrees of the analytic nature of the optimization problem in algorithm design. The deterministic global optimization method uses the analytic nature of the optimization problem to generate a deterministic finite or infinite point sequence to converge (or its subsequence converge) to the global optimum, which can theoretically guarantee the globality of the optimal solution. This type of method needs to design specific algorithms based on the analytic nature of different optimization problems. Since the converging point sequence may be infinite, it cannot guarantee that any type of global optimization problem can be solved in a limited time, and the amount of calculation increases rapidly with the increase of the problem scale. Although this method has been applied to simpler nonlinear dynamic system models in the fields of biology and chemistry (Esposito and Floudas, 2000; Miro *et al.*, 2012), it is not suitable for complex models (Li, 2012; Athias *et al.*, 2000).

The stochastic global optimization method cannot theoretically guarantee the globality of the optimal solution, and the degree of deviation between the obtained optimal solution and the global optimization solution is unpredictable, but the method relies on the probability search principle and is relatively effective in the acceptable calculation cost to locate the adjacent location to the global optimal solution so as to obtain an approximate global optimal solution (or quasi-global optimal solution) (Li, 2012; Athias *et al.*, 2000; Tashkova *et al.*, 2012). This method is a direct search which does not require knowing the analytic properties of the problem, nor does it require the problem to have explicit analytic expressions. It has an irreplaceable role in the optimization of complex model parameters (Li, 2012; Athias *et al.*, 2000; Tashkova *et al.*, 2012). Stochastic methods are mainly composed of a class of meta-heuristic algorithms combining local search and stochastic methods. Most of them are proposed by imitating natural phenomena or social behaviors, which mainly include evolutionary algorithms, simulated annealing, swarm intelligence, tabu search (TS), and so on (Table 4.1). The current

Table 4.1. The development and application of meta-heuristic algorithms.

Name	Established	Founder	Meaning
Evolutionary Algorithm			
Genetic Algorithm	1967	[US] J H Holland	The bionic algorithm based on the Darwin theory of natural evolution and Mendel theory of genetic variation
Evolution Strategies	1965	[Germany] I Recenberg and H P Schwefel	Emphasis on behavioral changes at the individual level, rather than genetic algorithms emphasizing the manipulation of chromosomes
Evolutionary Programming	1960s	[US] L J Fogel	Emphasizing changes in the behavior of the cluster
Genetic Programming	1992	[US] Koza	Having a tree topology of individual structure, not a character string of genetic algorithms that can represent independent computer programs, algorithmic, or logical relationships between variables
Simulated Annealing	1982	Kirkpatrick	Simulating the physical process of cooling of metals
Tabu/Taboo Search	1986	Glover	Simulating the intelligent process with memory
Swarm Intelligence			
Ant Colony Optimization	1991	[Italty] Dorigo	Biological community behavior, the process of an ant colony searching for food
Particle Swarm Optimization	1995	Kennedy and Eberhart	Migration and cluster behavior in the foraging of birds
Harmony Search	2001	[Korean] Z W Geem	Simulating the process of musicians adjusting the pitch of a musical instrument to a harmonious level

stochastic global optimization method has been widely used in river hydrology and water quality models, but there are few reports on the application of marine ecosystem dynamics model parameter optimization problems. The methods that have been tried in a marine ecosystem dynamics model are mainly genetic algorithms (Weber *et al.*, 2005; Oschlies and Schartau, 2005; Schartau and Oschlies,

2003; Ward *et al.*, 2010), followed by simulated annealing (Matear, 1995; Hurtt and Armstrong, 1996; 1999; Kavanagh *et al.*, 2004).

As the model to be optimized becomes more nonlinear and its structure becomes more complex, the commonly used meta-heuristic algorithms such as genetic algorithms and simulated annealing gradually exhibit various problems such as poor local search ability, local precocious convergence, and an unsatisfactory convergence rate. Therefore, the algorithm needs to be improved correspondingly. For example, a micro-genetic algorithm based on small populations and re-initialization operations can overcome the shortcomings of general genetic algorithms that require large-group operations when the dimension of the search space is high to avoid premature convergence that causes large computational resource and time consumption. Therefore, the micro-genetic algorithm is a genetic algorithm that has been commonly used in the optimization of marine ecosystem dynamics models (Weber *et al.*, 2005; Oschlies and Schartau, 2005; Schartau and Oschlies, 2003; Ward *et al.*, 2010). Since there is no complete theory to explain the convergence and convergence rate of different algorithms, the comparison of the skill of different algorithms often depends on numerical experiments. As the optimization efficiency of this kind of algorithm depends on the nature of the optimal solution and number of calculations, the results of different case studies are different. Recent studies show that micro-genetic algorithms are better than simulated annealing, and evolutionary strategies are better than genetic algorithms. For example, Athias *et al.* (2000) employed complex nonlinear marine biogeochemical models to simulate ocean particle system cycles, which indicated that micro-genetic algorithms had lower computational cost, faster convergence speed, and a more robust optimal solution than simulated annealing. After investigating the optimization of nonlinear dynamic biochemical pathway model parameters by different algorithms, Moles *et al.* (2003) considered an evolutionary strategy to be the most competitive stochastic optimization algorithm. Huret *et al.* (2007) applied evolutionary strategies to the phytoplankton production model in the Bay of Biscay and achieved good results. In addition to being determined by the nature of the model, the optimization efficiency of the algorithm is greatly affected by the amount of calculation needed. For example, Zhang *et al.* (2009), after investigating the parameter optimization of complex hydrological models by using a genetic algorithm, particle swarm optimization, differential evolution, artificial immune system optimization and other algorithms, concluded that a genetic algorithm was optimal when computing resources were sufficient, whereas particle swarm optimization was optimal when computational cost was limited.

Current research on random global optimization methods mainly focuses on the implementation, improvement, and application of the algorithm, but their theoretical research has always been a challenging topic. There is no complete theory to clarify its convergence or properly characterize its convergence speed, so it cannot accurately evaluate various algorithms or improve their skill. The evaluation of algorithm skill is primarily based on a large number of numerical experiments. Due to the high computational cost of the marine ecosystem dynamics model and late development of parameter optimization research of this model, there are few studies on the skill of different algorithms, which leads to the uncertainty of the optimization efficiency of different algorithms for marine ecosystem dynamics models. This topic is also the future research direction of the author's research group.

4.3 Model Skill Evaluation for Model Validation

4.3.1 *Different methods for estimating the goodness-of-fit of assimilated models*

Model verification generally analyzes and estimates the goodness-of-fit of the assimilated model (i.e., optimized model) to validate the effect of model calibration. Here, the goodness-of-fit refers to the degree of matching between model simulations and observations, which is the skill evaluation of the assimilated model. There is no other established method for model verification. Usually, the modeler tries to illustrate the accuracy of the model simulation from different perspectives. The most common form is a visual comparison using graphical analysis (Gregg *et al.*, 2009; Stow *et al.*, 2009). According to the meaning of each coordinate in the figure, the matching degree can be analyzed from different angles such as through the comparison of time-series plots of simulated and observed data (Arhonditsis and Brett, 2004; Stow *et al.*, 2009).

In addition, binary diagrams and residual distribution maps of simulation data and observation data are also common methods (Stow *et al.*, 2009). Also, the spatial map comparison method is used to investigate the spatial matching between simulation and observation (Rose *et al.*, 2009). Graphical analysis is generally a qualitative description of the degree of fit. Quantitative methods mainly refer to statistical analysis of simple metrics such as the correlation coefficient and root mean squared error (Gregg *et al.*, 2009; Stow *et al.*, 2009).

Because of the cost of calculations, the data available, and the purpose of modeling, most models in general only calibrate a few variables of interest. There are different degrees of correlation between multiple state variables in the marine ecosystem dynamics model. If the variables of interest to the modeler are optimized, other variables are affected, resulting in calibration bias (Arhonditsis and Brett, 2004). Modelers are reminded to consider the verification of unassimilated variables in order to examine the response of unassisted variables to parameter optimization (or data assimilation) (Gregg *et al.*, 2009). During verification, the errors between simulation and observation of all variables are compared simultaneously, and the mutual influence of each spatiotemporal variable can be considered, which is helpful to analyze the global matching characteristics of the model. By reference to the analysis method of individual variables, multi-comparison can be used by multiple statistical analyses and the cost function calculation of multivariate error. The latter is a more complex measure (Stow *et al.*, 2009). This multivariate comparison method is more specific in revealing the calibration effect of the assimilated model obtained by using the multi-objective optimization algorithm.

Due to insufficient attention to model verification in the past, model verification is generally based on visual comparison of time series plots, but quantitative verification applications are rare. After reviewing 153 aquatic biogeochemical models published from 1990 to 2002, Arhonditsis and Brett (2004) found that only 30.1% of the studies carried out model fit calculations; others only showed comparison charts of basic simulation and observation data. The subsequent analysis showed that there was almost no correlation between model skill level and the citation rate of the paper (Arhonditsis *et al.*, 2006).

In addition, in the marine field, Stow *et al.* (2009) reviewed 142 ocean-system biophysical coupling models published between 2000 and 2007 and found that only 31.7% of them quantitatively described the goodness-of-fit. As managers and decision-makers have higher requirements for models, model verification becomes more necessary and important.

4.3.2 *Different validation data for skill evaluation of assimilated models*

In addition to validation methods, the different sets of observations used for validation (validation data) can also reflect different skills of the model. According to whether the validation data is within the space-time range covered by the calibration data (that is, the observation data used for calibration, also known as assimilation data), different levels of the assimilated model's skills can be evaluated.

Replicative and predictive performances are the two levels of skill of a model. These two skills generally refer to a model's ability to replicate and predict patterns using the data provided (Arhonditsis and Brett, 2004). A model with good replicative skill is able to replicate observations, and can achieve high accuracy in the space-time range covered by the calibration data. This mainly depends on the skill of the optimization algorithm and calibration data. The data used for replicative calibration can be a sub-data set from the same data set as the validation data, or it can be a data set obtained using different observation methods in the same space-time range as the validation data, such as satellite remote sensing data and in situ observations. A model's level of predictivity refers to the prediction capabilities of the assimilated model for the area or time range beyond the coverage of the validation data, which is a data set that is independent of the validation data. The verification result is determined by the optimizing skill and the overall performance of the model. In the case of predictive verification, when the independent data sets used for verification are extracted from different research areas or different physical environments (in general, the model is built for a specific area), the skill of the model obtained by verification is also called the transferability or portability of the model, which is used to examine whether the structure of this type of optimized model can be extrapolated to different ecosystems (Arhonditsis and Brett, 2004; Friedrichs *et al.*, 2007). Models with good predictive capabilities are considered to have high reliability and robustness, which can provide scientific support for decision making — all crucial requirements for a basic tool for marine environmental management. That is why improving a model's skill is becoming a hot topic of current research (Gregg *et al.*, 2009).

A model with good predictive performance should generally also have good replicative performance, but not vice versa. Overfitting, equifinality, and other effects endow the model to exhibit high replicative performance but poor predictive capabilities performance. Overfitting refers to a situation in which the model error can be relatively arbitrarily small, such that though the model is highly replicative, it achieves this by making the free parameters (relative to the control parameters) take excessively large or excessively small values that do not conform to reality, which will inevitably reduce the predictive performance. Equifinality is a relatively widespread phenomenon in basin hydrological modeling (Beven and Binley, 1992) which also appears in the study of marine ecosystem dynamics parameter optimization. In particular, when different optimization algorithms are used at the same time, the model accuracies are almost the same even when using different parameter sets (Ward *et al.*, 2010).

Predictive verification can determine whether the assimilated model is overfitting, which can also choose a scientific and reasonable equivalent parameter set. Although the most direct reason for the emergence of phenomena such as overfitting and equifinality is that the termination criteria of optimization algorithms (especially random global optimization algorithms) do not have a sound theoretical basis; this is mainly due to the diversity and complexity of the uncertainty of marine ecosystem dynamics models in the final analysis. It can be seen that model predictive verification is an important content of marine ecosystem dynamics model skill evaluation.

4.4 Review on Skill Enhancement of Marine Ecosystem Dynamics Models

4.4.1 *The complexity of marine ecosystem dynamics models requires global treatment*

The internal relationships of marine ecosystems are complex, which are also impacted by global changes and human activities. Marine ecosystem dynamics models generally involve physical, chemical, and biological processes simultaneously, so that multiple parameters and state variables of marine ecosystem dynamics models often have different degrees of nonlinear correlation. This characteristic of the marine ecosystem dynamics model requires a global perspective to understand and improve the model skill on the whole when dealing with model parameter optimization and skill evaluation.

Specifically, the global processing of model parameters includes two aspects. First, to reduce the solution space for model parameter optimization, the parameters with a large degree of influence and low correlation with each other have been proven to be the best set of parameters for optimization. Therefore, in addition to analyzing the degree of influence of the parameters on the model results, the correlation among the various parameters should be considered during the selection of control parameters. The method of local sensitivity analysis combined with multivariate statistics with low calculation cost can obtain the control parameter set to a certain extent, but its validity has yet to be tested. On the other hand, the influence of each parameter and its interaction on the model results is considered simultaneously in the global sensitivity analysis method. Since this protocol can analyze the sensitivity of the parameters in the global category, it should be the basic method for selecting the control parameters.

Second, the complexity of the marine ecosystem dynamics model often leads to the objective function of the optimization problem belonging to a multi-peak function, which in turn makes the optimization problem exhibit multiple extremalities and requires the optimization algorithm to have a global search capability for the solution space. However, current global algorithm research has not been able to effectively implement global optimization capability in complex systems. Nevertheless, it is presently an effective method to achieve global optimization of complex problems through the use of a random search to eliminate local convergence and obtain an approximate global optimal solution. Most of the current marine ecosystem dynamics model parameter optimization uses variational adjoint and other algorithms that are local optimization algorithms by design. Although these methods are able to improve the accuracy of the model to a large extent, the model's predictive ability and model application are limited to a certain extent. Stochastic global optimization algorithms based on the design concept of imitating natural phenomena or social behavior have been exploratively applied in the optimization of marine ecosystem dynamics models. However, since such algorithms have a relatively late origin, their advantages in global optimization have not yet been widely recognized. Because the evaluation of algorithm skill is based on a large number of numerical experiments, the comparison and improvement of such global optimization algorithms in the field of marine ecosystem dynamics models should be reinforced.

In addition, regarding the state variables of marine ecosystem dynamics models, the observation data available to most researchers are not sufficient to support models with relatively high complexity. In recent years, efforts to increase the complexity of models have proven to insignificantly promote the improvement of model skill. The current commonly used marine ecosystem dynamics models are mainly based on lower-trophic-level models such as classic nutrient-phytoplankton-zooplankton (NPZ), to which organic debris, bacteria, dissolved organic matters, algae, fish, and other creatures are added according to the interests of the modeler. It can be seen that the number of state variables in the marine ecosystem dynamics model is relatively small, far less than the number of parameters. Therefore, the correlation among the various state variables is far less complicated than the parameters, and consideration of a single state variable, which is a local processing protocol, can be used to obtain a better understanding. However, with the increasing complexity of the practical problems that need to be applied to marine ecosystem dynamics models, research attention on marine ecosystems has gradually increased from lower to higher trophic levels, and different functional groups have gradually been divided into many different types for lower trophic level models. This trend makes the number of state variables of the marine ecosystem dynamics model gradually increase with a more complicated relationship. Hence, global processing based on multi-objective optimization and multi-dimensional comparison become necessary.

4.4.2 *Full use of observation data to improve model skill*

Observation data is a critical part of model research and application. Due to the high cost of observation data, most modelers do not have sufficient data for model calibration and validation, which is one of the reasons for the relatively slow development of marine ecosystem dynamics model skill research. Learning how to make full use of limited observation data to effectively reveal and improve a model skill is an issue that modelers must consider.

In addition to the indispensable observation data used to set the external physical drive of the model, the initial conditions, and boundary conditions of various state variables, model calibration requires the modeler to have at least one data set that contains different time and space within a certain research scope. When the modeler has only one set of data, if it is used for data calibration, there is no data available for model verification. As a result, it is impossible to provide a quantitative description to explain the skill of the assimilated model. Thus, a subset of the data set should be reserved for model verification to evaluate the replicative skill of the assimilated model. When adopting this approach, the modeler should reasonably allocate two subsets to ensure the validity of corresponding calibration and verification processes. Gregg *et al.* (2009) suggest that the entire data set be used for model calibration to obtain the optimal parameter set under the available data, which is then divided into two subsets of the same size for calibration and verification, so as to compare their effects.

When the modeler has two independent data sets that are used for calibration and verification separately, the modeler should examine multiple measures at the same time and combine them with graphical and quantitative methods to ensure the comprehensive reveal of the skill of the assimilated model. In addition to using the two data sets for calibration and verification separately, they can also be flexibly used to optimize and evaluate model skill with different calibration strategies. For example, two sets of data are used for calibration respectively to obtain two sets of optimized parameter sets, and then another set is used for verification. This method is generally called cross-validation. Alternatively, two independent sets are used at the same time for model calibration to obtain a set of parameters. This method can be used to obtain the common parameter set of different regional ecosystems (Friedrichs *et al.*, 2009). In addition, understanding the climatic conditions and physical environment at the time of obtaining available observation data before modeling and analyzing data from existing observations facilitate the understanding of the real world that the model needs to reproduce, thus guiding the model structure design.

Under normal circumstances, the modeler obtains a certain amount of observation data. By comparing the observed data with simulation data, the skill of the optimization algorithm and model to improve the goodness-of-fit is analyzed. In addition, by designing a certain strategy, the skill of the optimization algorithm and model can also be explained only by analyzing and using model-simulated data. This also provides a way for modelers to study model optimization algorithms and model skill without available observations. A more typical application is parameter sensitivity analysis, by which the degree of influence of model parameters is obtained through analysis of simulated data using different parameter values. In addition, the skill of the parameter optimization algorithm can also be analyzed by comparing the simulation results of an assimilated model with an unassimilated model or free-run model. This method generally combines actual observation data to investigate the capability of different optimization algorithms to improve the accuracy of the model. The typical application in the field of data assimilation is to use the model simulation data as the assimilation data for model calibration, which is referred to as identical twin experiments. Through the identical twin test, the skill of different optimization algorithms can be directly compared and tested without observing the data, and the influence of assimilation data changes on the optimization effect can be also investigated to guide the design of observation sampling strategies (Friedrichs, 2002).

4.4.3 *Investigate the uncertainty of the model to quantitatively analyze the influencing factors of model optimization*

The current technical measures to improve the skill of the model have basically enabled the model to meet the application requirements in terms of "reproducing real natural systems". Since the marine ecosystem dynamics model will be affected by many factors in the actual application process, the reliability and robustness of the model during application are important concerns for model users (i.e., managers and decision-makers). This concern puts forward high requirements on the predictive skill of the model. Because model skill is affected by many factors, errors in various aspects will cause model uncertainty, which will lead to predictive uncertainty. By analyzing the influence of different factors on the model's predictive skill model, uncertainty analysis can provide model users with quantitative model confidence, which can help promote the model application and is a critical part of model skill evaluation.

According to the factors affecting model skill, the sources of model uncertainty include the following four aspects:

(1) Model parameter uncertainty, which originates from parameter optimization algorithms and assimilation data, and the spatiotemporal heterogeneity of the parameters.
(2) Model structure uncertainty, which mainly originates from insufficient knowledge of actual problems and conceptualization and simplification of natural processes during modeling.
(3) Uncertainty of model solving, namely, the approximate error of the numerical algorithm used to solve the model by the computer.
(4) Uncertainty of input data such as external driving, initial conditions, and boundary condition comes from data observation errors and the randomness of natural processes (Gregg *et al.*, 2009).

Previous studies show that the uncertainty of marine ecosystem dynamics models is mainly caused by observation errors and model parameters. Therefore, the uncertainty analysis of marine ecosystem dynamics models mainly focuses on the uncertainty of model inputs and parameters, and the predictive uncertainty caused by the uncertainty of model inputs and parameters, and so on.

Current research on uncertainty analysis of marine ecosystem dynamics models overlaps with calibration and verification (Wallhead *et al.*, 2009). For example, according to the contribution rate of the parameters and their interactions obtained from the quantitative global sensitivity analysis to the uncertainty of the model, the uncertainty of the model parameters is described (Gibson and Spitz, 2011; Morris *et al.*, 2014). Based on the idea of sensitivity analysis, the uncertainty of the input data to model prediction is analyzed by setting up a certain degree of perturbation of the input data (Friedrichs *et al.*, 2009). In addition, the Bayesian method characterized by considering prior information is regarded as a good choice for model uncertainty analysis (Ward *et al.*, 2010). This type of method was first proposed in the field of basin hydrology and water quality models and has been widely used. For example, the earliest proposed generalized likelihood uncertainty estimation (GLUE) (Beven *et al.*, 1992), the widely used Markov chain Monte Carlo (MCMC) (Kuezera and Parent, 1998), and so on. Currently, these methods have also been tried in the marine field, but it is basically an application of simple models. Stow and Scavia (2009) combined Bayesian stratification with a simple process-based model to study hypoxia in the Chesapeake Bay and quantified the predictive uncertainty caused by parameters and model inputs. The Bayesian method combines *priori* and *posteriori* information, which can comprehensively analyze the uncertainty of the model and guide the reduction of the uncertainty of the model prediction. However, the application of the Bayesian method in marine ecosystem dynamics modeling has yet to be explored (Ward *et al.*, 2010).

4.5 Summary

The improvement and evaluation of model skill is the premise of model application. Although current research on marine ecosystem dynamics models has achieved some results, the improvement of model skill has not yet been effectively addressed from a wider perspective. In addition, current means of model skill evaluation are not sufficiently comprehensive, and observation data have not been fully utilized to optimize the model, which causes the model's predictive skill to be unreliable, and its uncertainty lacks quantitative description, thus limiting the application of marine ecosystem dynamics models. This chapter summarizes the methods and strategies for improving and evaluating the skill of marine ecosystem dynamics models, and attempts to construct a logical framework to inspire research ideas rather than cover all related issues. The purpose of this chapter is to make modelers pay more attention to the study of model skill in the process of model research, so that they can form a technical framework for improving the skill of marine ecosystem dynamics models in a scientific and standardized way.

Practical model research should take into account scholarly advancement and regulatory practice. A global model optimization scheme can significantly improve the skill of marine ecosystem dynamics models in theory. When the modeler cannot support relatively large calculation costs or is not familiar with the algorithm, the inversion modeling method based on the inverse problem method and the traditional local processing method can be regarded as an effective method. This protocol prompts the modeler to carry out further research or use the assimilated model for decision-making with low risk. In addition, most of the content of model validation is mainly to serve the practical application of the model, for scientific researchers only need to have an objective understanding of model performance.

From this study, it can be concluded that whether the study of the marine ecosystem dynamics model requires in-depth research on model skill should mainly depend on the modeling purpose of the modeler rather than the personal interest and ability of the modeler. Currently, there are relatively few studies on the skill of marine ecosystem dynamics models, and scientific and effective model skill improvement techniques are based on a large number of numerical experiments. Therefore, modelers should strengthen their research on model skill as much as possible to promote the standardization of dynamic model skill improvement techniques in the field of marine ecology.

Acknowledgment

The reference (Shen *et al.*, 2016) was reproduced with permission from Springer Nature.

References

Arhonditsis G B, Brett M T. 2004. Evaluation of the current state of mechanistic aquatic biogeochemical modeling. *Marine Ecology Progress Series*. **271**: 13–26.

Arhonditsis G B, Adams-Vanharn B A, Nielsen, L, *et al.* 2006. Evaluation of the current state of mechanistic aquatic biogeochemical modeling: Citation analysis and future perspectives. *Environmental Science and Technology*. **40**(21): 6547–6554.

Athias V, Mazzega P, Jeandel C. 2000. Selecting a global optimization method to estimate the oceanic particle cycling rate constants. *Journal of Marine Research*. **58**: 675–707.

Bennett A F, McIntosh P C. 1982. Open modeling problem: Tidal theory. *Journal of Physical Oceanography*. **12**(10): 1004–1018.

Beven K J, Binley A. 1992. The future of distributed models: Model calibration and uncertainty predictive. *Hydrological Processes*. **6**(3): 279–298.

Bouttier F, Courtier P. 1999. Data assimilation concepts and methods. *Meteorological Training Course Lecture Series.* ECMWF.

Chu P C, Ivanov L M, Margolina T M. 2007. On-linear sensitivity of marine biological models to parameter variations. *Ecosystem Modelling.* **206**(3): 369–382.

Chu-Agor M L, Munoz-Carpena R, Kiker G, *et al.* 2011. Exploring vulnerability of coastal habitats to sea level rise through global sensitivity and uncertainty analyses. *Environmental Modelling & Software.* **26**(5): 593–604.

Cossarini G, Solidoro C. 2008. Global sensitivity analysis of a trophodynamic model of the Gulf of Trieste. *Ecosystem Modelling.* **212**: 16–27.

Cukier R I, Fortuin C M, Shuler K E, *et al.* 1973. Study of the sensitivity of coupled reaction systems to uncertainties in rate coefficients. I Theory. *Journal of Chemical Physics.* **59**(8): 3873–3878.

Epsteines E S. 1969. Stochastic dynamic prediction. *J Tellus.* **21**: 739–759.

Esposito W R, Floudas C A. 2000. Deterministic global optimization in nonlinear optimal control problems. *Journal of Global Optimization.* **17**(1–4): 97–126.

Evensen G. 1994. Sequential data assimilation with a nonlinear quasigeos trophic model using Monte Carlo methods statistics. *Journal of Geophysical Research.* **99**: 1043–1062.

Fennel K, Losch M, Schroter J, *et al.* 2001. Testing a marine ecosystem model: Sensitivity analysis and parameter optimization. *Journal of Marine Systems.* **28**: 45–63.

Flynn K J. 2005. Castles built on sand: Dysfunctionality in plankton models and the inadequacy of dialogue between biologists and modelers. *Journal of Plankton Research* **2**(12): 1205–1210.

Friedrichs M A M. 2002. Assimilation of JGOFS EqPac and SeaWiFS data into a marine ecosystem model of the Central Equatorial Pacific Ocean. *Deep Sea Research Part II. Topical Studies in Oceanography.* **49**(1–3): 289–319. DOI: http://dx.doi.org/10.1016/S0967-0645(01)00104-7.

Friedrichs M A M, Carr M-E, Barber R T, *et al.* 2009. Assessing the uncertainties of model estimates of primary productivity in the tropical Pacific Ocean. *Journal of Marine Systems.* **76**(1–2): 113–133. DOI: http://dx.doi.org/10.1016/j.jmarsys.2008.05.010.

Friedrichs M A M, Dusenberry J A, *et al.* 2007. Assessment of skill and portability in regional Marine biogeochemical models: The role of multiple planktonic groups. *Journal of Geophysical Research.* **112**: C08001. DOI: 10.1029/2006JC003852.

Gibson G A, Spitz Y H. 2011. Impacts of biological parameterization, initial conditions, and environmental forcing parameter sensitivity and uncertainty in a marine ecosystem model for the Bering Sea. *Journal of Marine Systems.* **88**(2): 214–231.

Gregg WW, Friedrichs MAM, Robinson AR, *et al.* 2009. Skill assessment in ocean biological data assimilation. *Journal of Marine Systems.* **76** (1–2): 16–33. DOI: http://dx.doi.org/10.1016/j.jmarsys 2008.05.006.

Hood R R, Laws E A, Armstrong R A, *et al.* 2006. Pelagic functional group modelling: Progress, challenges and prospects. *Deep-Sea Research II.* **53**: 459–512.

Huret M, Gohin F, Delmas D, *et al.* 2007. Use of SeaWiFS data for light availability and parameter estimation of phytoplankton production model of the Bay of Biscay. *Journal of Marine Systems.* **65**(1): 509–531.

Hurtt G C, Armstrong R A. 1996. A pelagic ecosystem model calibrated with BATS data. *Deep-Sea Research II.* **43**: 653–683.

Hurtt G C, Armstrong R A. 1999. A pelagic ecosystem model calibrated with BATS and OWSI data. *Deep-Sea Research II.* **46**: 27–61.

Ishizaka J. 1990. Coupling of coastal zone color scanner data to a physical-biological model of the southeastern United States continental-shelf ecosystem. 3. Nutrient and phytoplankton fluxes and CZCS data assimilation. *Journal of Geophysical Research.* **95**: 20201–20212.

Jone S R H. 1965. Optimal estimation of initial conditions for numerical prediction. *Journal of the Atmospheric Sciences.* **22**: 658–663.

Kalman R E. I960. Approach to linear filtering and predict ion problems. *Transactions of the ASME. Series D, Journal of Basic Engineering.* **82D**: 34–45.

Kavanagh P, Newlands N, Christensen V, *et al.* 2004. Automated parameter optimization for Ecopath ecosystem models. *Ecosystem Modelling.* **172**(2): 141–149.

Kuezera G, Parent E. 1998. Monte Carlo assessment of parameter uncertainty in conceptual catchment models: The Metropolis Algorithm. *Journal of Hydrology.* **211**(1-4): 69–85.

Lawson L M, Spitz Y H, Hofmann E E, *et al.* 1995. A data assimilation technique applied to a predator-prey model. *Bulletin of Mathematical Biology.* **57**(4): 593–617.

Li L. 2012. *Global Optimization Algorithm and Its Application.* Qufu: Qufu Normal University.

Matear R J. 1995. Parameter estimation and analysis of ecosystem models using simulated annealing: Station P. *Journal of Marine Research.* **53**: 571–607.

McKay M D, Beckman R J, Conover W J. 1979. A comparison of three methods for selecting values of input variables in the analysis of output from a computer code. *Technometrics.* **21**(2): 239–245.

Miro A, Pozo C, Guillen-Gosalbez G, *et al.* 2012. Deterministic global optimization algorithm based on outer approximation for the parameter estimation of nonlinear dynamic biological systems. *BMC Bioinformatics.* **13**(1): 90.

Moles C G, Mendes P, Banga J R. 2003. Parameter estimation in biochemical pathways: A comparison of global optimization methods. *Genome Research.* **13**(11): 2467–2474.

Morris M D. 1991. Factorial sampling plans for preliminary computational experiments. *Technometrics.* **33**(2): 161–174.

Morris D J, Speirs D C, Cameron A I, *et al.* 2014. Global sensitivity analysis of model of the North Sea: Factors affecting the biomass of fish and benthos. *Ecosystem Modelling.* **273**: 251–263.

Oschlies A, Schartau M. 2005. Basin-scale skill of a locally optimized marine ecosystem model. *Journal of Marine Research.* **63**(2): 335–358.

Pahlow M, Oschlies A. 2009. Chain model of phytoplankton P, N and light colimitation. *Marine Ecology Progress Series.* **376**: 69–83.

Peng S T, Qin X B, Shi H H. 2012. Distribution and controlling factors of phytoplankton assemblages in closed bay during spring and summer. *Marine Pollution Bulletin.* **64**: 941–948.

Rose K A, Roth B M, Smith E P. 2009. Skill assessment of spatial maps for oceanographic modeling. *Journal of Marine Systems.* **76**(1–2): 34–48.

Saltelli A, Tarantola S, Chan K P S. 1999. A quantitative model-independent method for global sensitivity analysis of model output. *Technometrics.* **41**: 39–56.

Sasaki Y. 1970. Some basic formalisms in numerical variational analysis. *Monthly Weather Review.* **98**(12): 875–883.

Schartau M, Oschlies A. 2003. Simultaneous data-based optimization of a 2D-ecosystem model at three locations in the North Atlantic: Part I — Method and parameter estimates. *Journal of Marine Research.* **61**(6): 765–793.

Shen C, Shi H, Liu Y, *et al.* 2016. Discussion of skill improvement in marine ecosystem dynamic models based on parameter optimization and skill assessment. *Chinese Journal of Oceanology and Limnology.* **34**(4): 683–696.

Shi H. 2010. *Construction and Application of Several Marine Ecological Dynamics Models.* Qingdao: Institute of Oceanology, Chinese Academy of Sciences.

Shi H, Cheng C, Li F, *et al.* 2014. Sensitivity analysis of the bio-physical coupling model of the Jiaozhou Bay. *Acta Ecologica Sinica.* **34**(1): 41–49.

Sobol I M. 1993. Sensitivity estimates for nonlinear mathematical models. *Mathematical Modelling Computational Experiments.* **1**(4): 407–414.

Solidoro C, Crise A, Crispi G, *et al.* 2003. An a priori approach to assimilation of ecosystem data in marine ecosystem models. *Journal of Marine Systems.* **40**: 79–97.

Stow C A, Scavia D. 2009. Modeling hypoxia in the Chesapeake Bay: Ensemble estimation using a Bayesian hierarchimarmecal model. *Journal of Marine Systems*. **76**(1–2): 244–250.

Stow C A, Jolliff J, McGillicuddy Jr D J, *et al.* 2009. Skill assessment for coupled biological/physical models of marine systems. *Journal of Marine Systems*. **76**(1–2): 4–15. DOI: http://dx.doi.org/10.1016/j.jmarsys.2008.03.011.

Tashkova K, Silc J, Atanasova N, *et al.* 2012. Parameter estimation in a nonlinear dynamic model of an aquatic system with meta-heuristic optimization. *Ecosystem Modelling*. **226**(2): 36–61.

Wallhead P J, Martin A P, Srokosz M A. 2009. Skill assessment via cross-validation and Monte Carlo simulation: An application to Georges Bank plankton models. *Journal of Marine Systems*. **76**: 134–150.

Wang H, Liu G, Wan L. 2007. Application and research progress of data assimilation in marine ecological model. *Advances in Earth Science*. **22**(10): 989–996.

Ward B A, Friedrichs M A M, Anderson T R, *et al.* 2010. Parameter optimisation techniques and the problem of underdetermination in marine biogeochemical models. *Journal of Marine Systems*. **81**: 34–43.

Weber L, Volker C, Schartau M, *et al.* 2005. Modeling the speciation and biogeochemistry of iron at the Bermuda Atecounlantic Time-Series study site. *Global Biogeochemical Cycles*. **19**(1): GB1019. DOI: 10.1029/2004GBC00-2340.

Xu C, Hu Y, Chang Y, *et al.* 2004. Sensitivity analysis of ecological models. *Chinese Journal of Applied Ecology*. **15**(6): 1056–1062.

Zhang X, Srinivasan R, Zhao K. 2009. Evaluation of global optimization algorithms for parameter calibration of a computationally intensive hydrologic model. *Hydrological Processes*. **23**: 430–441.

Zheng W, Shi H H, Fang G H, *et al.* 2012. Global sensitivity analysis of a marine ecosystem dynamic model of the Sanggou Bay. *Ecosystem Modelling*. **247**: 83–94.

Chapter 5

Construction and Application of Coastal *Ulva Prolifera* Growth Model

In recent years, the pressure of eutrophication on the coastal areas in China has been continuously increasing. This frequently causes marine ecological disasters, such as red and green tides. Since the large-scale outbreak of *Ulva prolifera* (*U. prolifera*) in the Yellow Sea of China in June 2007, green tide disasters dominated by *U. prolifera* have occurred every year, and the occurrence and biomass of green tides have been significantly increasing (Liu *et al.*, 2009; Gao *et al.*, 2014). Therefore, the study of the growth of *U. prolifera* can help with early warnings and the prevention of green tide disasters.

At present, studies on *U. prolifera* are mainly conducted through remote sensing and monitoring and biological experiments. For example, Liu *et al.* (2009) monitored the floating path of *U. prolifera* by remote sensing. Hu (2009) proposed a global-scale remote sensing and monitoring system of planktonic algae. Pang *et al.* (2010) and Wang *et al.* (2010) studied the origin of *U. prolifera* outbreaks by a molecular biology method. Li *et al.* (2009) and Huo *et al.* (2014) investigated the growth mechanism of *U. prolifera* by biological experiments. Yet the *U. prolifera* has barely been studied by numerical simulation, which can be used to explore the growth mechanism and environmental affecting mechanism of *U. prolifera* from the perspective of ecological dynamics at lower costs, as compared with field and experimental studies. Currently, the numerical simulations of macroalgae mainly focus on the biomass changes of mature algae (Boyer and Fong, 2005; Ren *et al.*, 2014), barely involving the early life stages of *U. prolifera*, possibly because mature plants can cause more severe environmental damages. However, Martins *et al.* (2008) pointed out that the biomass of spores with lengths less than 10 cm played a more significant role in the outbreak of *U. prolifera*. Therefore, the study of the biomass changes of both spores and mature plants can help us understand the outbreak mechanism of *U. prolifera* from the perspective of a life cycle (Liu *et al.*, 2016). Studies have shown that the uncertainty of a model is mainly attributed to the uncertainty of parameters, and thus parameter optimization by sensitivity analysis can significantly improve the model accuracy (Xu *et al.*, 2004; Shen *et al.*, 2016). Parameter sensitivity analysis evaluates the degree of influence of each parameter on the simulation results. The analysis can identify the parameters that greatly affect the model for optimization (Zheng *et al.*, 2012; Shi *et al.*, 2014), and can be used as a reference for exploring the growth mechanism and influencing factors of *U. prolifera* (Liu *et al.*, 2016).

In this chapter, a population dynamics model for the growth of *U. prolifera* in the offshores of China is established in STELLA, a software for system dynamics modeling. The growths of spores and mature *U. prolifera* are numerically simulated. The main parameters of the model are screened by global sensitivity analysis, and the growth mechanism and its influencing factors of *U. prolifera* are discussed.

5.1 Model Formulation

The basic structure of the growth model is referred to the numerical simulation results of macroalgae reported by Martins *et al.* (2007; 2008), with comprehensive considerations of the limiting factors including nutrient, light, temperature, and salinity, as well as the predation process and growth and reproduction of *U. prolifera*. The model assumes that the nitrogen and phosphorus in *U. prolifera* are the nutritive salts that the growth relies on (Björnsäter and Wheeler, 1990). Figure 5.1 shows a conceptual model for the growth of *U. prolifera* (Liu *et al.*, 2016).

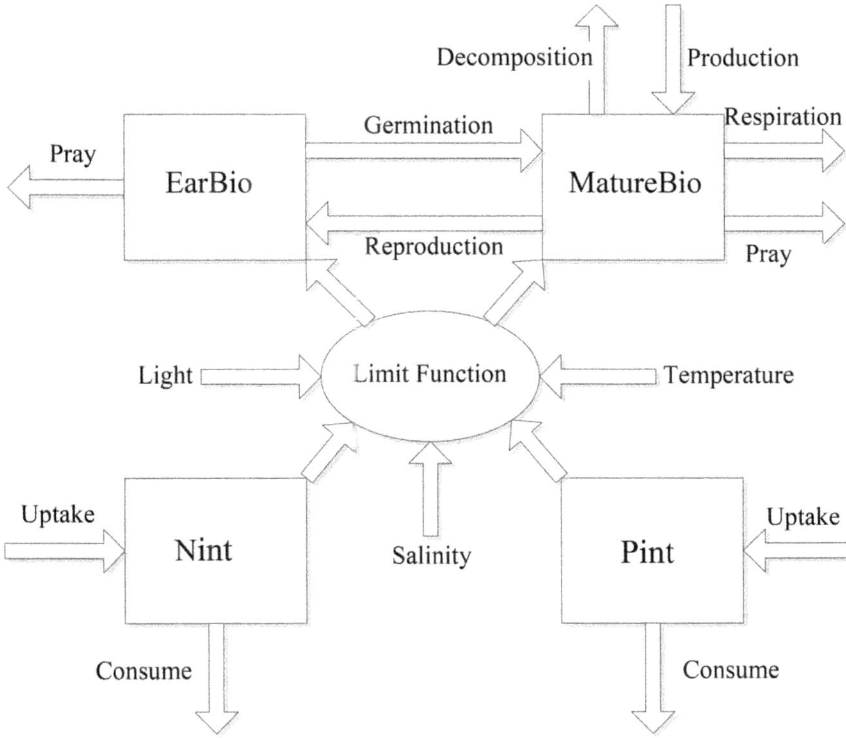

Fig. 5.1. Conceptual model for the growth of *U. prolifera.*

The state variables of the model include the spore biomass (in carbon) (EarBio, mg/m^2), mature biomass (MatureBio, mg/m^2), and the internal nitrogen concentration of *U. prolifera* (N_{int}, μmol N/mg C) and internal phosphorus concentration of *U. prolifera* (P_{int}, μmol P/mg C). The differential equation for the changes of these variables over time (t, unit d) can be expressed as:

$$\frac{\partial P}{\partial t} = -u\frac{\partial P}{\partial x} - v\frac{\partial P}{\partial y} - w\frac{\partial P}{\partial z} + \frac{\partial}{\partial x}\left(A_h\frac{\partial P}{\partial x}\right) + \frac{\partial}{\partial y}\left(A_h\frac{\partial P}{\partial y}\right) + \frac{\partial}{\partial z}\left(A_v\frac{\partial P}{\partial z}\right) + [\mu_{PPT\_G}(1 - r_{PPT\_E}) - \mu_{PPT\_D}]P - \mu_{PPT\_Z}Z, \tag{5.1}$$

$$\begin{aligned}\frac{dMatureBio}{dt} &= \text{GrossProduction } (T) + \text{Germination}\\ &\quad - \text{Respiration } (T) - \text{Mature Predation } (T)\\ &\quad - \text{Decomposition } (T) - \text{Reproduction } (T), \end{aligned} \tag{5.2}$$

$$\frac{dN_{\text{int}}}{dt} = N_{\text{uptake}} - N_{\text{consumption}}, \tag{5.3}$$

$$\frac{dP_{\text{int}}}{dt} = P_{\text{uptake}} - P_{\text{consumption}}, \tag{5.4}$$

where Reproduction (T) is the reproduction function of *U. prolifera*, Germination (T) is the spore germination function, SporePredation (T) is the preying function of zooplankton on spores, GrossProduction (T) is the total production of mature *U. prolifera*, Respiration (T) is the consumption function of cellular respiration, MaturePredation (T) is the preying function of zooplankton on mature *U. prolifera*, Decomposition (T) is the decomposition function of mature *U. prolifera*, N_{uptake} is the nitrogen uptake of *U. prolifera*, $N_{\text{consumption}}$ is the nitrogen consumption function of *U. prolifera*, P_{uptake} is the phosphorus uptake function of *U. prolifera*, $P_{\text{consumption}}$ is the phosphorus consumption function of *U. prolifera*, and T is the sea surface temperature. Table 5.1 lists the main parameters of the model.

5.1.1 *Spore biomass*

Spore biomass change is affected by three processes: reproduction of mature *U. prolifera*, spore germination, and preying on spores.

(1) *Mature propagule*

A portion of mature *U. prolifera* forms propagules for reproduction (Niesenbaum, 1988). The reproduction biomass function is defined as:

$$\text{Reproduction } (T) = \text{MatureBio} \cdot \text{Reprod\_rate} \cdot \theta^{\text{T-20}}. \tag{5.5}$$

Table 5.1. Main parameters of the *U. prolifera* growth model and their values.

Parameter	Value	Range	Definition	Unit	Reference
V_{maxn}	11.12	7.784 ~ 14.45	Maximum nitrogen uptake	μmol N/(mg C · d)	(Arhonditsis and Brett, 2005)
V_{maxp}	0.29	0.203 ~ 0.377	Maximum phosphorus uptake	μmol P/(mg C · d)	(Arhonditsis and Brett, 2005)
$\mu_{\max}$	1.2	0.8 ~ 1.2	Maximum growth rate	d^{-1}	(Martins *et al.*, 2007)
$R_{\max 20}$	0.08	0.02 ~ 0.1	Maximum respiration rate at 20°C	d^{-1}	(Martins *et al.*, 2007)
Z_{maxa}	0.02	0.014 ~ 0.026	Maximum preying rate on mature *U. prolifera*	d^{-1}	(Martins *et al.*, 2007)
$dec_{\max}$	0.025	0.025 ~ 0.065	Maximum decomposition rate	d^{-1}	(Paalme *et al.*, 2002)
S_{opt}	18	15 ~ 25	Optimal salinity for the germination and growth of *U. prolifera*	Psu	(Martins *et al.*, 1999) (Cong, 2011)
Z_{maxs}	0.14	0.06 ~ 0.14	Maximum preying rate on spores	d^{-1}	(Lotze and Worm, 2000)
θ	1.047	1.01 ~ 1.2	Empiric parameters of respiration, decomposition and reproduction	—	(Martins *et al.*, 2007)
Reprod_rate	0.01	0.01 ~ 0.06	Biomass loss rate caused by the formation of propagule from mature *U. prolifera*	d^{-1}	(Paalme *et al.*, 2002)
$G_{\max}$	0.964	0.958 ~ 0.964	Maximum germination rate	d^{-1}	(Chen *et al.*, 2011)
Is	320	320 ~ 400	Optimal light intensity for photosynthesis	μmol/m^2/s	(Cong, 2011)
T_{opt}	22	18 ~ 23	Optimal temperature for *U. prolifera* growth	°C	(Gao *et al.*, 2014)

(*Continued*)

Table 5.1. (*Continued*)

Parameter	Value	Range	Definition	Unit	Reference
k_{qN}	8	$5.6 \sim 10.4$	Half saturation coefficient of nitrogen needed for the growth of *U. prolifera*	μmol N/(mg C)	(Zheng *et al.*, 2012)
k_{qP}	0.6	$0.42 \sim 0.78$	Half saturation coefficient of phosphorous needed for the growth of *U. prolifera*	μmol P/(mg C)	(Zheng *et al.*, 2012)
k_{N}	3.214		Half saturation coefficient of nitrogen uptake	μmol N/L	(Arhonditsis and Brett, 2005)
K_{P}	0.322		Half saturation coefficient of phosphorous uptake	μmol P/L	(Arhonditsis and Brett, 2005)
N_{req}	0.45		Free nitrogen concentration for the growth of *U. prolifera*	d^{-1}	(Martins *et al.*, 2007)
P_{req}	0.25		Free phosphorous concentration for the growth of *U. prolifera*	d^{-1}	(Martins *et al.*, 2007)
N_{imin}	5.7		Minimum internal nitrogen concentration	d^{-1}	(Arhonditsis and Brett, 2005)
N_{imax}	12.85		Maximum internal nitrogen	d^{-1}	(Arhonditsis and Brett, 2005)
P_{imin}	0.258		Minimum internal phosphorus	d^{-1}	(Arhonditsis and Brett, 2005)
P_{imax}	0.87		Maximum internal phosphorus	d^{-1}	(Arhonditsis and Brett, 2005)

(2) *Spore germination*
The spore germination of *U. prolifera* is defined as (Martins *et al.*, 2007):

$$\begin{aligned}\text{Germination }(T) &= \text{EarBio}\cdot \text{G}_{\text{max}}\cdot \text{Lightlimit }(I)\cdot \text{Tlimit }(T)\\ &\quad\cdot \text{Salinityinfluence }(S)\cdot \text{NPlimit }(N,P). \qquad (5.6)\end{aligned}$$

(a) The light limit for germination is defined as:

$$\text{Lightlimit }(I) = \frac{I}{Is}\exp\left(1-\frac{I}{Is}\right), \qquad (5.7)$$

where I represents the photosynthetic photon flux density ($\mu\text{E}/(\text{m}^2\cdot\text{s})$).

(b) The temperature limit is defined as (Martins *et al.*, 2007):

$$\text{Tlimit }(T) = \exp\left[-2.3\left(\frac{T-T_{\text{opt}}}{T_x-T_{\text{opt}}}\right)^2\right], \qquad (5.8)$$

where T is the sea surface temperature (°C). $T_x = T_{\text{min}}$, if $T \le T_{\text{opt}}$ and $T_x = T_{\text{max}}$, if $T > T_{\text{opt}}$. Considering the changing characteristics of the coastal sea surface temperature, the temperature function can be calculated with Eq. (5.9).

$$T = \frac{T_{\text{max}}+T_{\text{min}}}{2} - \delta\cos\left[\frac{2\pi(t-90)}{365}\right], \qquad (5.9)$$

where t is the number of days counted from January 1 and

$$\delta = \frac{T_{\text{max}}-T_{\text{min}}}{2},$$

where T_{min} is the low tolerable temperature limit of *U. prolifera* and T_{max} is the upper tolerable temperature limit.

(c) Based on the experimental results reported by Martins *et al.* (1999), the salinity limit for spore germination is defined as follows:

$$\text{if } S \geq 5, \quad \text{Salinityinfluence(S)} = 1 - \left(\frac{S - S_{\text{opt}}}{S_x - S_{\text{opt}}}\right)^m, \tag{5.10}$$

$$\text{if } S < 5, \quad \text{Salinityinfluence(S)} = \frac{S - S_{\min}}{S_{\text{opt}} - S_{\min}}, \tag{5.11}$$

$$\text{if } S < S_{\text{opt}}, \quad S_x = S_{\min} \quad \text{and} \quad m = 2.5, \tag{5.12}$$

$$\text{if } S \geq S_{\text{opt}}, \quad S_x = S_{\max} \quad \text{and} \quad m = 2, \tag{5.13}$$

where S is the environment salinity (psu).

(3) The nitrogen and phosphorus limits for spore germination are defined following Liebig's law of the minimum. To make the model more suitable for the growth of *U. prolifera* in the coastal waters of China, the limit functions of nitrogen and phosphorus are modified based on the effects of nutrients on the growth of *U. prolifera* reported by Wu (2013) and the ocean numerical simulation results as follows (Björnsäter *et al.*, 1990; Wu, 2013).

$$\text{If } 12 \leq N_{\text{int}}/P_{\text{int}} \leq 16, \ \text{NPlimit}(N_{\text{int}}, P_{\text{int}}) = 1, \tag{5.14}$$

$$\text{if } N_{\text{int}}/P_{\text{int}} < 12, \ \text{NPlimit}(N_{\text{int}}, P_{\text{int}}) = N\text{limit}(N_{\text{int}}), \tag{5.15}$$

$$\text{if } N_{\text{int}}/P_{\text{int}} > 16, \ \text{NPlimit}(N_{\text{int}}, P_{\text{int}}) = P\text{limit}(P_{\text{int}}), \tag{5.16}$$

$$N\text{limit}(N_{\text{int}}) = \frac{N_{\text{int}} - N_{\text{imin}}}{k_{\text{qN}} + N_{\text{int}} - N_{\text{imin}}}, \tag{5.17}$$

$$P\text{limit}(P_{\text{int}}) = \frac{P_{\text{int}} - P_{\text{imin}}}{k_{\text{qp}} + P_{\text{int}} - P_{\text{imin}}}, \tag{5.18}$$

where N_{imin} is the minimum internal nitrogen concentration of mature *U. prolifera* in μmol N/mg C, P_{imin} is the minimum internal phosphorous concentration of mature *U. prolifera* in μmol P/mg C, k_{qN} is the half saturation coefficient of the nitrogen needed for the growth of *U. prolifera*, and k_{qP} is the half saturation coefficient of the phosphorous needed for the growth of *U. prolifera*.

(4) Preying of zooplankton on *U. prolifera* spore

The preying function of zooplankton on *U. prolifera* spores is defined as:

$$\text{SporePredation}\ (T) = \text{EarBio} \cdot Z_{\text{maxs}} \cdot \text{Tlimit}\ (T). \tag{5.19}$$

5.1.2 *Mature biomass*

The model assumes that the biomass of mature *U. prolifera* is decided by the following six processes: gross production, spore germination, predation by zooplankton, decomposition, and reproduction. The definitions of spore germination, reproduction, and spore biomass can be found above. Other functions are defined below:

(1) Gross production of mature *U. prolifera*

The gross production of mature *U. prolifera* is defined as (Martins and Marques, 2002):

$$\begin{aligned}\text{Gross Production}\ (T) &\\ = \text{MatureBio} \cdot \mu_{\max} \cdot \text{Lightlimit}\ (I) \cdot \text{Tlimit}\ (T) &\\ \cdot \text{Salinityinfluence}\ (S) \cdot \text{NPlimit}\ (N_{\text{int}}, P_{\text{int}}). &\end{aligned} \tag{5.20}$$

(2) Respiration consumption of mature *U. prolifera*

The respiration consumption function is defined as:

$$\text{Respiration}\ (T) = \text{MatureBio} \cdot R_{\max 20} \cdot \theta^{\text{T-20}}. \tag{5.21}$$

(3) Preying of zooplankton on mature *U. prolifera*

The preying function of zooplankton is defined as:

$$\text{MaturePredation}\ (T) = \text{Z}_{\text{maxa}} \cdot \text{T}_{\text{limit}}\ (T) - \text{MatureBio}. \tag{5.22}$$

(4) Decomposition of mature *U. prolifera*

Decomposition is one of the major causes for the biomass loss of mature *U. prolifera.* In particular, during the rapid growth season, the light shading of the surface layer can cause exponential decreases of growth rate (Paalme *et al.*, 2002). The *U. prolifera* on the water surface can perform normal photosynthesis, but those under are subjected to decomposition due to light shading (Cong, 2011). The decomposition function is defined as:

$$\text{Decomposition}\ (T) = \text{MatureBio} \cdot \text{dec}_{\max} \cdot \theta^{\text{T-20}}. \tag{5.23}$$

5.1.3 *Internal nitrogen concentration*

The nitrogen uptake function for mature *U. prolifera* is defined as (Wu, 2013):

$$N_{\text{uptake}} = \frac{N_{\text{imax}} - N_{\text{int}}}{N_{\text{imax}} - N_{\text{imin}}} \cdot \frac{V_{\text{maxN}} \cdot N}{K_N + N}, \tag{5.24}$$

where N is the nitrogen concentration in the water environment in μ mol/L, and N_{imax} is the maximum internal nitrogen concentration of mature *U. prolifera* is in μ mol N/mg C.

The nitrogen consumption function is defined as:

$$N_{\text{consume}} = N_{\text{int}} \cdot N_{\text{req}}. \tag{5.25}$$

5.1.4 *Internal phosphorous concentration*

The phosphorous uptake function of mature *U. prolifera* is defined as (Wu, 2013):

$$P_{\text{uptake}} = \frac{P_{\text{imax}} - P_{\text{int}}}{P_{\text{imax}} - P_{\text{imin}}} \cdot \frac{V_{\text{maxP}} \cdot P}{K_{\text{p}} + P}, \tag{5.26}$$

where P is the phosphorous concentration in the water environment in μmol/L, and P_{imax} is the maximum phosphorous concentration of mature *U. prolifera* is, μmol P/mg C.

The phosphorous consumption function is defined as:

$$P_{\text{consume}} = P_{\text{int}} \cdot P_{\text{req}}. \tag{5.27}$$

5.2 Simulation Results

The growth model is run in STELLA using the equations and parameters set above. The numerical solution is implemented by the fourth-order Runge–Kutta method. Figure 5.2 shows the main architecture of the model. In STELLA, the dynamic changes of state variables are regarded as "water tanks", and the changes of various ecological processes are regarded as the dynamic changes of "water tank capacity". These dynamic changing processes reflect the dynamic changes of the growth model and can be used for the dynamic monitoring of the model.

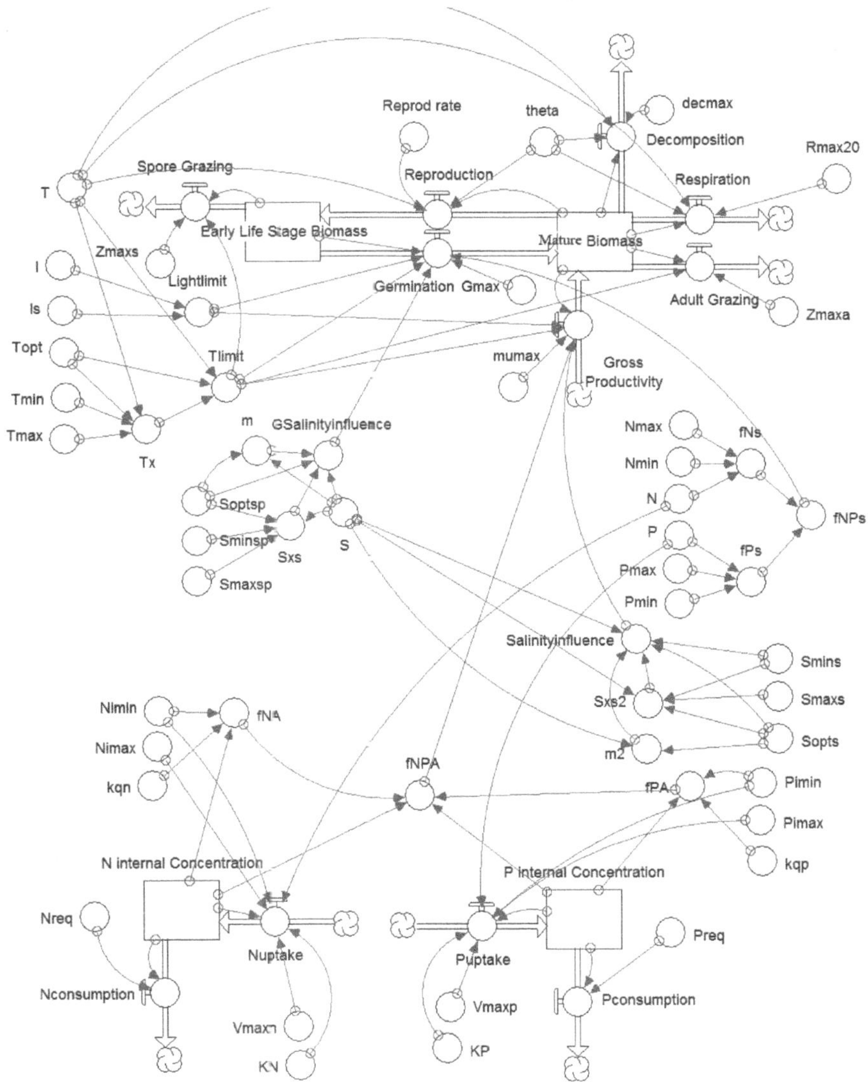

Fig. 5.2. Schematic diagram of the *U. prolifera* growth model in the system dynamics modeling software STELLA.

The large-scale reproduction of *U. prolifera* mainly occurs from the end of April to early May in the coastal waters in China. Therefore, the simulation time is set to 200 days from May 1. Both initial values of spores and mature biomasses are set to $50\,\mathrm{mg\,C/m^2}$.

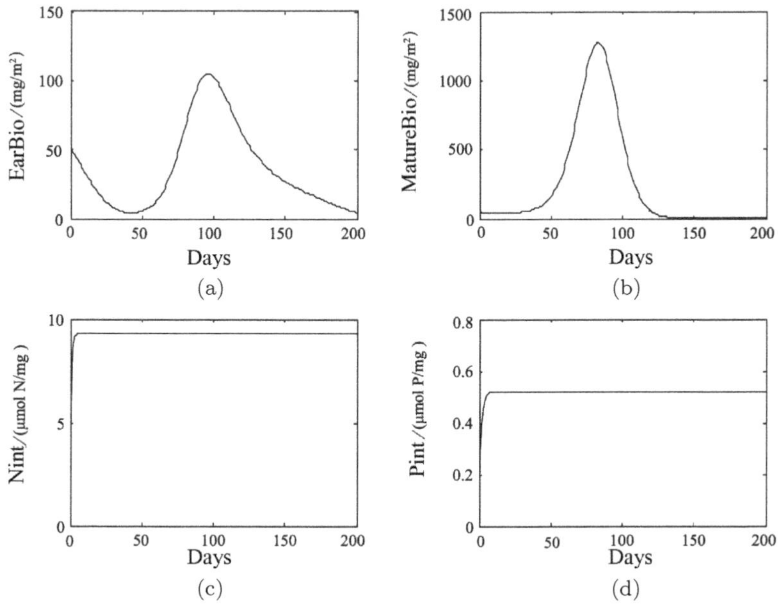

Fig. 5.3. Simulation results of the *U. prolifera* growth process with the starting date set to May 1. (a) Spore biomass, (b) mature biomass, (c) internal nitrogen concentration, and (d) internal phosphorus concentration.

The initial internal nitrogen and phosphorous concentrations are set to 5.7 μmol N/mg C and 0.258 μmol P/mg C (Martins *et al.*, 2007), respectively. The sea surface temperature on May 1 is set as the initial temperature. The initial nitrogen and phosphorous concentrations in the environment are set to 24 μmol N/L and 2 μmol P/L, respectively. The initial photosynthetic photon flux density is set to 245 μmol E/($m^2 \cdot s$) and the initial salinity is 34.

Figure 5.3 shows the numerical simulation results of the growth of *U. prolifera*. The spore biomass shows a fluctuating trend, decreasing monotonically first, reaching the minimum value of 4.23 mg C/m^2 on day 42 (June 11), then increasing monotonically and reaching the maximum value of 104.97 mg C/m^2 on day 96 (August 4), and then decreasing again (Fig. 5.3(a)). The mature biomass gradually decreases first, increases sharply until reaching the peak value of 1273.29 mg C/m^2 on day 83 (July 22), and then decreases to 1.04 mg C/m^2 on day 177 (October 24) (Fig. 5.3(b)). Both internal nitrogen and phosphorous concentrations show monotonically increasing trends in the first 5 days, and become stable and remain at 9.28 μmol N/mg C and 0.51 μmol P/mg C, respectively, thereafter (Fig. 5.3(c)).

5.3 Global Sensitivity Analysis of Key Influencing Parameters

Parameter sensitivity analysis is an effective method to determine the degree of influence of a parameter on the modeling result. It is a key step of the parameter optimization and model calibration of marine ecological dynamics models, and an important basis for understanding the key influencing parameters of the model and related processes (Xu *et al.*, 2004; Shi *et al.*, 2014; Liu *et al.*, 2016; Shen *et al.*, 2016). Parameter sensitivity analysis can be conducted by either global or local sensitivity analysis that is differentiated by whether the interactions between parameters are involved or not. Local sensitivity analysis only examines the local small disturbances of a single parameter. The analysis is simple and operable without including the interactions among parameters. In addition, the analysis results are affected by the initial parameter values significantly (Shi *et al.*, 2014). Global sensitivity analysis tests the effects of multiple parameters on the model simultaneously in the specific ranges. It analyzes the sensitivity of each parameter of the model and the interaction of each parameter with other parameters as a whole, and thus can fully reveal the key influence parameters of the model (Cossarini and Solidoro, 2008; Zheng *et al.*, 2012; Morris *et al.*, 2014; Shen *et al.*, 2016). Compared with local sensitivity analysis, global sensitivity analysis is more complex with intensive computation and thus is barely applied to marine ecosystem dynamics modeling. The Morris method (Morris, 1991; Liu, 2012) and the Sobol method (Sobol, 1993) are the most commonly used global sensitivity analysis algorithms. The Sobol method usually requires frequent and intensive calculations of Monte Carlo integrals (Liu, 2012). In contrast, the Morris method is based on the sampling method proposed by Morris (1991), which can rank parameter sensitivity levels with small amounts of calculation and qualitatively describe the parameter interactions (Cossarini and Solidoro, 2008). In this section, the key parameters that affect the growth model of *U. prolifera* are determined by the Morris method for global sensitivity analysis.

5.3.1 Morris sampling based global sensitivity analysis

The Morris method is a global sensitivity analysis algorithm based on statistical sampling. It improves the reliability of sensitivity analysis by repeated experiments (Morris, 1991). The method gives the initial parameter values with a cut set matrix randomly. The randomness and sampling times of parameter generation ensure that the initial values of the parameter set are highly representative, and the mean and standard deviation analysis of the sensitivity further reduce the randomness (Morris, 1991). The 15 variable parameters listed in Table 5.1 are subjected to global sensitivity analysis by the Morris method following the steps described below. The number of repeated experiments is set to 10,000.

(1) Interval discretization and elementary effect calculation

The parameter values are first mapped over the interval $[0, 1]$, and the interval is discretized to $\left\{0, \frac{1}{p-1}, \frac{2}{p-1}, \ldots, 1\right\}$. The elementary effect (EE) of the i-th parameter can be expressed as:

$$EE_i(x) = |[y(x_1, x_2, \ldots, x_{i-1}, x_i + \delta, x_{i+1}, \ldots, x_k) - y(x)]/\delta|, \quad (5.28)$$

where k is the total number of parameters, $y(x)$ is the output of the model with the corresponding parameter, and $\delta = \frac{m}{p-1}, m = 1, 2, \ldots, p-1$ is the parameter change.

EE is then calculated with a parameter that does not equal to x_i and can change the value of δ (different from the δ of the last step). The process is repeated until all parameters are varied once and the corresponding EEs are calculated. This process forms a track in the parameter space as shown in Fig. 5.4.

(2) Sampling

The Euclidean vector of the k coordinate basis x^* is defined and the elements are sampled from $\left\{0, \frac{1}{p-1}, \frac{2}{p-1}, \ldots, 1 - \frac{1}{p-1}\right\}$.

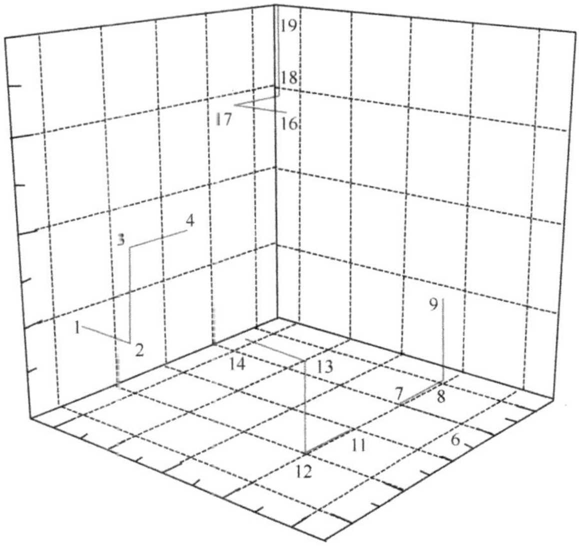

Fig. 5.4. Schematic diagram of the Morris sampling.

The $(k+1) \times k$ matrix B is defined as:

$$B = \begin{pmatrix} 0 & 0 & 0 & \cdots & 0 \\ 1 & 0 & 0 & \cdots & 0 \\ 1 & 1 & 0 & \cdots & 0 \\ 1 & 1 & 1 & \cdots & 0 \\ \vdots & \vdots & \vdots & \vdots & \vdots \\ 1 & 1 & 1 & 1 & 1 \end{pmatrix}. \tag{5.29}$$

The $k \times k$ diagonal matrix is defined as D^* and the diagonal elements of D^* are equally assigned as -1 or 1. All elements of the $(k+1) \times k$ matrix $J_{k+1,k}$ are 1

The $(k+1) \times k$ matrix J^* is defined as:

$$J^* = \frac{1}{2}[(2B - J)D^* + J]. \tag{5.30}$$

The $k \times k$ random matrix is defined as P^*, and there is only 1 in each row and the remaining elements are 0.

The matrix B^* is defined as:

$$B^* = (J_{k+1,k}x^* + \delta J^*)P^*. \tag{5.31}$$

It can be concluded, based on the randomness of matrixes D^*, x^* and P^*, B^* is a random array.

(3) Sensitivity analysis

The matrix for the whole test is defined as: $X = \begin{pmatrix} B_1^* \\ B_2^* \\ \vdots \\ B_{k+1}^* \end{pmatrix}$.

There are only two different rows in column j, e.g., $B(j) = \begin{pmatrix} x_1 & \cdots & x_{j1} & \cdots & x_m \\ x_1 & \cdots & x_{j2} & \cdots & x_m \end{pmatrix}$, where $z_{j1} - z_{j2} = \delta$.

The EE of the j-th parameter can be calculated using $B(j)$ as the parameter input as:

$$EE_j = \frac{y(z_1, \ldots, z_{j1}, \ldots, z_m) - y(z_1, \ldots, z_{j2}, \ldots, z_m)}{\delta y(z_1, \ldots, z_{j2}, \ldots, z_m)}. \tag{5.32}$$

Therefore, the sensitivities of all parameters in one sampling can be calculated using matrix X. For the r times sampling, the calculation is repeated r times, and the mean and standard deviation of the EE of each parameter are calculated. The larger the mean value, the more significant the effect of the parameter. The larger the standard deviation, the stronger the interaction of the parameter with other parameters and the higher the nonlinearity of the model (Morris, 1991; Saltelli *et al.*, 2000; Cossarin and Solidoro, 2008; Liu, 2012; Morris *et al.*, 2014).

5.3.2 *Results of parameter global sensitivity analysis*

Figures 5.5 and 5.6 show the plots of the mean and standard deviation of the normalized EEs, respectively. The larger the value, the larger the mean or standard deviation of the corresponding EE.

The plots in Fig. 5.5 suggest that for the spore biomass, T_{opt} is most sensitive, I_{s}, $G_{\max}$, and Z_{maxs} are sensitive, and *Reprod_rate* is not sensitive (Fig. 5.5(a)). For the mature biomass, T_{opt} is the most sensitive parameter, $G_{\max}$, k_{qN}, and V_{maxN} are sensitive, and

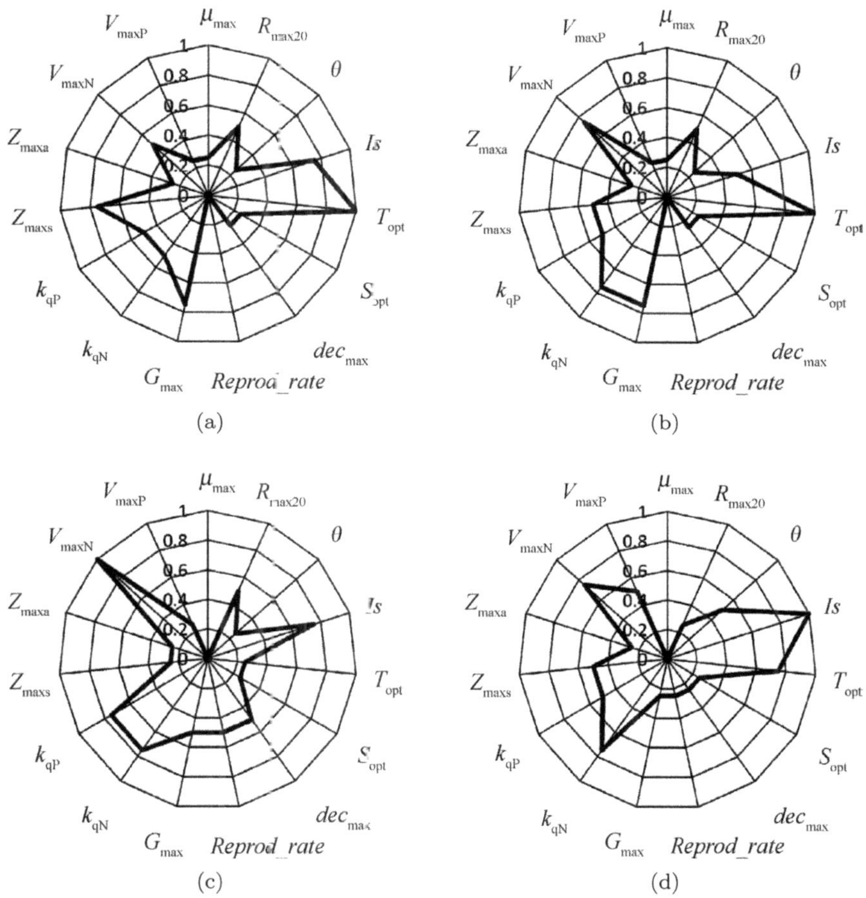

Fig. 5.5. Mean EE value of each parameter. (a) Spore biomass, (b) Mature biomass, (c) Internal nitrogen concentration, and (d) Internal phosphorus concentration.

Reprod_rate is not sensitive (Fig. 5.5(b)). For the internal nitrogen concentration, V_{maxN} is most sensitive, k_{qN}, k_{qP}, and I_{s} are sensitive, and μ_{max} is not sensitive (Fig. 5.5(c)). For internal phosphorus concentration, I_{s} is most sensitive, T_{opt}, V_{maxP}, and k_{qN} are sensitive, and μ_{max} is not sensitive (Fig. 5.5(d)).

The standard deviation plots in Fig. 5.6 indicate that for the spore biomass, T_{opt} has the strongest correlations with other parameters, and I_{s}, G_{max}, and Z_{maxs} correlate with other parameters strongly

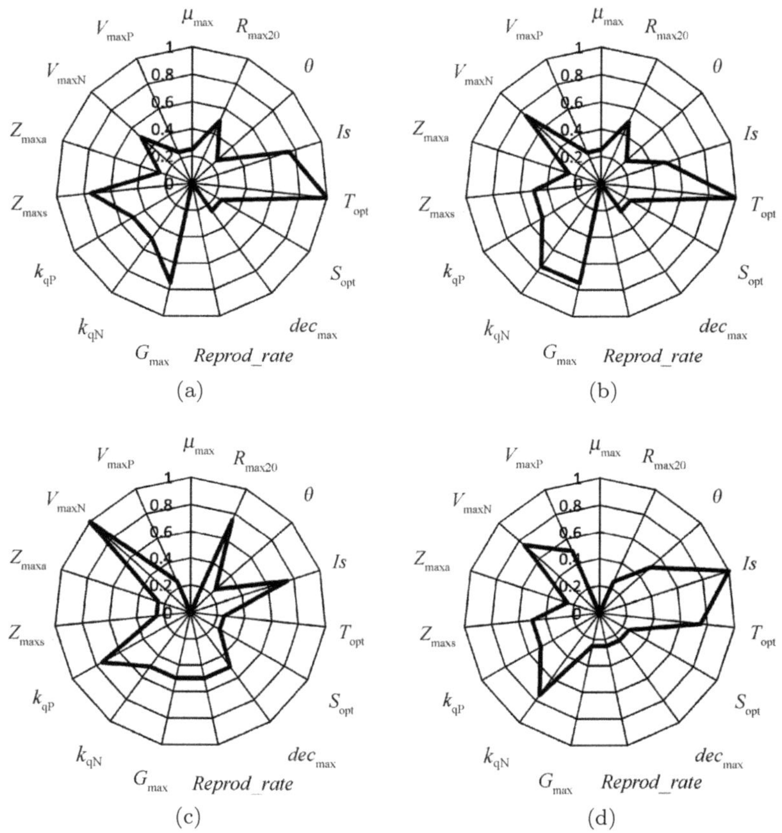

Fig. 5.6. Standard deviation of EE of each parameter. (a) Spore biomass, (b) mature biomass, (c) internal nitrogen concentration, and (d) internal phosphorus concentration.

(Fig. 5.6(a)). For the mature biomass, T_{opt} has the strongest correlations with other parameters, and G_{max}, k_{qN}, and V_{maxN} correlate with other parameters strongly (Fig. 5.6(b)). For the internal nitrogen concentration, V_{maxN} correlates with other parameters most strongly, and k_{qP}, I_{s}, and R_{max20} correlate with other parameters strongly (Fig. 5.6(c)). For the internal phosphorous concentration, I_{s} has the strongest correlations with other parameters, and T_{opt}, k_{qN}, and V_{maxN} correlate with other parameters strongly (Fig. 5.6(d)).

Based on the ranks of the mean EE of each state variable, it can be seen that T_{opt} is the most sensitive parameter for more than two state variables, and T_{opt}, I_{s}, G_{max}, k_{qN}, and V_{maxN} are the most sensitive or sensitive parameters for more than two state variables, and μ_{max} and *Reprod_rate* are the parameters not sensitive to the above state variables.

5.3.3 *Parameters to be optimized*

Parameter sensitivity analysis is a key step in establishing an ecological model (Xu *et al.*, 2004). It evaluates the influence of each parameter on the model, which can establish parameter optimization strategies to optimize the parameters with high sensitivities and improve the model accuracy. The sensitivity analysis reveals that T_{opt} is the most sensitive parameter of the growth model of *U. prolifera*, and I_s, G_{max}, k_{qN}, and V_{maxN} are the sensitive parameters. Therefore, these five parameters are optimized first. μ_{max} and *Reprod_rate* are insensitive parameters, and thus the empirical values are adopted. The remaining parameters can adopt empirical data or be optimized as needed.

5.3.4 *Parameters affecting the growth of U. prolifera*

Sensitivity analysis results suggest that among the 15 main parameters involved in the growth model, there are only 1 very sensitive parameter, 4 sensitive parameters, 8 less sensitive parameters, and 2 insensitive parameters. T_{opt}, I_s, G_{max}, k_{qN}, and V_{maxN} are very sensitive and sensitive parameters, which mainly affect the germination and growth of *U. prolifera*. T_{opt} is most sensitive, indicating that temperature affects the growth of *U. prolifera* the most. This is consistent with the conclusions reported by Ren *et al.* (2014). I_s, G_{max}, k_{qN}, and V_{maxN} are sensitive parameters, indicating that the nitrogen concentration, light, and germination rate also affect *U. prolifera*, and the growth of *U. prolifera* is mainly limited by nitrogen concentration and light. The experiments conducted by Abreu *et al.* (2011) show that the growth of macroalgae requires large amounts of nitrogen, and fast-growing marine algae species have higher demands on nitrogen. The simulation here also gives the same conclusion. Light intensity affects the photosynthesis rate of *U. prolifera*, and can also indirectly affect nutrient absorption (Cong, 2011). Temperature, light, and nitrogen concentration affect the maximum germination rate of *U. prolifera* (Song *et al.*, 2014), and thus G_{max} is also a sensitive parameter.

T_{opt} is the most sensitive parameter, I_s, G_{max}, and Z_{maxs} are sensitive parameters, and *Reprod_rate* is an insensitive parameter for the spore biomass. Ren *et al.* (2014) found that the upper and lower limits of temperature tolerance range significantly affected the growth of *U. prolifera.* The parameter sensitivity analysis conducted here suggests that the optimal temperature for the growth of *U. prolifera* also has significant impacts on spore germination and growth. G_{max} affects the spore germination process and thus shows influences on the mature biomass (Song *et al.*, 2014). The high sensitivity of I_s indicates that the light affects the photosynthesis rate and nutrient absorption process of spores, and thus impacts the growth of *U. prolifera.* Z_{maxs} also shows obvious effects on the mature biomass due to its impacts on the spore survival rate. The narrow fluctuation range of *Reprod_rate* suggests that it has no obvious effect on the spore biomass.

T_{opt} is the most sensitive parameter, G_{max}, k_{qN}, and V_{maxN} are sensitive parameters, and *Reprod_rate* is an insensitive parameter. T_{opt}, I_s, and *Reprod_rate* exhibit similar effects on the spores and mature biomass. k_{qN} and V_{maxN} affect the mature biomass more significantly, suggesting the growth of *U. prolifera* is mainly limited by nitrogen. k_{qN}, k_{qP}, V_{maxN}, and V_{maxN} are insensitive for the spore, indicating that nutrients have no significant effects on spores.

5.3.5 *Advantages of global sensitivity analysis*

The advantages of global sensitivity analysis over local sensitivity analysis are demonstrated using T_{opt} as an example. As shown in Table 5.2, no significant differences are found between the results obtained by the two methods, although the magnitudes of the results

Table 5.2. Comparison of the global and local sensitivity analysis results of T_{opt}.

Initial parameter values		Local sensitivity analysis		Global sensitivity analysis	
T_{opt}	I_s	Mature	Spore	Mature	Spore
22.0	320	10.75	1.22	321.79	138.14
22.0	304	12.51	1.44	321.79	138.14
23.1	320	7.89	0.96	321.79	138.14

are different. However, the local sensitivity analysis only can analyze one parameter at a time and the parameter sensitivity depends on the initial value selection, and the initial values of other parameters affect the sensitivity of the analyzed parameter (Liu, 2012; Shi *et al.*, 2014). Global sensitivity analysis is based on the sampling in the entire parameter value range and thus can analyze the sensitivities of all parameters simultaneously. It overcomes the limitations of local sensitivity analysis and is particularly important in the parameter optimization of complex ecosystem dynamics models. In addition, the global sensitivity analysis by the Morris method can characterize the interactions among parameters using the standard deviations of their EEs. As demonstrated in Figs. 5.5 and 5.6, the degrees of discretization of the parameters with higher sensitivities are also higher, and they are affected by other parameters more significantly. Therefore, the sensitivity and the degree of discretization of EE are correlated to certain degrees. In this chapter, the correlations of the means and standard deviations of the EEs of spore biomass, mature biomass, internal nitrogen concentration, and internal phosphorus concentration are analyzed, and the corresponding Pearson correlation coefficients are determined to be 0.678, 0.808, 0.967, and 0.984, respectively, suggesting the significant correlations between the parameters with high sensitivities and other parameters, consistent with the results reported by Cossarini (2008).

5.4 Summary

In this chapter, a growth model of *U. prolifera* including both spores and mature plants is constructed, and parameter global sensitivity analysis is conducted using the Morris sampling method. The results show that T_{opt} is the most sensitive parameter, I_{s}, G_{max}, k_{qN}, and V_{maxN} are sensitive parameters, and μ_{max} and *Reprod_rate* are insensitive parameters. Therefore, the growth of *U. prolifera* is mainly affected by temperature, light, and nitrogen. Unlike local sensitivity analysis, global sensitivity analysis can analyze the influences of all parameters on the model in the value range of each parameter simultaneously, as well as those of the interactions among the parameters, which gives highly referential and reliable results. The parameters with higher sensitivities often show stronger correlations with other parameters, and their EEs fluctuate in larger ranges.

References

Abreu M H, Pereira R, Buschmann A H, Sousa-Pinto I, Yarish C. 2011. Nitrogen uptake responses of Gracilaria vermiculophylla (Ohmi) Papenfuss under combined and single addition of nitrate and ammonium. *Journal of Experimental Marine Biology and Ecology*. **407**(2): 190–199.

Arhonditsis G B, Brett M T. 2005. Eutrophication model for Lake Washington (USA): Part II — Model calibration and system dynamics analysis. *Ecological Modelling*. **187**(2–3): 179–200.

Björnsäter B R, Wheeler P A. 1990. Effect of nitrogen and phosphorus supply on growth and tissue composition of Ulva fenestrata and Enteromorpha intestinalis (Ulvales, Chlorophyta). *Journal of Phycology*. **26**(4): 603–611.

Boyer K E, Fong P. 2005. Macroalgal-mediated transfers of water column nitrogen to intertidal sediments and salt marsh plants. *Journal of Experimental Marine Biology and Ecology*. **321**(1): 59–69.

Chen Q F, He P M, Feng Z H, *et al.* 2011. Reproduction of spores/gametes of floating green tide algae *Ulva prolifera*. *Journal of Fishery Sciences of China*. **18**(5): 1069–1076.

Cong S S. 2011. *Experimental Study of the Effects of Environmental Factors on Growth, Survival Conditions, and Nutrient Absorption of Enteromorpha Prolifera*. Qingdao: Ocean University of China.

Cossarini G, Solidoro C. 2008. Global sensitivity analysis of a trophodynamic model of the Gulf of Trieste. *Ecological Modelling*. **212**(1–2): 16–27.

Gao S, Fan S L, Han X R, *et al.* 2014. Relations of Enteromorpha prolifera blooms with temperature, salinity, dissolved oxygen and pH in the Southern Yellow Sea. *China Environmental Science*. **34**(1): 213–218.

Hu C M. 2009. A novel ocean color index to detect floating algae in the global oceans. *Remote Sensing of Environment*. **113**(10): 2118–2129.

Huo Y Z, Hua L, Wu H L, *et al.* 2014. Abundance and distribution of Ulva microscopic propagules associated with a green tide in the southern coast of the Yellow Sea. *Harmful Algae*. **39**: 357–364.

Li R X, Wu X W, Wei Q S, *et al.* 2009. Growth of Enteromorpha prolifera under different nutrient conditions. *Advances in Marine Science*. **27**(2): 211–216.

Liu S B. 2012. *Study on Parameter Estimation and Sensitivity Analysis Algorithms Based on Hydrological Model and Ecological Model.* Lanzhou: Lanzhou University.

Liu D Y, Keesing J K, Xing Q G, Shi P. 2009. World's largest macroalgal bloom caused by expansion of seaweed aquaculture in China. *Marine Pollution Bulletin.* **58**(6): 888–895.

Liu Y Z, Shen C C, Shi H H, *et al.* 2016. Factors influencing Ulva prolifera growth revealed by model based on global sensitivity analysis. *Acta Ecologica Sinica.* **36**(13).

Lotze H K, Worm B. 2000. Variable and complementary effects of herbivores on different life stages of bloom-forming macroalgae. *Marine Ecology Progress Series.* **200**: 167–175.

Martins I, Marques J C. 2002. A model for the growth of opportunistic macroalgae (Enteromorpha sp.) in tidal estuaries. *Estuarine, Coastal and Shelf Science.* **55**(2): 247–257.

Martins I, Marcotegui A, Marques J C. 2008. Impacts of macroalgal spores on the dynamics of adult macroalgae in a eutrophic estuary: High versus low hydrodynamic seasons and long-term simulations for global warming scenarios. *Marine Pollution Bulletin.* **56**(5): 984–998.

Martins I, Lopes R J, Lillebø A I, *et al.* 2007. Significant variations in the productivity of green macroalgae in a mesotidal estuary: Implications to the nutrient loading of the system and the adjacent coastal area. *Marine Pollution Bulletin.* **54**(6): 678–690.

Martins I, Oliveira J M, Flindt M R, Marques J C. 1999. The effect of salinity on the growth rate of the macroalgae Enteromorpha intestinalis (Chlorophyta) in the Mondego estuary (west Portugal). *Acta Oecologica.* **20**(4): 259–265.

Morris M D. 1991. Factorial sampling plans for preliminary computational experiments. *Technometrics.* **33**(2): 161–174.

Morris D J, Speirs D C, Cameron A I, Heath M R. 2014. Global sensitivity analysis of an end-to-end marine ecosystem model of the North Sea: Factors affecting the biomass of fish and benthos. *Ecological Modelling.* **273**: 251–263.

Niesenbaum R A. 1988. The ecology of sporulation by the macroalga Ulva lactuca L. (chlorophyceae). *Aquatic Botany.* **32**(1): 155–166.

Paalme T, Kukk H, Kotta J, Orav H. 2002. "In vitro" and "in situ" decomposition of nuisance macroalgae Cladophora glomerata and Pilayella littoralis. *Hydrobiologia.* **475–476**(1): 469–476.

Pang S J, Liu F, Shan T F, *et al.* 2010. Tracking the algal origin of the Ulva bloom in the Yellow Sea by a combination of molecular, morphological and physiological analyses. *Marine Environmental Research.* **69**(4): 207–215.

Ren J S, Barr N G, Scheuer K, *et al.* 2014. A dynamic growth model of macroalgae: Application in an estuary recovering from treated wastewater and earthquake-driven eutrophication. *Estuarine, Coastal and Shelf Science.* **148**: 59–69.

Saltelli A, Tarantola S, Campolongo F. 2000. Sensitivity analysis as an ingredient of modeling. *Statistical Science.* **15**(4): 377–395.

Shen C, Shi H, Liu Y, Li F, Ding D. 2016. Discussion of skill improvement in marine ecosystem dynamic models based on parameter optimization and skill assessment. *Chinese Journal of Oceanology and Limnology.* **34**: 683–696. DOI: 10.1007/s00343-016-5068-3.

Shi H H, Shen C C, Li F, *et al.* 2014. Parameter sensitivity analysis of a coupled biological-physical model in Jiaozhou Bay. *Acta Ecologica Sinica.* **34**(1): 41–49.

Sobol I M. 1993. Sensitivity estimates for nonlinear mathematical models. *Mathematical Modelling and Computational Experiments.* **14**(4): 407–414.

Song W, Peng K Q, Xiao J, *et al.* 2014. Effects of temperature on the germination of green algae micro-propagules in coastal waters of the Subei Shoal, China. *Estuarine, Coastal and Shelf Science.* DOI: 10.1016/j.ecss.2014.08.007.

Wang J F, Jiang P, Cui Y L, *et al.* Molecular analysis of green-tide-forming macroalgae in the Yellow Sea. *Aquatic Botany.* **93**(1): 25–31.

Wu T. 2013. *Preliminary Study on the Influence of Nutrients on the Growth of Ulva Prolifera and Its Absorption of Different Nitrogen Species.* Qingdao: Ocean University of China.

Xu C G, Hu Y M, Chang Y, *et al.* 2004. Sensitivity analysis in ecological modeling. *Chinese Journal of Applied Ecology.* **15**(6): 1056–1062.

Zheng W, Shi H H, Fang G H, *et al.* 2012. Global sensitivity analysis of a marine ecosystem dynamic model of the Sanggou Bay. *Ecological Modelling.* **247**: 83–94.

Chapter 6

Construction and Applications of Planktonic Ecosystem Model in the Coastal Kelp Culturing Area

Kelp is a large brown marine algae plant that grows in low-temperature seawater and is widely used in food processing, medicine, and chemicals. In China, kelp is a product that is important for China's economy seaweed with the largest aquaculture area and production in the world. In this chapter, with the nutrient-phytoplankton-zooplankton-detritus (NPZD) model as the basic structure and sea surface temperature and light intensity as the impact of external environment on marine ecology, kelp was added as a large algae component to construct a phytoplankton ecosystem model in the kelp culture area, while Sanggou Bay, as a typical kelp culture area in Northern China, was used to set all parameters. The effects of marine aquaculture activities on planktonic ecosystems were analyzed with different kelp culture densities and terrestrial nutrient supplementation. By means of the local sensitivity analysis and global sensitivity analysis, the later including the Morris and Sobol methods, the key influence parameters of the kelp culturing model were investigated.

6.1 Construction of Planktonic Ecosystem Model of the Coastal Kelp Culturing Area

6.1.1 *Construction of the NPZKD model*

Based on phytoplankton (P), zooplankton (Z), kelp (K), dissolved inorganic nitrogen (DIN), dissolved inorganic phosphorus (DIP), organic debris (D), light intensity (I), and temperature (T), the phytoplankton ecosystem model in the kelp culturing area can be named as the NPZKD model, and its differential equation over time is expressed as:

$$
\begin{cases}
\dfrac{dP}{dt} &= U_m \min\left(\dfrac{DIN}{K_N+DIN}, \dfrac{DIP}{K_P+DIP}\right) f(I)g(T)P \\
& \quad -G_m h(T)Z(1-e^{-2P}) - M_p P \\
\dfrac{dZ}{dt} &= \gamma G_m h(T)Z(1-e^{-\lambda P}) - M_Z Z \\
\dfrac{dK}{dt} &= U_k Q(T) \min\left(\dfrac{DIN}{Kk_N+DIN}, \dfrac{DIP}{Kk_P+DIP}\right) K - M_K K \\
\dfrac{dDIN}{dt} &= \alpha_1 \left(-U_m \min\left(\dfrac{DIN}{K_N+DIN}, \dfrac{DIP}{K_P+DIP}\right) f(I)g(T)P\right. \\
& \quad \left. +\theta G_m h(T)Z(1-e^{-2P}) + eD\right) \\
& \quad -\beta_1 U_k Q(T) \min\left(\dfrac{DIN}{Kk_N+DIN}, \dfrac{DIP}{Kk_P+DIP}\right) K \\
& \quad +N_C \alpha_1/\alpha_2 \left(0.003 + 0.001\cos\left(2\pi\dfrac{t-180}{365}\right)\right) \\
\dfrac{dDIP}{dt} &= \alpha_2 \left(-U_m \min\left(\dfrac{DIN}{K_N+DIN}, \dfrac{DIP}{K_P+DIP}\right) f(I)g(T)P\right. \\
& \quad \left. +\theta G_m h(T)Z(1-e^{-\lambda P}\right) + eD) \\
& \quad -\beta_2 U_k Q(T) \min\left(\dfrac{DIN}{Kk_N+DIN}, \dfrac{DIP}{Kk_P+DIP}\right) K \\
& \quad +N_C \left(0.003 + 0.001 \times \cos\left(2\pi\dfrac{t-180}{365}\right)\right) \\
\dfrac{dD}{dt} &= (1-\gamma-\theta) G_m h(T)Z(1-e^{-2P}) \\
& \quad -eD + M_P P + M_Z Z + M_K K
\end{cases}
\tag{6.1}
$$

where $g(T)$ and $h(T)$ are Q_{10} law of photosynthesis and respiration (Fransz *et al.*, 1985).

$$g(T) = Qg_{10}^{\left(\frac{T-10}{T}\right)}, \quad h(T) = Qh_{10}^{\left(\frac{T-10}{T}\right)} \tag{6.2}$$

$Q(T)$ is the temperature limit function of kelp growth (Duarte *et al.*, 2003), which can be defined as:

$$Q(T) = \frac{2.0 \cdot (1+\beta) \cdot x_t}{x_t^2 + 2.0 \cdot \beta \cdot x_t + 1.0}, \quad x_t = \frac{T - T_{leth}}{T_{opt} - T_{leth}}. \tag{6.3}$$

$f(I)$ is the light intensity limit function (Steele, 1962). According to the measured data, the light intensity of the seawater surface is fitted as the following formula:

$$I_S = 200.38 - 116.47 \times \cos\left(2\pi \cdot \frac{t-60}{365}\right). \tag{6.4}$$

In particular, $t = 0$ is defined as the initial time of kelp culturing.

6.1.2 *Setting of the NPZKD model*

As an example, the coastal kelp culturing cycle in Sanggou Bay is usually from early November to May the following year, so the simulation time of the NPZKD model is set to 0–200 days (Zheng *et al.*, 2012). The initial conditions of the model were set according to the survey data of Sanggou Bay in November. The phytoplankton (P), zooplankton (Z), dissolved inorganic nitrogen (DIN), dissolved inorganic phosphorus (DIP), and organic debris (D) were set to 3.51, 4.14, 8.96, 0.45, and 16.76 μmol/L, respectively. Kelp (K) could be set according to a different culture density. Parameters and values of the model are shown in Table 6.1.

6.2 Scenario Simulation of the Impact of Kelp Culturing on the Planktonic Ecosystem

In order to analyze the impact of kelp culture density on phytoplankton ecosystems, the kelp culture density was designed to be at three

Table 6.1. Parameters and their values of the phytoplankton ecosystem model.

Parameter	Value	Unit	Definition of the parameter	Reference
U_m	1.8	d^{-1}	The maximum growth rate of phytoplankton	(Wang *et al.*, 2006)
k_N	8	μmol/L	Half-saturation constant of phytoplankton DIN adsorption	(Wang *et al.*, 2006)
k_P	0.6	μ mol/L	Half-saturation constant of phytoplankton DIP adsorption	(Wang, *et al.*, 2006)
Kk_N	8	μmol/L	Half-saturation constant of kelp DIN absorption	(Wang, *et al.*, 2006)
Kk_P	0.6	μmol/L	Half-saturation constant of kelp DIP absorption	(Wang, *et al.*, 2006)
U_k	0.04	d^{-1}	The maximum growth rate of kelp	(Mao *et al.*, 1993)
G_m	0.6	d^{-1}	The maximum intaking ratio of zooplankton	This study
λ	0.1	μmol/L	Ivlev uptaking constant	(Wang *et al.*, 2006)
γ	0.2	d^{-1}	Growth factor of zooplankton	(Gao *et al.*, 1998)
θ	0.12	d^{-1}	Excretion coefficient of zooplankton	(Wang *et al.*, 2006)
I_0	180	W/m^2	Optimal light intensity	(Wu *et al.*, 2009)
Qg_{10}	2.08	—	Temperature dependence coefficient of growth	(Fransz *et al.*, 1985)
Qh_{10}	3.1	—	Temperature dependence coefficient of food intaking	(Fransz *et al.*, 1985)
M_P	0.1	d^{-1}	Mortality of phytoplankton	(McCreary *et al.*, 1996)
M_Z	0.2	d^{-1}	Mortality of zooplankton	(Fransz *et al.*, 1985)
M_K	5×10^{-4}	d^{-1}	Mortality of kelp	This study
e	0.05	d^{-1}	Remineralization rate of debris	(McCreary *et al.*, 1996)
β	3	—	Temperature regulation parameters for kelp growth	(Andersen and Nival, 1989)
T_{opt}	13	°C	Optimum temperature for kelp growth	(Petrell *et al.*, 1993)
T_{ieth}	25	°C	Deadly temperature for kelp growth	(Petrell *et al.*, 1993)
α_1	16/106	mol/mol	Nitrogen to carbon ratio of phytoplankton	(Redfield, 1958)
α_2	1/106	mol/mol	Phosphorous to carbon ratio of phytoplankton	(Redfield, 1958)
β_1	1/6.5	mol/mol	Nitrogen to carbon ratio of kelp	(Huang *et al.*, 2005)
β_2	1/215	mol/mol	Phosphorus to carbon ratio of kelp	(Huang *et al.*, 2005)
N_c	0/1	—	Without terrestrial input/with terrestrial input	—

different levels: high, medium, and low, in which the initial values of kelp were set to 20, 10, and 5 μmol/L, respectively. Numerical simulation was carried out with and without terrestrial nutrition with salt supplement.

6.2.1 *Scenario simulation of kelp culture density with terrestrial nutrient supplementation*

Scenario simulation results showed (Fig. 6.1) that the growth of phytoplankton was slower at a higher culture density. Therefore, kelp was always in a favorable position in the feeding competition with phytoplankton, regardless of the culture density. As zooplankton mainly feed on phytoplankton, its growth trend had a relatively heavy dependence on phytoplankton as organic debris. In addition, there was a small peak of phytoplankton biomass in spring, which was consistent with the actual observation results of Sanggou Bay, and the peak got smaller with the increase of culture density (Fig. 6.1). In the later period of kelp growth, photosynthesis significantly increased its nutrient uptake, making DIN decrease gradually and then stabilize. The main reason is that the phytoplankton biomass and kelp are at a relatively high value and consume a large amount of DIP, and DIP gradually becomes the main limiting factor.

6.2.2 *Scenario simulation of kelp culture density without terrestrial nutrient supplementation*

Regardless of the input of exogenous nutrients, the growth of phytoplankton and kelp showed a similar trend (Fig. 6.2). In addition, the growth rate of phytoplankton and kelp had been reduced to some degree, no matter the culture density. Especially under the conditions at low culture density, the peak value of phytoplankton in spring was significantly lower than that with the supplement of land-based nutrients (Figs. 6.1 and 6.2). This happens because supplementary exogenous nutrients are absent, thus limiting effects of nutrient salts are more significant on the growth of phytoplankton and kelp.

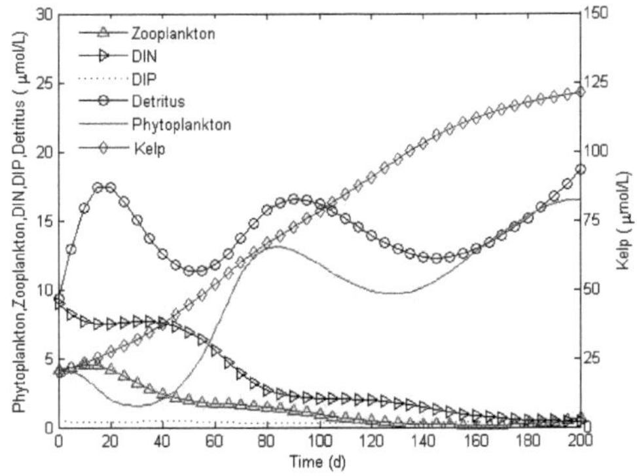

(a)

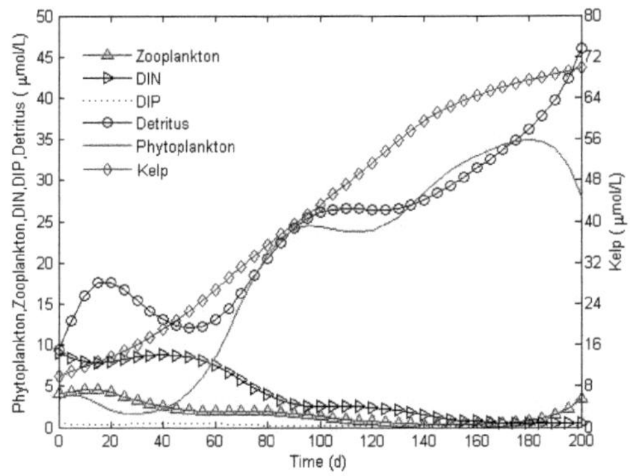

(b)

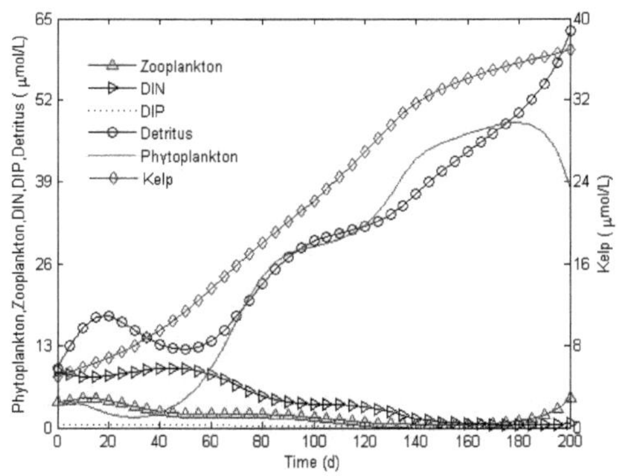

(c)

Fig. 6.1. NPZKD model results at different culture densities with terrestrial nutrition: (a) high culture density (initial value of kelp was 20 μmol/L), (b) medium culture density (initial value of kelp was 10 μmol/L), and (c) low culture density (the initial value of kelp is 5 μmol/L).

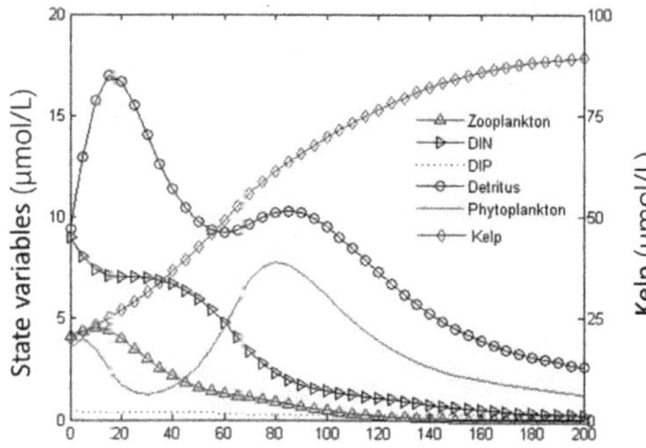

(a)

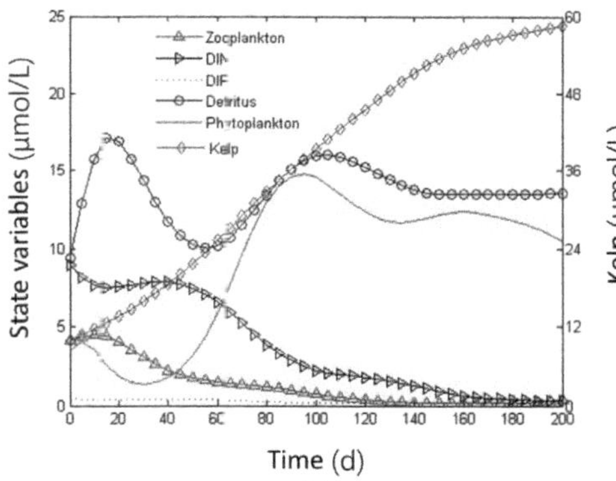

(b)

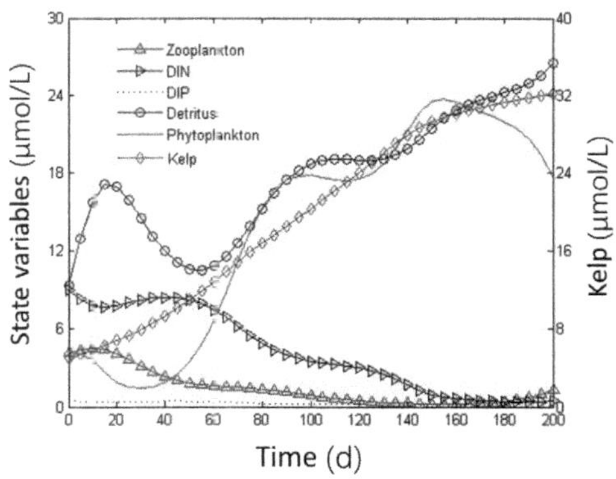

(c)

Fig. 6.2. NPZKD model results at different culture densities without terrestrial nutrition: (a) high culture density (initial value of kelp was 20 μmol/L), (b) medium culture density (initial value of kelp was 10 μmol/L), and (c) low culture density (the initial value of kelp was 5 μmol/L).

6.3 Analysis of Key Influence Parameters of Kelp Culturing According to the Local Sensitivity Analysis

Sensitivity analysis is an important basis for model optimization. Because the dynamic system is usually very sensitive to the initial value, setting the initial value will also affect the accuracy of the model. In this chapter, based on the local sensitivity analysis method, the response of the NPZKD model to initial values and changes in parameters will be discussed in the case of medium culturing density.

6.3.1 *Analysis method of local sensitivity*

The local sensitivity is generally defined as the ratio of the change degree of the model state variable to the change degree of the parameter (the initial value is also a model parameter), so the sensitivity of the i-th state variable (F_i) to the j-th parameter (x_j) is defined as:

$$SA_{ij} = \frac{\Delta F_i/F_i}{\Delta x_j/x_j}, \tag{6.5}$$

where

$\Delta F_i/F_i$ is the changing ratio of a state variable in the model,
$\Delta x_j/x_j$ is the changing ratio of a parameter that is related to F_i.

6.3.2 *Response of model state variables to changes in initial values*

(1) *With terrestrial nutrition supplements*
According to the definition of local sensitivity, the sensitivity of each state variable to its initial value was calculated with the supplement of terrestrial nutrients. The results were shown in Table 6.2.

With terrestrial nutrient supplementation, the absolute value of the sensitivity coefficient of each state variable of the model to the initial value was in the following order as: (from small to large) Z_0, DIP_0, DIN_0, D_0, P_0 and K_0. The sensitivity of zooplankton was the largest because its nutritional level was the highest in this model. This was exacerbated by the increase of the nutritional level due to the cumulative effect of interference of the initial value change of the state variable on itself.

Table 6.2. Local sensitivity of model state variables to initial values (with terrestrial nutrition supplements).

Initial value	Phytoplankton	Zooplankton	Kelp	DIN	Detritus	DIP	Average
P_0 (Phytoplankton)	0.02	−0.32	−0.04	−0.10	0.05	−0.11	−0.08
Z_0 (Zooplankton)	0.13	1.42	0.06	0.19	0.16	0.11	0.34
K_0 (Kelp)	−0.45	−0.33	0.92	−0.13	−0.37	0.46	0.02
DIN_0 (DIN)	0.43	0.13	0.07	0.85	0.34	−1.02	0.13
D_0 (Detritus)	0.08	0.21	0.05	0.03	0.19	0.08	0.11
DIP_0 (DIP)	−0.33	−0.35	0.34	−0.30	−0.06	1.64	0.16

Table 6.3. Local sensitivity of model state variables to initial values (without terrestrial nutrition supplements).

Initial value	Phytoplankton	Zooplankton	Kelp	DIN	Detritus	DIP	Average
P_0 (Phytoplankton)	−0.03	−0.28	0.01	−0.04	0.05	−0.01	−0.05
Z_0 (Zooplankton)	0.34	1.33	0.01	0.04	0.25	−0.03	0.32
K_0 (Kelp)	−0.70	−0.68	0.88	−0.28	−0.54	0.37	−0.16
DIN_0 (DIN)	0.78	0.60	0.13	1.24	0.60	−1.12	0.37
D_0 (Detritus)	0.18	0.37	0.06	0.04	0.26	0.10	0.17
DIP_0 (DIP)	−0.50	−0.18	0.36	−0.41	−0.09	1.98	0.19

(2) *Without terrestrial nutrition supplements*

The response of the model state variable to the initial value change without the input of terrestrial nutrients and calculated sensitivity matrix is listed in Table 6.3.

As listed in Table 6.3, the model was sensitive the most to the initial values of nutrients (DIN, DIP), mainly due to the material basis of nutrients in planktonic ecosystems and kelp culturing. The initial sensitivity of zooplankton was due to the fact that it did not participate in the kelp growth cycle and had no direct feeding relationship with nutrients. The change in its initial value had a relatively large impact on the zooplankton growth equation (the sensitivity to itself reached 1.33).

As shown in Tables 6.2 and 6.3, after the terrestrial nutrition supplements were eliminated, the sensitivity of kelp was improved to some degree because kelp consumed a large amount of nutrients in later cultivation, and a nutrient limitation could be easily formed with the initial value increase but without terrestrial supplements. The sensitivity improvement of DIN and DIP was also closely

Table 6.4. Sensitivity of model state variables to parameters (with terrestrial nutrition supplements).

Parameter	Phytoplankton	Zooplankton	Kelp	DIN	Detritus	DIP	Average
U_m	0.65	0.61	−0.60	−0.87	0.92	−1.21	−0.08
k_N	−0.28	−0.10	0.16	0.53	−0.21	0.41	0.08
k_P	−0.23	−0.46	0.25	0.20	−0.43	0.47	−0.03
Kk_N	0.17	0.05	−0.16	0.04	0.12	−0.19	0.00
Kk_P	0.45	0.29	−0.56	0.14	0.37	−0.45	0.04
G_m	−0.82	2.25	0.26	0.75	−0.27	0.53	0.45
λ	−0.28	0.47	0.10	0.10	−0.18	0.11	0.05
γ	−0.83	1.82	0.19	0.46	−0.35	0.37	0.28
θ	0.00	0.01	0.01	0.01	−0.04	0.02	0.00
I_0	−0.59	−1.01	0.49	0.67	−0.84	0.99	−0.05
Qg_{10}	−0.86	−1.22	0.41	0.62	−0.69	0.96	−0.13
Qh_{10}	0.45	−0.55	−0.13	−0.23	0.17	−0.31	−0.10
M_P	−0.71	−0.69	0.27	0.58	−0.15	0.63	−0.01
M_Z	1.15	−1.44	−0.30	−0.36	0.62	−0.54	−0.15
M_K	0.04	0.02	−0.05	0.01	0.04	−0.01	0.01
e	0.22	0.36	0.10	0.08	−0.54	0.17	0.07
β	−0.01	0.00	0.01	0.00	−0.01	0.01	0.00
T_{opt}	0.09	0.03	−0.09	0.02	0.06	−0.09	0.00
T_{leth}	−0.04	−0.02	0.05	−0.01	−0.03	0.05	0.00
U_k	−1.04	−0.68	1.24	−0.28	−0.85	1.08	−0.09
α_1	−0.44	−0.17	−0.06	−0.18	−0.35	1.03	−0.03
α^2	−0.05	−0.19	−0.01	0.08	−0.07	−0.56	−0.13
β_1	0.01	0.00	0.00	0.02	0.00	−0.03	0.00
β_2	−0.35	−0.07	−0.07	−0.03	−0.29	−0.09	−0.15

related — the nutrients easily limited the growth of phytoplankton and kelp at the late operation of the model.

6.3.3 *Response of model state variables to parameter changes*

(1) *With terrestrial nutrition supplements*

With terrestrial nutrition supplements, the sensitivity of each state variable to the model parameters was calculated (Table 6.4). Based on these results, the response strength of the state variables to parameter changes could be analyzed in detail.

According to the average sensitivity of each state variable, the maximum zooplankton intake rate (G_m), zooplankton growth coefficient (λ), phytoplankton phosphorus/carbon ratio (α_1), zooplankton mortality (M_Z), and kelp phosphorus/carbon ratio (β_2) were listed as the five most sensitive parameters. Therefore, it could be found that zooplankton growth parameters had a great impact on the reliability of the model. The parameter sensitivity ranking is shown in Table 6.5.

(2) *Without terrestrial nutrition supplements*
Similarly, with the absence of terrestrial nutrients, the sensitivity of each state variable to model parameters was calculated (Table 6.6). Based on these results, the response strength of the state variables to parameter changes could be analyzed in detail.

According to the average sensitivity of each state variable, parameters, such as the maximum growth rate of kelp (U_k), Qg_{10}, optimal light intensity, half-saturation constant of kelp DIP absorption (Kk_p), and phytoplankton maximum growth rate (U_m) were listed as the five most sensitive parameters. The parameter sensitivity ranking is shown in Table 6.7.

Compared with that with terrestrial input, the choice of kelp growth parameters became more sensitive. The lethal temperature of kelp growth (T_{leth}), temperature adjustment parameter of kelp growth (β), zooplankton excretion coefficient (θ), and other parameters were the most insensitive parameters, with or without external input.

Parameter optimization is a very complicated task, especially for the ecological dynamics model, because it has a many parameters, the equation is complicated, there is a nonlinear relationship between the state variables, and parameters often affect each other simultaneously when they are optimized. Although existing optimization methods, such as accompanying assimilation, least squares, etc., can improve the accuracy of the model through optimization, it is difficult to ensure that the optimized parameters have practical significance due to the interaction between parameters. We believe that after the sensitivity of each parameter is sorted in order, the main parameters with high sensitivity can be optimized. If those main parameters are obtained through experiments, we do not need to optimize them further.

Table 6.5. The average sensitivity ranking of state variables to parameters (with terrestrial nutrition supplements).

Order	1	2	3	4	5	6	7	8
Parameter	The maximum intaking of zooplankton G_m	Zooplankton growth factor λ	Phosphorus to carbon ratio of phytoplankton α_2	Mortality of zooplankton M_Z	Phosphorus to carbon ratio of kelp β_2	Correlation coefficient of temperature to growth Qg_{10}	Correlation coefficient of temperature to food intaking Qh_{10}	Maximum growth rate of kelp U_k

Order	9	10	11	12	13	14	15	16
Parameter	Half-saturation constant of phytoplankton DIN adsorption k_N	Maximum phytoplankton growth rate U_m	Remineralization rate of debris e	Ivlev uptake constant λ	Optimal light intensity I_0	Half-saturation constant of kelp DIP absorption Kk_P	Half-saturation constant of phytoplankton DIP adsorption k_P	Nitrogen to carbon ratio of kelp β_1

Order	17	18	19	20	21	22	23	24
Parameter	Mortality of phytoplankton M_P	Mortality of kelp M_K	Half-saturation constant of Kelp DIN absorption Kk_N	Optimal temperature for kelp growth T_{opt}	Deadly temperature for kelp growth T_{leth}	Temperature regulation parameters for kelp growth β	Excretion coefficient of zooplankton θ	Nitrogen to carbon ratio of phytoplankton α_1

Table 6.6. Sensitivity of model state variables to parameters (without terrestrial nutrition supplements).

Parameter	Phytoplankton	Zooplankton	Kelp	DIN	Detritus	DIP	Average
U_m	0.94	1.37	−0.46	−0.65	1.18	−1.01	0.23
k_N	−0.36	−0.27	0.10	0.42	−0.27	0.30	−0.02
k_P	−0.41	−0.85	0.22	0.12	−0.59	0.43	−0.18
Kk_N	0.26	0.20	−0.12	0.10	0.17	−0.14	0.08
Kk_P	0.76	0.75	−0.53	0.30	0.59	−0.39	0.25
G_m	−0.55	1.18	0.16	0.16	−0.35	0.24	0.14
λ	−0.33	0.38	0.10	0.03	−0.26	0.09	0.00
γ	−0.77	0.91	0.16	0.12	−0.50	0.29	0.04
θ	−0.01	0.01	0.01	0.00	−0.04	0.02	0.00
I_0	−0.87	−1.48	0.38	0.47	−1.03	0.81	−0.29
Qg_{10}	−1.27	−1.56	0.33	0.51	−0.95	0.88	−0.34
Qh_{10}	0.50	−0.30	−0.10	−0.15	0.26	−0.28	−0.01
M_P	−0.99	−1.15	0.22	0.52	−0.37	0.59	−0.20
M_Z	1.44	−1.18	−0.30	−0.21	0.91	−0.57	0.02
M_K	0.06	0.06	−0.04	0.02	0.07	0.00	0.03
e	0.22	0.50	0.09	0.06	−0.61	0.16	0.07
β	−0.02	−0.01	0.01	−0.01	−0.01	0.01	−0.01
T_{opt}	0.13	0.11	−0.07	0.05	0.10	−0.07	0.04
T_{leth}	−0.07	−0.06	0.04	−0.03	−0.05	0.04	−0.02
U_k	−1.53	−1.32	1.02	−0.63	−1.16	0.78	−0.47
$alpha_1$	−0.67	−0.50	−0.11	−0.36	−0.51	0.97	−0.20
α_2	−0.09	−0.25	−0.01	0.12	−0.11	−0.59	−0.15
β_1	−0.10	−0.08	−0.02	−0.07	−0.09	0.16	−0.03
β_2	0.05	0.16	−0.03	0.09	0.02	−0.25	0.01

Of course, parameter optimization and sensitivity analysis are often carried out according to the main purpose of modeling. The specific analysis and parameter optimization of the sensitivity of the state variables will not be repeated here due to the limited space.

6.4 Morris Sampling-based Global Sensitivity Analysis on Key Impact Parameters of Kelp Farming

6.4.1 *Analysis settings of parameter global sensitivity*

The key parameters affecting kelp culture were analyzed with the global sensitivity analysis method based on Morris sampling. In order to avoid the interference of the initial values of the state variables

Table 6.7. Average sensitivity ranking of state variables to parameters (without terrestrial nutrition supplements).

Order	1	2	3	4	5	6	7	8
Parameter	The maximum growth rate of kelp U_k	Correlation coefficient of temperature to growth Qg_{10}	Optimal light intensity I_0	Half-saturation constant of kelp DIP absorption Kk_P	Maximum phytoplankton growth rate U_m	Nitrogen to carbon ratio of phytoplankton α_1	Mortality of phytoplankton M_P	DIP Half-saturation constant of phytoplankton DIP absorption k_P

Order	9	10	11	12	13	14	15	16
Parameter	Phosphorus-carbon ratio of phytoplankton α_2	Maximum intake rate G_m	Half-saturation constant of kelp DIN absorption Kk_N	Remineralization rate of debris e	Optimal temperature for kelp growth T_{opt}	Growth factor of zooplankton γ	Nitrogen to carbon ratio of kelp β_1	Mortality of kelp M_K

Order	17	18	19	20	21	22	23	24
Parameter	Deadly temperature for kelp growth T_{leth}	Mortality of zooplankton M_Z	Half-saturation constant of phytoplankton DIN adsorption k_N	Correlation coefficient between temperature and food intake Qh_{10}	Phosphorus to carbon ratio of kelp β_2	Temperature regulation parameters for kelp growth β	Ivlev uptake constant λ	Excretion coefficient of zooplankton θ

and selection of parameters at the initial construction period of the model, the global sensitivity analysis was performed in spring (March to May, 90 days in total) after which the model ran smoothly, and the medium density of kelp culture was selected. With the terrestrial input, the model could basically simulate the evolution of the ecosystem after debugging (Fig. 6.1). In the global sensitivity analysis, the parameter range should not be set too large. Generally, the experiment should be started from a relatively small range. If the accuracy requirements could be satisfied, the range would not be expanded. If the parameter range was set too large, it would be easy to make values of some parameters out of the reasonable range. In the global sensitivity analysis, all model parameters (24 in total) changed at the same time, the parameter change range was the default parameter, and all parameters were divided into uniform ten intervals within the range ($P = 11$), meaning that the range of each parameter change was 2% of the original default value. A total of 10,000 simulations were performed. The parameter set and corresponding sensitivity were recorded each time, based on which the average sensitivity and standard deviation were calculated.

6.4.2 *Analysis results of parameter global sensitivity*

(1) *Distribution map of sensitivity*

In order to avoid excessive dispersion of the sensitivity distribution, the calculated data was divided into 50 groups; each was repeated 200 times The sensitivity distribution of each state variable with respect to parameter changes was shown in Figs. 6.3 to 6.6.

As shown in Figs. 6.3 to 6.6, many sensitivity distributions contained both positive and negative numbers, meaning that the same state variable might be positively or negatively correlated to a model parameter under different parameter states. These results show the complexity of sensitivity analysis of the marine ecological dynamics model. Therefore, if only the average value was used to reflect the response sensitivity of each state variable to parameter changes, there might be some error generated.

It was found that the sensitivity of each state variable to the same parameter was also very different. As shown in Fig. 6.3(b),

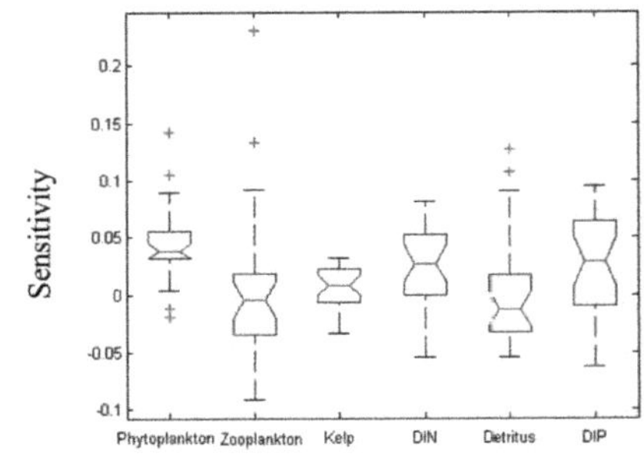

(1) Maximum growth rate of phytoplankton (U_m)

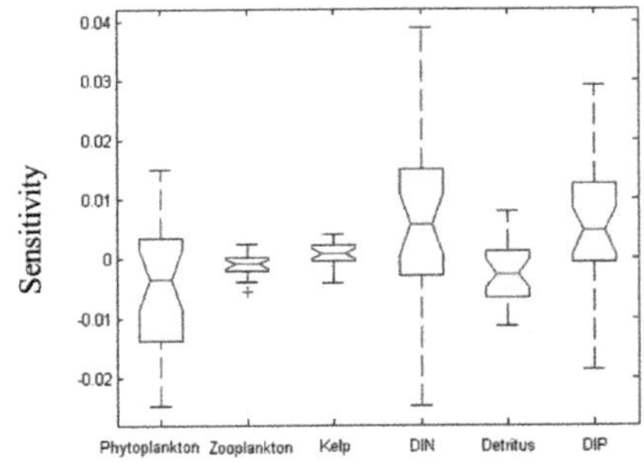

(2) Half-saturation constant of Phytoplankton DIN absorption (K_N)

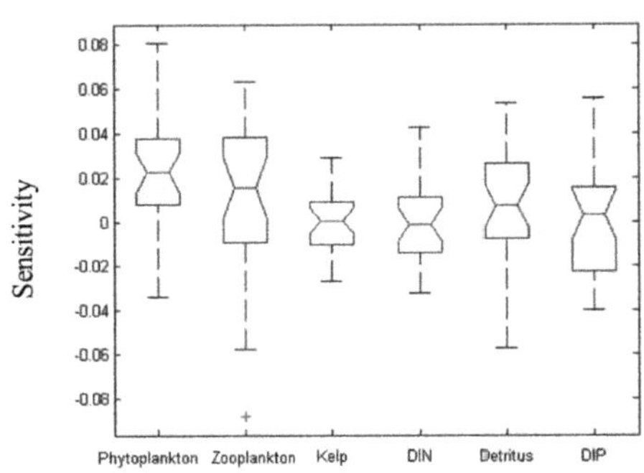

(3) Half-saturation constant of phytoplankton DIP absorption (K_P)

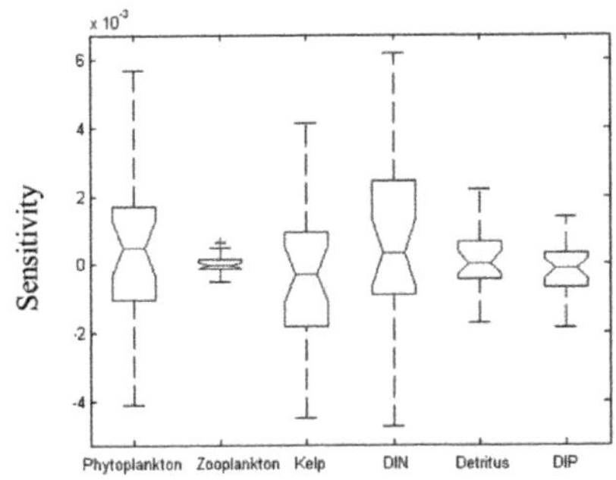

(4) Half-saturation constant of kelp DIN absorption (KK_N)

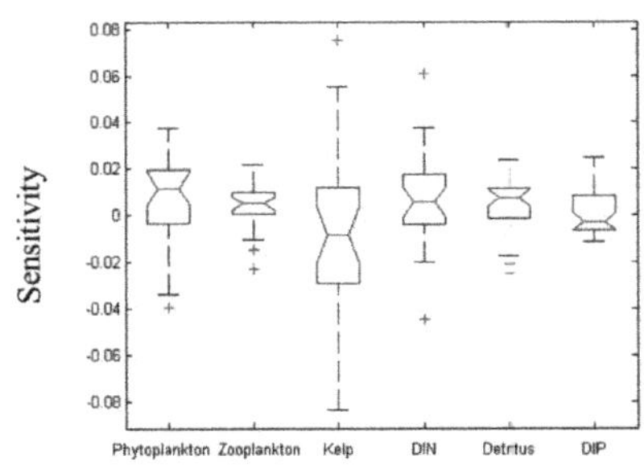

(5) Half-saturation constant of kelp DIP absorption (KK_P)

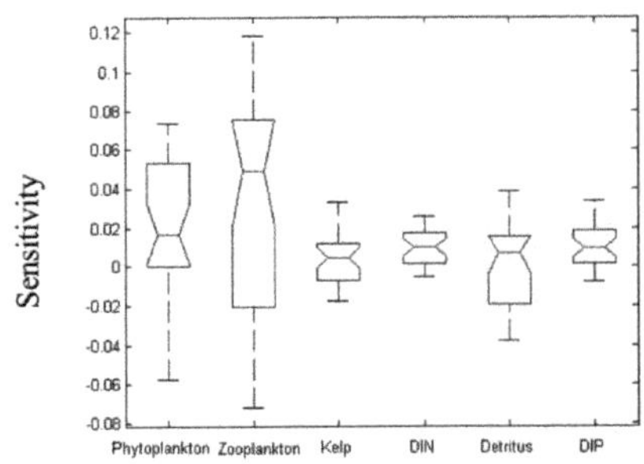

(6) Maximum feeding rate of zooplankton (G_m)

Fig. 6.3. Sensitivity distribution of each state variable related to 24 parameters.

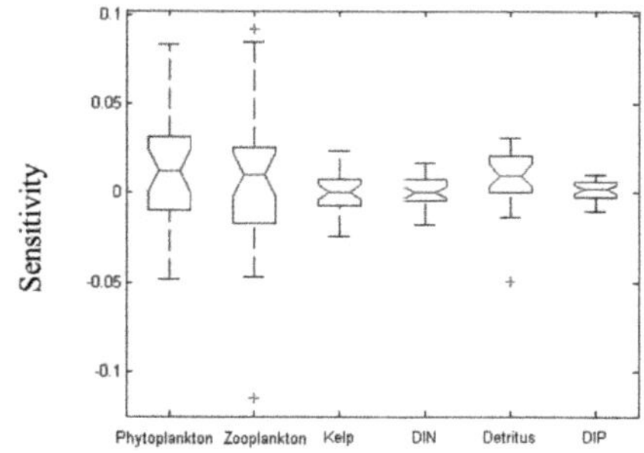

(7) Ivlev uptake constant λ

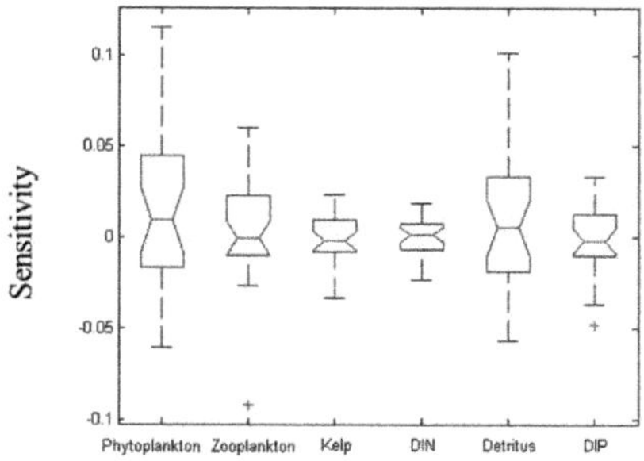

(8) Growth factor of zooplankton γ

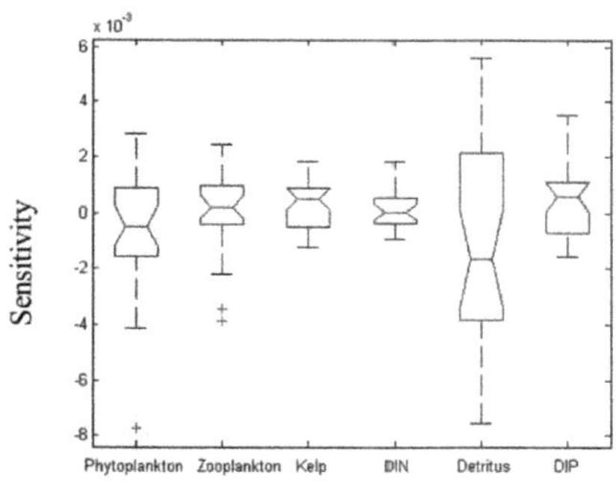

(9) Excretion coefficient of zooplankton θ

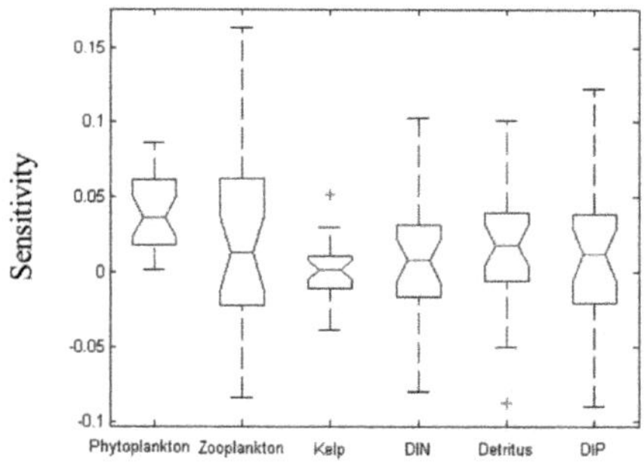

(10) Optimal light intensity I_0

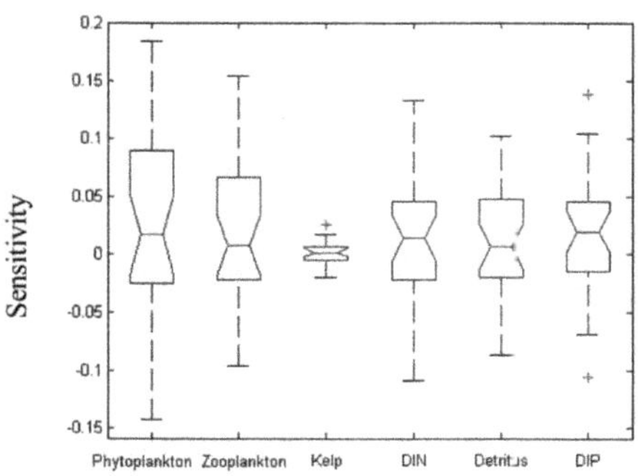

(11) Growth temperature dependence coefficient Q_{g10}

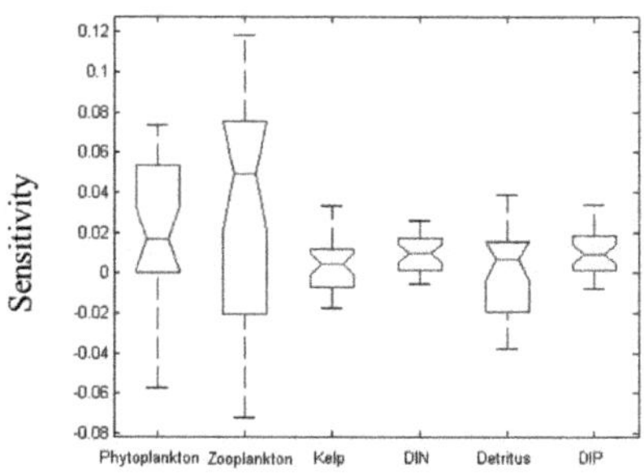

(12) Feeding temperature dependence coefficient Q_{h10}

Fig. 6.3. (*Continued*)

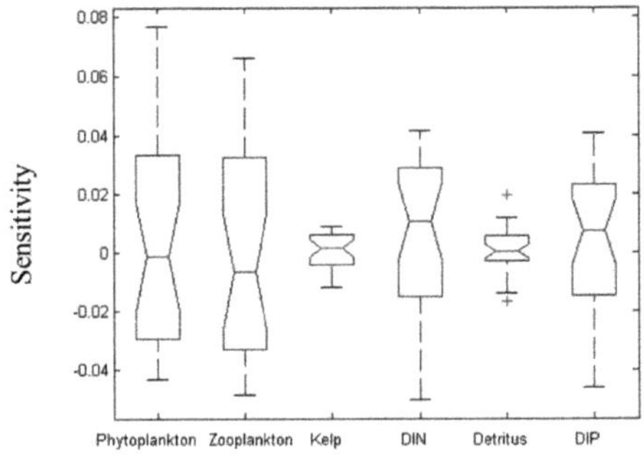

(13) Mortality of phytoplankton M_P

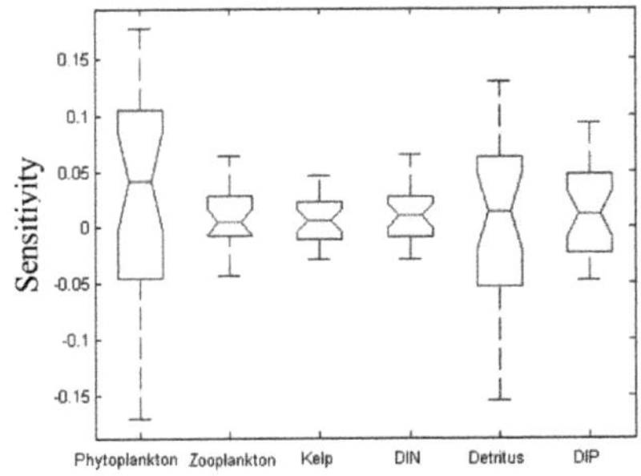

(14) Mortality of zooplankton mortality M_Z

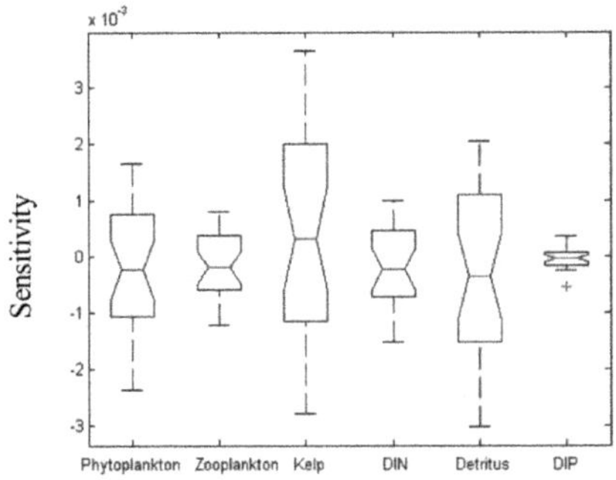

(15) Mortality of kelp M_K

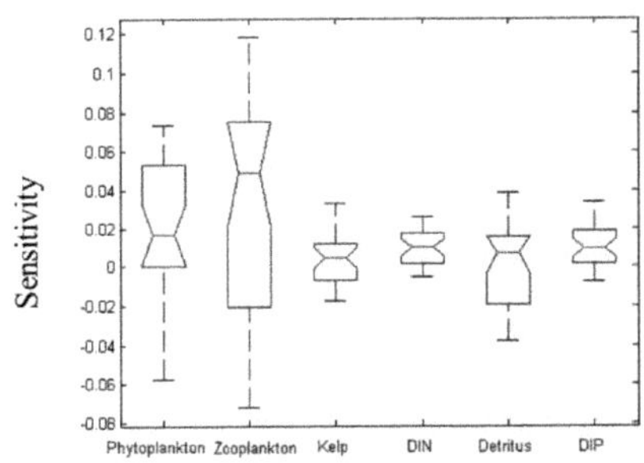

(16) Remineralization rate of debris e

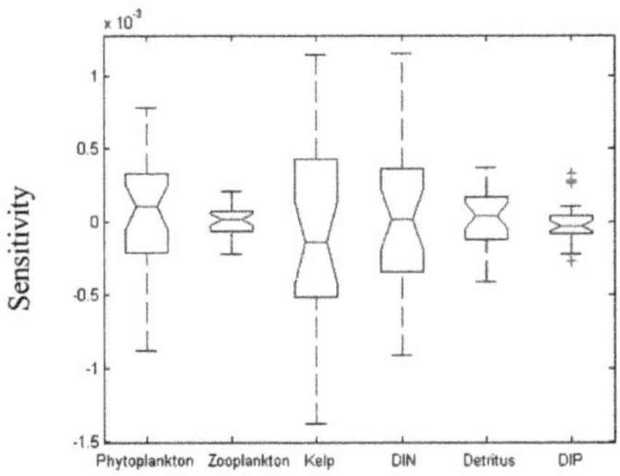

(17) Temperature regulation parameters for kelp growth β

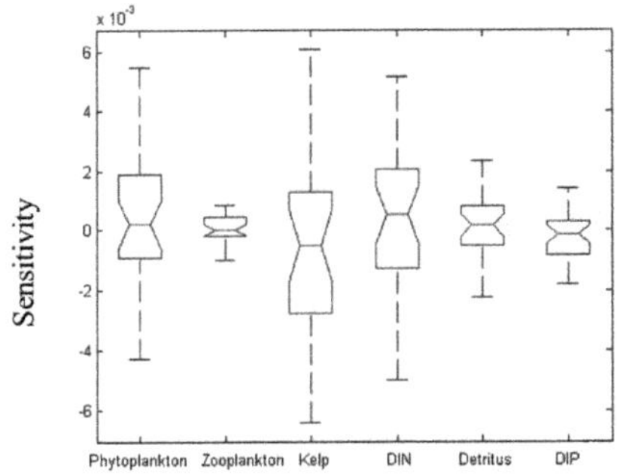

(18) Optimum temperature for kelp growth T_{opt}

Fig. 6.3. (*Continued*)

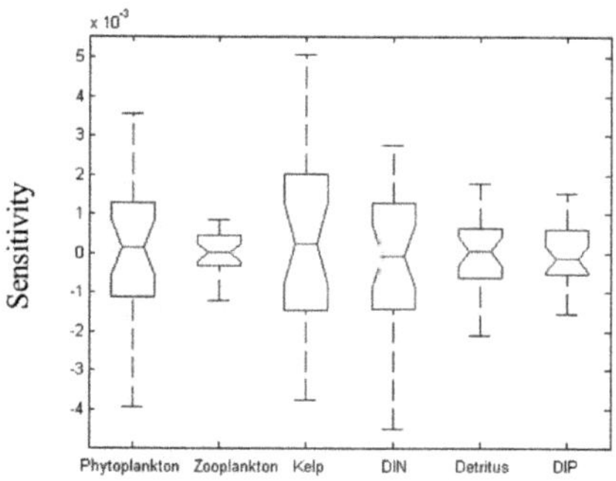

(19) Deadly temperature for kelp growth T_{leth}

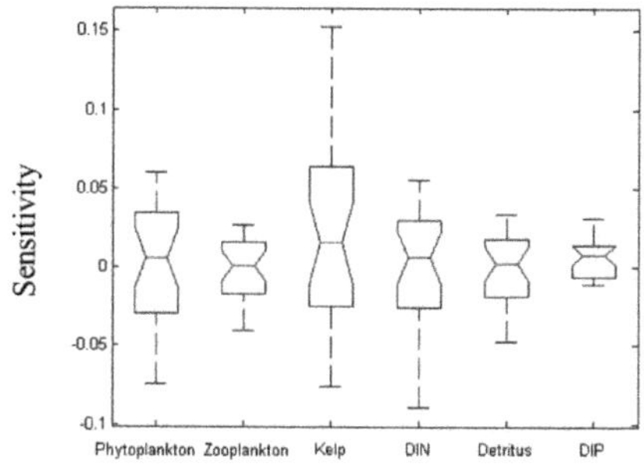

(20) Maximum growth rate of kelp U_k

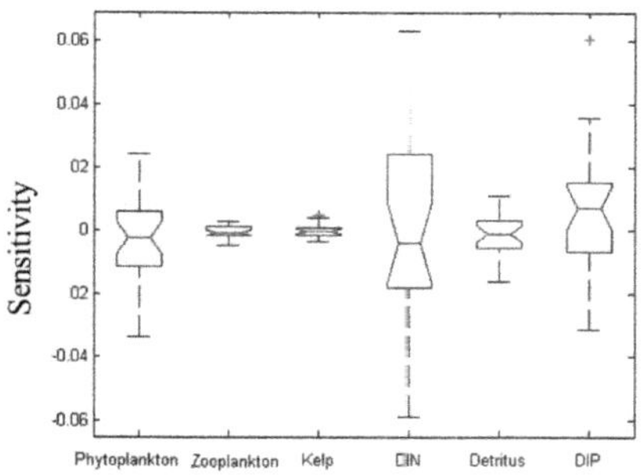

(21) Nitrogen to carbon ratio of kelp β_1

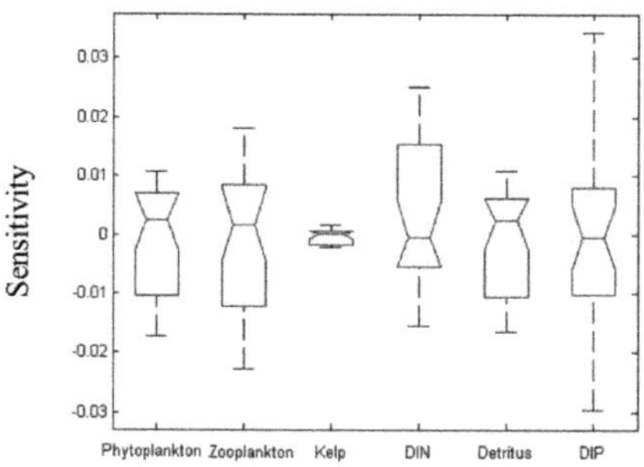

(22) Phosphorus to carbon ratio of kelp β_2

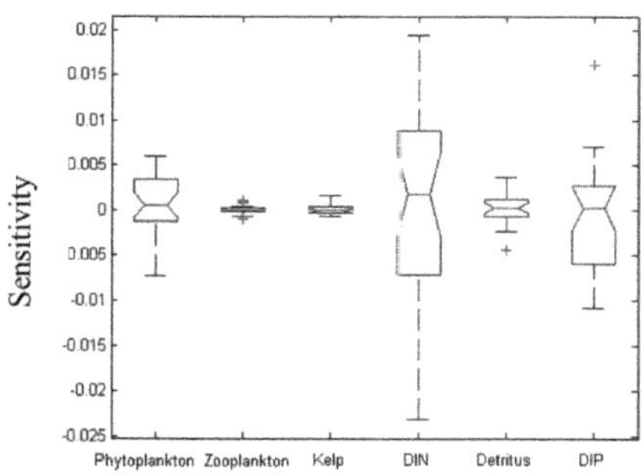

(23) Nitrogen to carbon ratio of phytoplankton α_1

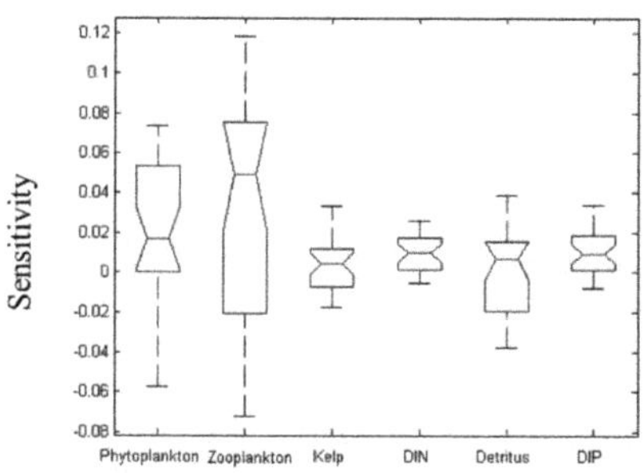

(24) Phosphorus to carbon ration of phytocarbon α_2

Fig. 6.3. (*Continued*)

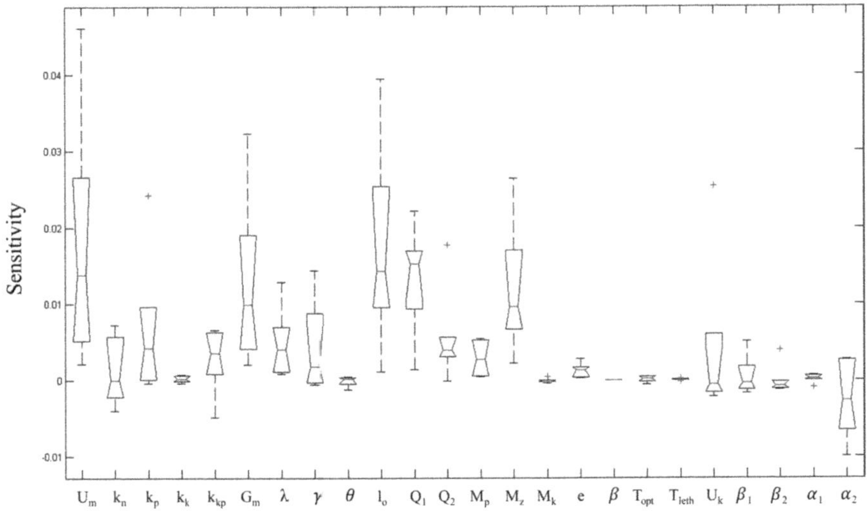

Fig. 6.4. Relationship between the mean sensitivity of each state variable and all model parameters.

the half-saturation constant of phytoplankton DIN absorption was more sensitive to phytoplankton, DIN, and DIP, and its sensitivity to zooplankton, kelp, and organic debris was less. Other parameters were similar. In addition, the sensitivity of state variables to different parameters also varied significantly, even in magnitude.

Therefore, the contribution of the parameters to the model stability as a whole body should also be needed. Further analysis on the average sensitivity of each state variable of the model to parameter changes was done and the results were shown in Fig. 6.4.

As shown in Fig. 6.4, parameters that were relatively sensitive to the model state variables included the maximum growth rate (G_m) of phytoplankton, optimal light intensity (I_0), growth temperature dependence coefficient (Qg_{10}), zooplankton mortality (M_Z), and the maximum feeding rate of zooplankton (G_m), etc. The results were different from those of the local sensitivity analysis (Table 6.4). The results of local sensitivity analysis showed that the maximum feeding rate of zooplankton (G_m), zooplankton growth coefficient (λ), zooplankton mortality (M_Z), and other zooplankton growth (death)

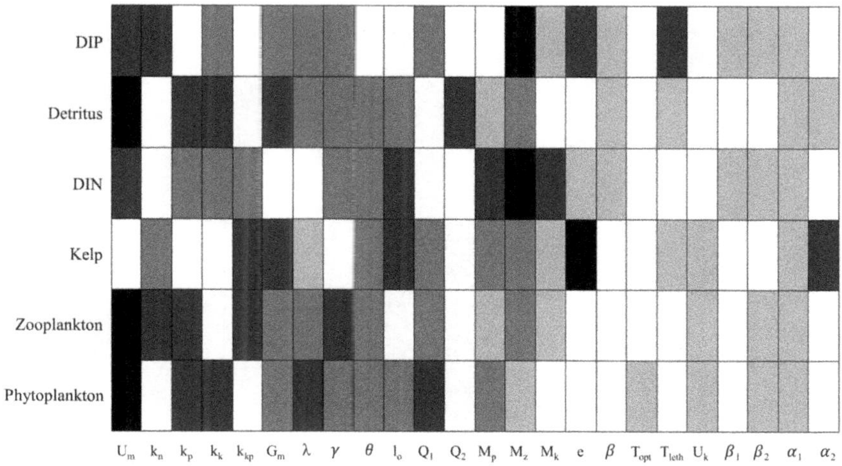

Fig. 6.5. Classification of mean sensitivity.

parameters were included in the top five most sensitive parameters, showing that zooplankton growth parameters had a great influence on the model reliability. However, the global sensitivity analysis further reflected the importance of phytoplankton growth parameters, such as maximum growth rate (U_m), optimal light intensity (I_0), and growth temperature dependence coefficient (Qg_{10}).

(2) *Classification of parameter sensitivity*

The model in this chapter involved a total of 24 parameters. In order to classify the model parameter sensitivity, the following ordering method was used: each state variable (such as phytoplankton, zooplankton, etc.) is used as the unit and the sensitivity (mean and standard deviation) of each parameter calculated: the parameter with the highest sensitivity (absolute value) was level 1, the 2nd to 4th ones were listed as level 2, the 5th to 10th ones were level 3, the 11th to 15th ones were level 4, and the rest were level 5. Each level was represented by a different color, in which a darker color indicated a higher level (closer to level 1). Sensitivity distribution mapping was made according to the mean and standard deviation of sensitivity (Figs. 6.5 and 6.6).

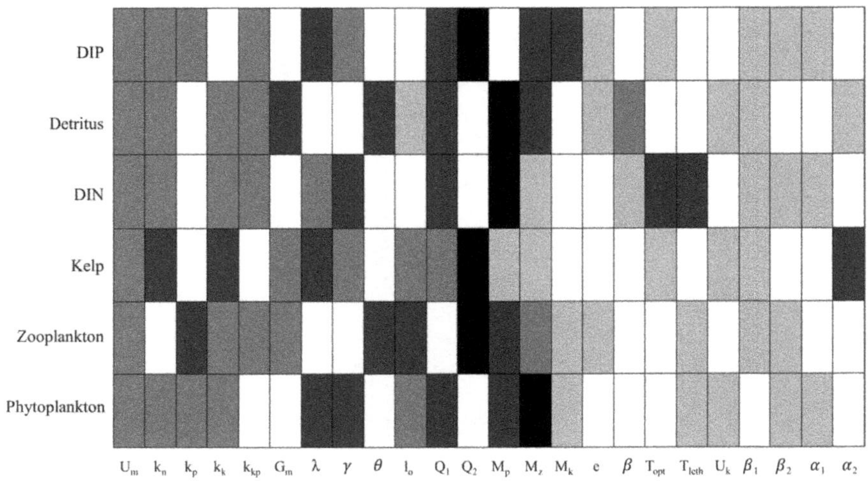

Fig. 6.6. Classification of sensitivity standard deviation.

As shown in Fig. 6.5, the most important parameters that affected each state variable could be clearly identified, and the degree of influence of each parameter on different state variables could also be obtained. According to the average sensitivity, the maximum growth rate of phytoplankton (U_m) on state variables other than kelp belonged to level 1 or level 2, and zooplankton mortality was also the most sensitive parameter of both DIN and DIP. If the status of an individual state variable was considered, it could be analyzed according to the horizontal axis of Fig. 6.5. For example, the most sensitive parameter of phytoplankton was the maximum growth rate of phytoplankton (U_m), followed by the growth temperature dependence coefficient (Q_1), zooplankton growth coefficient (λ), half-saturation constant of phytoplankton DIP adsorption, half-saturation constant of kelp DIN adsorption, etc. Other state variables could be analyzed similarly but are not described in detail here.

Table 6.8 lists the parameters that have relatively large impacts on the state variables, and Table 6.9 lists the parameters that have relatively small impacts on the state variables. Tables 6.8 and 6.9 can provide important references for parameter acquisition.

Table 6.8. Parameters having large impacts on state variables.

Impact	One having impacts on 3 or more state variables is listed as level 1 or 2	One having impacts on 2 or more state variables is listed as level 1 or 2
Parameter	U_m	U_m
	k_p	k_p
		M_Z
		K_n
		Kk_N
		Kk_P
		G_m
		I_o
		e

Table 6.9. Parameters having limited impacts on state variables.

Impact	One having impacts on 4 or more state variables is listed as level 5, and there is no level 1	One having impacts on 3 or more state variables is level 5, and there is no level 1 or 2
Parameter	Q_2	Q_2
	T_{opt}	T_{opt}
	β_1	β_1
	α_2	α_2
		β
		θ
		U_k

Similar conclusions could be obtained from grading parameters according to the standard deviation of the state variables' sensitivity to the parameters. This showed that the state variable had a strong correlation with the parameter-sensitive mean and variance. A larger average sensitivity had a larger fluctuation, which was mainly due to the cumulative effects from the high nonlinearity of the model. This result was consistent with the conclusions from Cossarini and Solidoro (2008).

6.5 Sobol Method-based Global Sensitivity Analysis On Key Influence Parameters of Kelp Farming

6.5.1 *Sobol method-based global sensitivity analysis method*

To reduce the uncertainty of the model, we must first identify the source of the uncertainty. The Sobol method uses the model variance to measure the uncertainty of the model. The main source of uncertainty is identified by dividing the model variance into different sub-items (Saltelli *et al.*, 2008), and quantifying the contribution percentage of each parameter (or parameter group) to the model output.

The key to sensitivity analysis is to reduce the cost of calculations. $y = f(x_1, x_2, \ldots, x_k)$ is usually a complex numerical model and takes a long period to run. In a general study, only first-order and total-order sensitivities are calculated (Homma and Saltelli, 1996; Saltelli *et al.*, 1999), and the latter describes the correlation between input variables. When the first-order sensitivity is calculated, the calculation cost is low and the dependence on k is small. When the total-order sensitivity is calculated, the variance decomposition is performed on a model with k parameters. The number of decompositions increases by2^k, and the calculation cost is higher (Saltelli *et al.*, 2010).

(1) *Basic concepts*
Domain $X = [x_1, x_2, \ldots, x_k]$ is limited to k dimensional cube as $I^k = \{X|0 \leq x_i \leq 1;\ i = 1, \ldots, k\}$.

(a) *Hoeffding decomposition and uniqueness theorem*
If x_i, $i = 1, \ldots, k$ are independent of each other and $f(X)$ is quadratic integrable on I^k, $f(X)$ can be decomposed into 2^k different function children with different dimensions as,

$$f(x_1, x_2, \ldots, x_k) = f_0 + \sum_i f_i(x_i) + \sum_i \sum_{j \succ 1} f_{ij}(x_i, x_j) + \cdots + f_{12\cdots k}(x_1, x_2, \ldots, x_k). \tag{6.6}$$

Based on this equation, Sobol (1993) proved its uniqueness conditions as:

$$\int_0^1 f_{i_1,i_2,\ldots,i_s}(x_{i_1}, x_{i_2}, \ldots, x_{i_s})dx_{i_w} = 0,$$
$$1 \leq i_1 \leq \cdots \leq i_s \leq k,\ i_w \in \{i_1, i_2, \ldots, i_s\}.$$

The function sub-items satisfy the following form:

$$f_M(x_M) = \sum_{N:N\subseteq M} (-1)^{|M|-|N|} E(f|x_N),$$
$$M = \{i_1, i_2, \ldots, i_s\},\ x_M = (x_{i_1}, x_{i_2}, \ldots, x_{i_s}) \tag{6.7}$$

where $E(f|x_N)$ is the expected conditions of $f(X)$ with the given x_N.

(b) *First-order sensitivity and total order sensitivity*

The first-order effect of parameter x_i is denoted as $V_{x_i}[E_{x\sim i}(y|x_i)]$, where $x_{\sim i}$ represents all parameters except x_i. $E_{x_{\sim i}}(y|x_i)$ represents the expected value with fixed x_i and varied other parameters. $V_{x_i}[E_{x\sim i}(y|x_i)]$ is the variance with all x_i values on this basis. The first-order sensitivity index is:

$$S_i = V_{x_i}[E_{x\sim i}(y|x_i)]/[V(y)]. \tag{6.8}$$

So, $V(y) = V_{x_i}[E_{x\sim i}(y|x_i)] + E_{x_i}[V_{x\sim i}(y|x_i)]$ can be obtained (Mood *et al.*, 1974). Thus, S_i is a standardized index. Here, $V_{x_i}[E_{x\sim i}(y|x_i)]$ measures the first-order effect of parameter x_i on the model output. Generally, $E_{x_i}[V_{x\sim i}(y|x_i)]$ is named as the remainder.

Another commonly used metric is the overall order sensitivity index of the parameter (Homma, 1996) as:

$$S_{T_i} = \frac{E_{x_{\sim i}}(Vx_i(y|x_{\sim i}))}{V(y)} = 1 - \frac{V_{x_{\sim i}}(Ex_i(y|x_{\sim i}))}{V(y)}. \tag{6.9}$$

Here, S_{T_i} is used to measure the total effect of the parameter x_i on the model output, meaning the sum of the first-order and higher-order effects of the parameter (the relationship with other parameters). Some believe that $V_{x_{\sim i}}(Ex_i(y|x_{\sim i}))$ represents the first-order effects of all non-x_i parameters on the model output, and thus, the sum of the variance contributions of all the related sub-items is given by $V(y) - V_{x_{\sim i}}(Ex_i(y|x_{\sim i}))$.

(c) *Variance decomposition and general order sensitivity*

According to Eq. (6.7), the following can be derived,

$$\begin{aligned} f_0 &= E(y), \\ f_i &= Ex_{\sim i}(y|x_i) - E(y), \\ f_{ij} &= Ex_{\sim ij}(y|x_i, x_j) - f_i - f_j - E(y), \\ &\cdots \end{aligned}$$

then:

$$\begin{aligned} V_i &= V(f_i(x_i)) = V_{x_i}(Ex_{\sim i}(y|x_i)), \\ V_{ij} &= V(f_{ij}(x_i, x_j)) = V_{x_i,x_j}(Ex_{\sim ij}(y|x_i, x_j)) \\ &\quad - V_{x_i}(Ex_{\sim i}(y|x_i)) - V_{x_j}(Ex_{\sim j}(y|x_j)) \\ &\cdots \end{aligned}$$

then:

$$V(y) = \sum_i V_i + \sum_i \sum_{j \succ 1} V_{ij} + \cdots + V_{1,2,\ldots,k},$$

then:

$$\sum_i S_i + \sum_i \sum_{j \succ 1} S_{ij} + \cdots + S_{1,2,\ldots,k} = 1,$$

in which

f_0 is the average level of f,
f_i is the individual effect of parameter x_i on f,
f_{ij} is the interaction of parameters x_i and x_j on f.

Assuming that all parameters x_i follow a uniform distribution over the interval [0,1], we can obtain:

$$V(y) = \int f^2(x)p(x)dx - \left(\int f(x)p(x)\right)^2,$$

where $p(x) = p(x_1) = \cdots = p(x_k)$,

$$V_M = \int \cdots \int f^2_{i_1,i_2,\ldots,i_s}(x_{i_1}, x_{i_2}, \ldots, x_{i_s})p(x_{i_1})p(x_{i_2}) \cdots p(x_{i_s})dx - \left(\int f(x)p(x)\right)^2 dx_{i_1} \cdots dx_{i_s}$$

$S_M = V_M/V$, $M = \{i_1, i_2, \ldots, i_s\} \subseteq I^k$ is the global sensitivity index of general order.

(2) *Monte Carlo algorithm*

(a) Algorithm principle

With $V(M) = \sum_{N \subseteq M} V_N$, Sobol proved $V(M) = \int f(x)f(x_M, x'_{\bar{M}})dxdx'_{\bar{M}} - f_0^2$ in 1993, in which $x = (x_M, x_{\bar{M}})$, $x' = (x'_M, x'_{\bar{M}})$ are two independent random vectors with I^k (Sobol, 1990). According to the law of large numbers, we can obtain:

$$\frac{1}{N}\sum_{i=1}^{N} f(x_i) \to f_0,$$

$$\frac{1}{N}\sum_{i=1}^{N} f(x_i)f(x_{iM}, x'_{i\bar{M}}) \to \int f(x)f(x_M, x'_{\bar{M}})\, dxdx'_{\bar{M}}.$$

Thus, $\frac{1}{N}\sum_{i=1}^{N} f(x_i)f(x_{iM}, x'_{i\bar{M}}) - \left(\frac{1}{N}\sum_{i=1}^{N} f(x_i)\right)^2$ can be used to estimate $V_{(M)}$, then to estimate V_M and S_M.

In 2007, Sobol (2007) further modified the equation as:

$$V_{(M)} = \int f(x)f(x_M, x'_{\bar{M}})dxdx'_{\bar{M}} - \int f(x)f(x')\, dxdx'$$
$$= \int f(x)(f(x_M, x'_{\bar{M}}) - f(x'))\, dxdx'$$

Thus, $\hat{V}_{(M)} = \frac{1}{N}\sum_{i=1}^{N} f(x_i)(f(x_{iM}, x'_{i\bar{M}}) - f(x'_i))$ can be used to estimate $V(M)$, then to obtain the estimated values of V_M and S_M, and $\hat{V}_{(M)}$ has a high convergence rate.

(b) Steps of the algorithm

(1) $2k$ random numbers $(\alpha_1, \ldots, \alpha_{2k})$ are generated to meet function $U(0,1)$, and lead to $x_1 = (x_{1M}, x_{1\bar{M}}) = (\alpha_1, \ldots, \alpha_k)$, $x_1' = (x_{1M}', x_{1\bar{M}}') = (\alpha_{m+1}', \ldots, \alpha_{2m}')$.
(2) $f(x_1), f(x_1'), f(x_1)f(x_{1M}, x_{1\bar{M}}')$ are calculated.
(3) Repeat (1) and (2) for N times, and $f(x_j), f(x_j'), f(x_j)$ $f(x_{jM}, x_{j\bar{M}}')$, $j = 1, 2, \ldots, N$ can be obtained.
(4) To $M \subseteq I^k$, $\hat{V}_{(M)} = \frac{1}{N}\sum_{j=1}^{N} f(x_j)[f(x_{jM}, x_{j\bar{M}}') - f(x_j')]$ is calculated.
(5) $\hat{S}_{(M)} = \hat{V}_{(M)}/\hat{V}_{I^k}$ is calculated, then $S_{(M)}$ can be estimated according to $\hat{S}_{(M)} = \sum_{N:N\subseteq M}(-1)^{|M|-|N|}\hat{S}_{(N)}$.

(3) *Pseudo Monte Carlo algorithm*
In order to improve the calculation efficiency of the MC algorithm and convergence rate of the estimator, a series of numbers are generated according to certain rules, which are pseudo-random numbers (the value is determined by the sample size when the rule is determined). These numbers are used to replace the random numbers to make the distribution of sampling points in the domain as uniform as possible, thus improving the accuracy of the estimation. The effect of the algorithm is mainly related to the selection of the sampling plan, statistics, and pseudo-random number generation methods (Wang, 2012).

(a) *Statistics*
Assuming that $A(a_{ji})$ and $B(b_{ji})$ are two independent $N \times k$ dimensional sampling matrices and N is the number of sampling, $i = 1, 2, \ldots, k$; $j = 1, 2, \ldots, N$, we can say $A_B^{(i)}$ $(B_A^{(i)})$ indicates that the matrices with the i column are from $B(A)$ and the remaining columns from $A(B)$. The existing statistical formulas for the sensitivity of each order were shown in Table 6.10. The selection of different statistics directly affects the convergence rate of the algorithm. It was found that the Satelli estimation performed better (Saltelli *et al.*, 2010, Wang, 2012).

(b) *Sampling method*
In order to estimate the global sensitivity, three sample matrices A, B, A_B^M (or B_A^M) are needed. The selection of different sample matrices will affect the result of the pseudo MC algorithm. $A(a_{ji})$

Table 6.10. Formulas and sources for calculating the sensitivity of each order.

First-order: Estimated V_i value when S_i is calculated	Reference
(a) $\frac{1}{N}\sum_{j=1}^{N} f(A)_j f(B_A^{(i)})_j - f_0^2$	(Sobol, 1993)
(b) $V(y) - \frac{1}{2N}\sum_{j=1}^{N} (f(B)_j f(A_B^{(i)})_j)^2$	(Jansen, 1999)
(c) $\frac{1}{N}\sum_{j=1}^{N} f(B)_j (f(A_B^{(i)})_j - f(A)_j)$	(Saltelli *et al.*, 2010)
Total order: Estimated value of V_{Ti} when S_{Ti} is calculated	
(d) $V(y) - \frac{1}{N}\sum_{j=1}^{N} (f(A)_j f(A_B^{(i)})_j)^2 + f_0^2$	(Homma, 1996)
(e) $\frac{1}{N}\sum_{j=1}^{N} f(A)_j (f(A)_j - f(A_B^{(i)})_j)$	(Sobol, 2007)
(f) $\frac{1}{2N}\sum_{j=1}^{N} (f(A)_j - f(A_B^{(i)})_j)^2$	(Saltelli *et al.*, 2010; Jansen, 1999)
General order: Estimated value when S_M is calculated	
(g) $\frac{1}{N}\sum_{j=1}^{N} f(A)_j f(B_A^{(M)})_j - \left(\frac{1}{N}\sum_{j=1}^{N} f(A)_j\right)^2$ *Note*: $\hat{S}_{(M)} = \sum_{N:N\subseteq M} (-1)^{\lvert M\rvert-\lvert N\rvert}\hat{V}_{(M)}/\hat{V}_{Ik}$	(Sobol, 1993)
(h) $V(y) - \frac{1}{2N}\sum_{j=1}^{N} (f(B)_j - f(A_B^M)_j)^2$ *Note*: $\hat{V}(y) = \frac{1}{N}\sum_{j=1}^{N} f(A)_j^2 - \left(\frac{1}{N}\sum_{j=1}^{N} f(A)_j\right)^2$	(Jansen, 1999)
(i) $\frac{1}{N}\sum_{j=1}^{N} f(A)_j (f(B_A^M)_j - f(B)_j)$	(Sobol, 2007)
(j) $\frac{1}{N}\sum_{j=1}^{N} f(B)_j (f(A_B^M)_j - f(A)_j)$	(Saltelli *et al.*, 2010)

Table 6.11. The first step of sampling in MC simulation.

Radial sampling	Targeted variable	Winding sampling
$a_{11}, a_{12}, a_{13}, \ldots, a_{1k}$		$a_{11}, a_{12}, a_{13}, \ldots, a_{1k}$
$b_{11}, a_{12}, a_{13}, \ldots, a_{1k}$	x_1	$b_{11}, a_{12}, a_{13}, \ldots, a_{1k}$
$a_{11}, b_{12}, a_{13}, \ldots, a_{1k}$	x_2	$b_{11}, b_{12}, a_{13}, \ldots, a_{1k}$
$a_{11}, a_{12}, b_{13}, \ldots, a_{1k}$	x_3	$b_{11}, b_{12}, b_{13}, \ldots, a_{1k}$
$\cdots$		$\cdots$
$a_{11}, a_{12}, a_{13}, \ldots, b_{1k}$	x_m	$b_{11}, b_{12}, b_{13}, \ldots, b_{1k}$

and $B(b_{ji})$ are the replacing matrix and replaced matrix. There are two common sampling plans: Radial Sampling (Saltelli, 2002) and Winding Sampling (Jansen, 1999) (Table 6.11). Studies have shown that the former plan is better (Wang, 2012).

(c) *Types of pseudo-random numbers*
There are three commonly used pseudo-random numbers: Halton pseudo-random numbers (Halton, 1960), Sobol pseudo-random numbers (Sobol, 1967; Niederreiter, 1988), and Faure pseudo-random numbers (Faure, 1982). Studies have shown that the error of Sobol pseudo-random numbers is relatively small (Wang, 2012).

(d) *Steps of the algorithm*
In the pseudo-MC algorithm in this chapter, a combination of Sobol pseudo-random numbers, Saltelli estimates, and Radial sampling is chosen for use. The algorithm is as follows:

(1) Sobol pseudo-random numbers are used to generate a $2k$ dimensional sequence with a length of N, of which each is cut from the middle into A and B. It should be noted that, in order to maintain independence between sequences and independence between A and B, a total of $2k$ dimensional sequences should be generated instead of two k dimensional sequences.
(2) $\hat{V}_{(M)} = \frac{1}{N}\sum_{j=1}^{N} f(B)_j(f(A_B^M)_j - f(A)_j)$ is calculated.
(3) $\hat{S}_{(M)} = \sum_{N:N\subseteq M}(-1)^{|M|-|N|}\hat{V}_{(M)}/\hat{V}_{I^k}$ is calculated.

6.5.2 *Analysis settings of parameter global sensitivity*

In this chapter, 24 model parameters were selected for global sensitivity analysis. The default value of each parameter was determined according to reported literature. A total of $\pm 20\%$ of the default value of each parameter (McKay, 1995; Chu-Agor *et al.*, 2011) was used as the domain, respectively. The Sobol method was used to analyze the global sensitivity of 24 parameters, and the pseudo-MC algorithm in the Sobol method was used with Sobol pseudorandom numbers, Saltelli estimated values, and Radial sampling.

6.5.3 *Analysis results of parameter global sensitivity*

(1) Analysis results of parameter global sensitivity
The first-order sensitivity S_i represents the influence of the i-th parameter on the model output after interaction with other parameters is excluded, and the total-order sensitivity S_{Ti} represents the total influence of the i-th parameter on the model output (including

the changes in model output due to the interaction with other parameters). Therefore, the interaction sensitivity $S_{Ti} - S_i$ represents the interaction strength between the i-th parameter and other parameters. The analysis results in this chapter are shown in Table 6.12, and Figs. 6.7 and 6.8.

As listed in Table 6.12, the average first-order sensitivity of each parameter to inorganic nitrogen, inorganic phosphorus, and organic debris is relatively high, while the average first-order sensitivity to phytoplankton, zooplankton, and kelp is relatively low. The average first-order sensitivity of all parameters to inorganic nitrogen was the highest at 6.26%. Among them, the first-order sensitivity of Kk_P, T_{opt}, β, α_1, Qg_{10}, Qh_{10}, e, and β_1 to inorganic nitrogen was negative, and the first-order sensitivity of other parameters to inorganic nitrogen was positive. λ (Ivlev uptake constant) had the highest first-order sensitivity to inorganic nitrogen at 17.03%, which was followed by that to inorganic phosphorus at 5.77%. The first-order sensitivity of k_N, Kk_N, G_m, λ, γ, θ, M_K, T_{leth}, T_{opt}, U_k, and α_2 to inorganic phosphorus was negative, and that of the remaining parameters was positive. The highest first-order sensitivity of e (Debris remineralization rate) to inorganic phosphorus was the highest at 31.01%.

The average first-order sensitivity of all parameters to organic debris is 4.05%. The first-order sensitivity of Kk_P, T_{leth}, β, Qh_{10} and e to organic debris was negative, and that of the remaining parameters was positive. The first-order sensitivity of M_Z (zooplankton mortality) to organic debris was the highest at 14.70%.

The average first-order sensitivity of all parameters to kelp was 3.62%, while the first-order sensitivity of k_P, G_m, λ, T_{leth}, β, α_1, Qh_{10}, and e to kelp was negative, and that of the remaining parameters was positive. The first-order sensitivity of U_k (the maximum growth rate of kelp) to kelp was the highest at 28.20%.

The average first-order sensitivity of all parameters to zooplankton was 3.59%, while the first-order sensitivity of k_N, G_m, λ, θ, M_K, T_{opt}, β, U_k, α_2, and e to zooplankton was negative, and that of the remaining parameters was positive. The first-order sensitivity of γ (plant zooplankton growth coefficient) to zooplankton was the highest at 16.74%.

The average first-order sensitivity of all parameters to phytoplankton was 3.18%, while the first-order sensitivity of k_N, Kk_N,

Table 6.12. Global sensitivity analysis results of the Sobol method.

Input variable	Phytoplankton	Zooplankton	Kelp	Inorganic Nitrogen	Inorganic Phosphorus	Organic Debris	Absolute Average
	Output variable						
	Sobol First-order Sensitivity (%), S_i						
U_m	4.26	11.51	6.03	15.76	12.75	13.87	10.70
k_N	−3.06	−0.86	1.43	4.60	−4.46	3.37	2.96
k_P	4.18	0.16	−1.31	4.00	4.16	3.08	2.81
Kk_N	−2.82	0.38	1.99	1.84	−3.19	2.18	2.07
Kk_P	4.93	1.56	5.93	−0.81	5.72	−1.44	3.40
G_m	9.91	−2.58	−3.33	7.27	−8.86	0.18	5.35
λ	−0.52	−2.93	−0.95	17.03	−6.98	5.59	5.67
M_P	6.82	4.10	0.26	9.06	2.07	7.23	4.92
M_Z	7.87	11.64	5.67	16.12	2.97	14.70	9.83
γ	3.35	16.74	5.07	8.54	−2.72	8.94	7.56
θ	0.46	−0.64	0.17	0.34	−1.40	0.34	0.56
I_0	1.63	16.69	8.43	4.37	18.12	13.52	10.46
M_K	−0.13	−0.04	0.32	0.15	−0.38	0.08	0.18
T_{leth}	−0.52	1.06	−0.98	2.25	−0.29	−0.42	0.92
T_{opt}	−0.24	−0.01	2.38	−1.69	−0.26	0.20	0.80
β	0.12	−0.02	−0.05	−0.15	0.04	−0.07	0.08
U_k	−0.34	−1.80	28.20	14.63	−1.92	4.94	8.64
α_1	0.39	0.36	−0.21	−5.14	0.68	1.50	1.38
α_2	−1.24	−0.71	0.07	2.95	−1.57	1.62	1.36
Qg_{10}	4.72	7.68	5.87	−6.74	17.43	4.14	7.76
Qh_{10}	6.94	1.15	−4.23	−1.53	3.28	−0.88	3.00
e	−8.09	−0.28	−1.64	−0.39	31.01	−2.37	7.30
β_1	−0.17	0.06	0.26	−14.46	0.26	0.00	2.53
β_2	3.74	3.26	2.24	10.41	7.97	6.60	5.70
Absolute Average	3.18	3.59	3.62	6.26	5.77	4.05	4.41
	Sobol Interaction Sensitivity (%), $S_{Ti} - S_i$						
U_m	4.33	7.04	6.62	3.23	8.44	7.03	6.12
k_N	3.70	0.95	−1.09	−3.47	5.15	−2.75	2.85
k_P	−1.36	5.17	4.64	2.95	2.16	3.37	3.28
Kk_N	3.32	−0.29	−1.40	−1.22	3.66	−1.87	1.96
Kk_P	−2.91	−1.30	5.61	3.78	−3.85	2.11	3.26
G_m	4.70	25.34	5.35	3.23	12.28	6.03	9.49
λ	2.75	7.51	2.33	−11.34	10.18	−2.79	6.15
M_P	7.45	−0.61	2.92	−4.00	−0.03	−1.45	2.74
M_Z	1.35	−0.50	−0.98	−4.36	4.34	−3.80	2.55
γ	5.59	6.80	−3.05	−0.83	5.18	−3.02	4.08
θ	−0.46	0.68	−0.17	−0.33	1.46	−0.33	0.57
I_0	5.55	1.56	1.97	11.12	−0.16	3.16	3.92
M_K	0.14	0.04	−0.27	−0.15	0.40	−0.08	0.18
T_{leth}	0.94	−1.01	1.46	−1.52	0.57	0.60	1.02
T_{opt}	0.36	0.01	−1.92	1.77	0.35	−0.14	0.76
β	−0.11	0.02	0.07	0.15	−0.04	0.07	0.08
U_k	8.88	2.99	8.23	−4.52	9.50	−0.88	5.83
α_1	0.78	−0.28	0.63	12.00	0.21	1.37	2.54
α_2	1.41	0.77	−0.01	−1.97	1.82	−0.62	1.10
Qg_{10}	3.97	0.95	3.59	23.07	−2.60	12.30	7.75
Qh_{10}	−1.72	4.95	5.65	6.70	−2.04	4.08	4.19
e	11.66	2.06	1.99	1.43	−20.67	3.00	6.80
β_1	0.57	−0.05	−0.14	21.86	0.03	0.79	3.91
β_2	8.15	−2.27	−1.55	−4.74	1.19	−3.52	3.57
Absolute Average	3.42	3.05	2.57	5.41	4.01	2.71	3.53

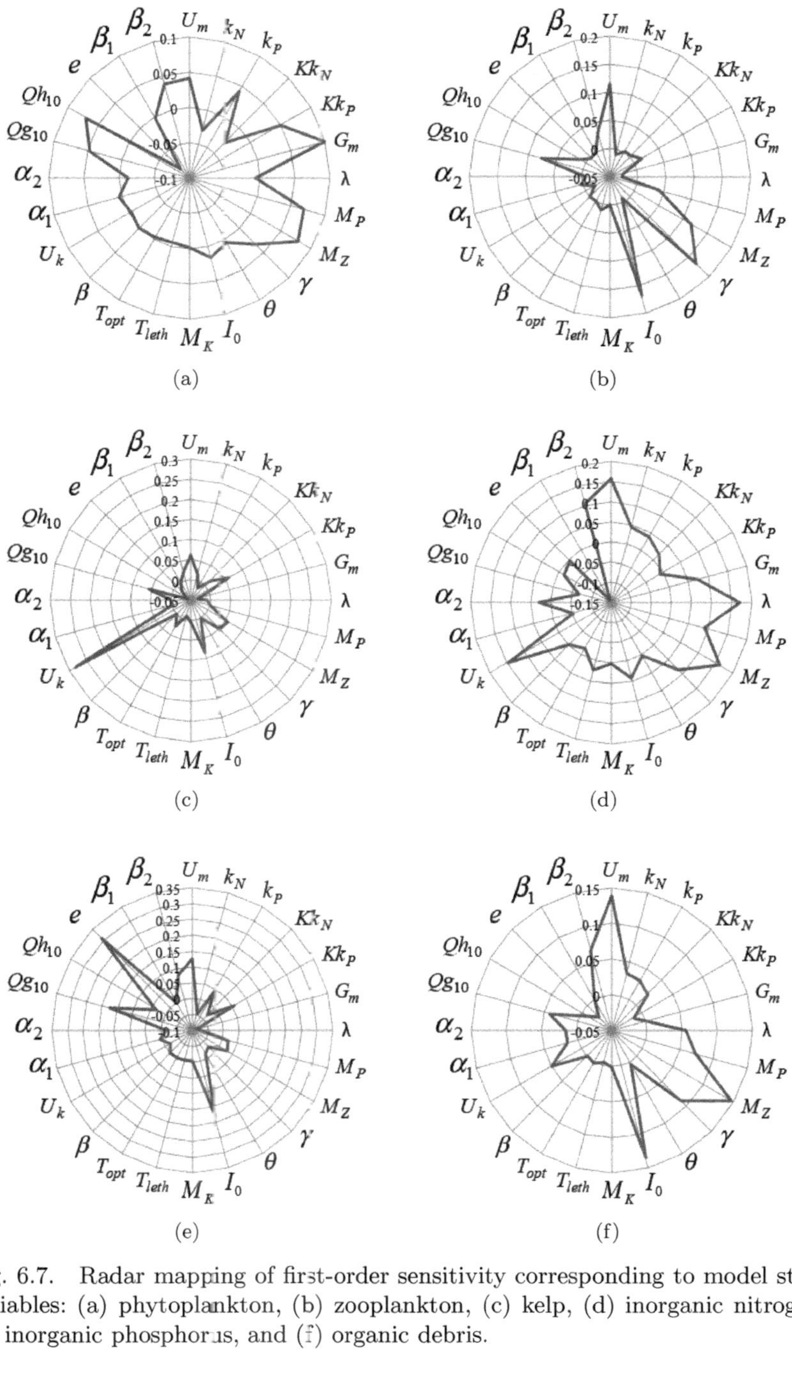

Fig. 6.7. Radar mapping of first-order sensitivity corresponding to model state variables: (a) phytoplankton, (b) zooplankton, (c) kelp, (d) inorganic nitrogen, (e) inorganic phosphorus, and (f) organic debris.

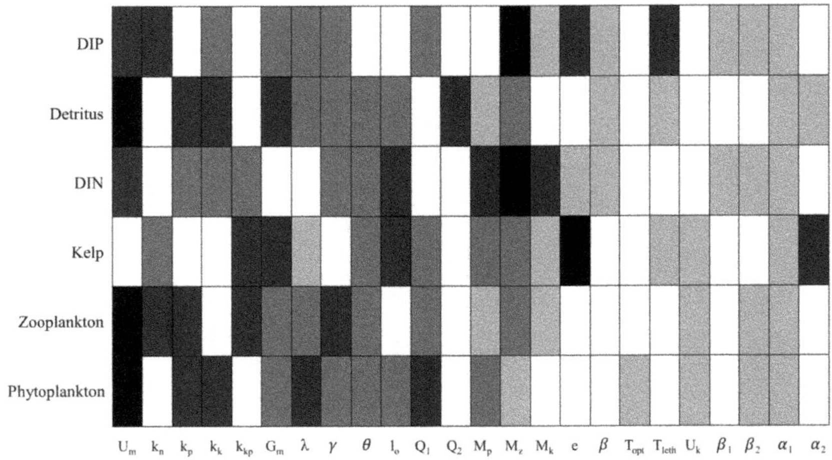

Fig. 6.8. Classification of sensitivity mean value of different parameter to state variables: (a) Phytoplankton, (b) Zooplankton, (c) Kelp, (d) Dissolved inorganic nitrogen (DIN), (e) Dissolved inorganic phosphorus (DIP), and (f) Detribus.

λ, M_K, T_{leth}, T_{opt}, U_k, α_2, e, and β_1 to phytoplankton was negative, and that of the remaining parameters was positive. The first-order sensitivity of G_m (maximum intake of zooplankton) to phytoplankton was the highest at 9.91%.

Differing from the results of the first-order sensitivity analysis, interactions related to various parameters had relatively great impacts on inorganic nitrogen and inorganic phosphorus, but relatively small impacts on kelp and organic debris. The average interaction sensitivity of all parameters to inorganic nitrogen was the highest at 5.41%, in which the interaction between Qg_{10} (growth temperature dependence coefficient) and other parameters had the largest effect on inorganic nitrogen at 23.07%, followed by that on inorganic phosphorus (4.01%), in which the interaction of e (debris remineralization rate) with other parameters had the greatest effect on inorganic phosphorus at −20.07%. The average interaction sensitivity of all parameters to phytoplankton was 2.88%, in which the interaction of e (debris remineralization rate) with other parameters had the largest impact on phytoplankton at 11.66%. The average interaction sensitivity of all parameters to zooplankton was 3.05%, in which the interaction between G_m (the maximum intake of zooplankton) and other parameters had the largest effect on zooplankton at 25.34 %. The average interaction sensitivity of all parameters to organic debris was 2.71%, in which the interaction between Qg_{10} (growth temperature dependence coefficient) and other parameters had the largest impact on organic debris at 12.30%. The average interaction sensitivity of all parameters to kelp was 2.57%, in which the interaction of U_k (the largest growth rate of kelp) and other parameters had the greatest impact on kelp at 8.23%.

(2) *Ranking of global sensitivity*

Because the model had six output variables, the absolute values of the sensitivity of each parameter to each output variable were averaged to measure the total impact of each parameter on the model. The absolute average of the first-order sensitivity and interaction sensitivity of each parameter were sorted by order and listed in Tables 6.13 and 6.14, where the first-order sensitivity of U_m, I_0, M_Z, U_k, Qg_{10}, γ, and e was relatively high, the interaction sensitivity of G_m, Qg_{10}, e, λ, U_m, and U_k was relatively high, and the first-order sensitivity and interaction sensitivity of Kk_N, α_1, α_2, T_{leth}, T_{opt}, θ, M_K, and β were relatively low.

Using the Sobol method, because the sensitivity of each parameter (group) could be quantified with more accurate results, optimization parameters were selected according to the global sensitivity analysis results. The rankings of the first-order sensitivity and interaction sensitivity of each parameter in Tables 6.13 and 6.14 were added and sorted by order, which was equivalent to sorting weights of their first-order and interaction sensitivity. The results are shown

Table 6.13. First-order sensitivity ranking of each parameter.

Parameter	Order	First-order S_i(%)	Parameter	Order	First-order S_i(%)	Parameter	Order	First-order S_i(%)
U_m	1	10.70	λ	9	5.67	Kk_N	17	2.07
I_0	2	10.46	G_m	10	5.35	α_1	18	1.38
M_Z	3	9.83	M_P	11	4.92	α_2	19	1.36
U_k	4	8.64	Kk_F	12	3.40	T_{leth}	20	0.92
Qg_{10}	5	7.76	Qh_{10}	13	3.00	T_{opt}	21	0.80
γ	6	7.56	k_N	14	2.96	θ	22	0.56
e	7	7.30	k_P	15	2.81	M_K	23	0.18
β_2	8	5.70	β_1	16	2.53	β	24	0.08

Table 6.14. Interaction sensitivity ranking of each parameter.

Parameter	Order	Interaction $S_{Ti} - S_i$ (%)	Parameter	Order	Interaction $S_{Ti} - S_i$ (%)	Parameter	Order	Interaction $S_{Ti} - S_i$ (%)
G_m	1	9.49	I_0	9	3.92	α_1	17	2.54
Qg_{10}	2	7.75	β_1	10	3.91	Kk_N	18	1.96
e	3	6.80	β_2	11	3.57	α_2	19	1.10
λ	4	6.15	k_P	12	3.28	T_{leth}	20	1.02
U_m	5	6.12	Kk_P	13	3.26	T_{opt}	21	0.76
U_k	6	5.83	k_N	14	2.85	θ	22	0.57
Qh_{10}	7	4.19	M_P	15	2.74	M_K	23	0.18
γ	8	4.08	M_Z	16	2.55	β	24	0.08

Table 6.15. Ranking of global sensitivity of all parameters.

Parameter	S_i order	$S_{Ti} - S_i$ order	Sum of ranks	Parameter	S_i order	$S_{Ti} - S_i$ Order	Sum of ranks
U_m	1	5	6	M_P	11	15	26
Qg_{10}	5	2	7	β_1	16	10	26
U_k	4	6	10	k_P	15	12	27
e	7	3	10	k_N	14	14	28
G_m	10	1	11	Kk_N	17	18	35
I_0	2	9	11	α_1	18	17	35
λ	9	4	13	α_2	19	19	38
γ	6	8	14	T_{leth}	20	20	40
M_Z	3	16	19	T_{opt}	21	21	42
β_2	8	11	19	θ	22	22	44
Qh_{10}	13	7	20	M_K	23	23	46
Kk_P	12	13	25	β	24	24	48

in Table 6.15. According to results of the Sobol global sensitivity analysis, the parameters that were determined to optimize were the maximum growth rate of phytoplankton (U_m), growth temperature dependence coefficient (Qg_{10}), maximum growth rate of kelp (U_k), debris remineralization rate (e), maximum intake rate of zooplankton (G_m), optimal light intensity (I_0), Ivlev uptake constant (λ), and zooplankton growth coefficient (γ). Parameters that were not needed to optimize were the half-saturation constant of kelp DIN absorption (Kk_N), nitrogen-carbon ratio of phytoplankton (α_1), phosphorus-carbon ratio of phytoplankton (α_2), deadly temperature for kelp growth (T_{leth}), the optimum temperature for kelp growth (T_{opt}), excretion coefficient of zooplankton (θ), kelp mortality (M_K), and kelp growth temperature adjust parameters (β), whose empirical data could be used directly with optimization. Other parameters could be chosen to optimize or use their empirical data directly according to different requirements of applications.

Acknowledgment

The reference (Zheng *et al.*, 2012) was reprinted from Ecological Modelling, 247, with permission from Elsevier.

References

Andersen V, Nival P. 1989. Modelling of phytoplankton population dynamics in an enclosed water column. *Journal of the Marine Biological Association UK.* **69**: 625–646.

Chu-Agor M L, Munoz-Carpena R, Kiker G, *et al.* 2011. Exploring vulnerability of coastal habitats to sea level rise through global sensitivity and uncertainty analyses. *Environmental Modelling & Software.* **26**(5): 593–604.

Cossarini G, Solidoro C. 2008. Global sensitivity analysis of a trophodynamic model of the Gulf of Trieste. *Ecological Modelling.* **212**: 16–27.

Duarte P, Meneses R, Hawkins A J S, *et al.* 2003. Mathematical modelling to assess the carrying capacity for multi-species culture within coastal waters. *Ecological Modelling.* **168**: 109–143.

Faure H. 1982. Discrepance de suits associees a un systeme de numeration (en dimensions). *Acta Arithmetica.* **41**: 337–351.

Fransz H G, Verhagen J H G. 1985. Modelling research on the production cycle of phytoplanton in the southern Bight of North Sea in relation to river-borne nutrient loads. *Netherlands Journal of Sea Research.* **19**: 241–250.

Gao H, Feng S, Guan Y. 1998. Modelling annual cycles of primary production in different regions of the Bohai Sea. *Fisheries Oceanography.* **7**(3/4): 258–264.

Halton J H. 1960. On the efficiency of certain quasi-random sequences of points in evaluating multi-dimensional integrals. *Numerical Mathematics.* **2**(1): 84–90.

Homma T, Saltelli A. 1996. Importance measures in global sensitivity analysis of model output. *Reliability Engineering and System Safety.* **52**(1): 1–17.

Huang D J, Huang X P, Yue W Z. 2005. Contents of TN, TP in macroalgae and its significance for remediation of coastal environment, *Journal of Oceanography in Taiwan Strait.* **4**(3): 316–321.

Jansen M J W. 1999. Analysis of variance designs for model output. *Computer Physics Communications.* **117**: 35–43.

Mao X H, Zhu M Y, Yang X L. 1993. The photosynthesis and productivity of benthic macrophytes in Sanggou Bay, *Acta Ecologica Sinica.* **13**(1): 25–29.

McCreary J P, Kohler K E, Hood R R, Olson D B. 1996. A four-component ecosystem model of biological in Arabian Sea. *Progress in Oceanography.* **37**: 193–240.

McKay M D. 1995. *Evaluating Prediction Uncertainty.* Los Alamos: US Nuclear Regulatory Commission and Los Alamos National Laboratory.

Mood A M F, Graybill A, Boes D C. 1974. *Introduction to the Theory of Statistics*, 3rd Edition. McGraw-Hill.

Niederreiter H. 1988. Quasi-Monte Carlo methods for multidimensional numerical integration. In: *Numerical Integration 3, International Series of Numerical Mathematics.* Birkhanser Basel, 157–171.

Petrell R J, Tabrizi K M, Harrison P J, Druehl L D. 1993. Mathematical model of Laminaria production near a British Columbian salmon sea cage farm. *Journal of Applied Phycology.* **5**:1–144.

Redfield A C. 1958. The biological control of chemical factors in the environment. *American Scientist.* **46**: 205–211.

Saltelli A. 2002. Making best use of model evaluations to compute sensitivity indices. *Computer Physics Communications.* **145**(2): 280–297.

Saltelli A, Annoni P, Azzini I, *et al.* 2010. Variance based sensitivity analysis of model output. Design and estimator for the total sensitivity index. *Computer Physics Communications.* **181**(2): 259–270.

Saltelli A, Ratto M, Andres T, *et al.* 2008. *Global Sensitivity Analysis: The Primer.* John Wiley and Sons.

Saltelli A, Tarantola S, Chan K. 1999. Quantitative model-independent method for global sensitivity analysis of model output. *Technometrics.* **41**(1): 39–56.

Sobol I M. 1967. On the distribution of points in a cube and the approximate evaluation of integrals. *Zhurnal Vychislitel'noi Matematiki i Matematicheskoi Fiziki.* **7**(4): 784–802.

Sobol I M. 1990. Sensitivity estimates for nonlinear mathematical models (in Russian), *Matematicheskoe Modelirovanie.* **2**(1): 112–118, *MMCE,* 1993, **1**(4): 407–414 (in English).

Sobol I M. 2007. Global sensitivity analysis indices for the investigation of nonlinear mathematical models (in Russian). *Matematicheskoe Modelirovanie.* **19**(11): 23–24.

Steele J H. 1962. Environmental control of photosynthesis in the sea. *Limnology and Oceanography.* **7**: 137150.

Wang X D. 2012. *Global Sensitivity Analysis for High-dimensional Complex Models*. Shanghai: East China Normal University.

Wang X L, Li K Q, Shi X Y. 2006. *Marine Environmental Capacity of Major Chemical Pollutants in Jiaozhou Bay*. Beijing: Science Press.

Wu R J, Zhang X L, Zhu M Y, Zhen Y F. 2009. Studies on the growth model of kelp culturing. *Marine Science Bulletin*. **28**(2): 34–40.

Zheng W, Shi H H, Fang G H, et al. 2012. Global sensitivity analysis of a marine ecosystem dynamic model of the Sanggou Bay. *Ecological Modelling*. **247**: 83–94.

Chapter 7

Construction and Applications of Coastal Biophysical Coupling Model

With the interdisciplinary and comprehensive development of various sciences, studies on marine ecosystems have gradually evolved from a single discipline (or specific research direction), such as biooceanology and marine chemistry, to multiple interdisciplinary comprehensive ones. At the end of the 1990s, the implementation of the Global Ocean Ecosystem Hydrodynamics Research Program (GLOBEC) marked an overall rise in studies on the hydrodynamics of the marine ecosystem. During this period, many three-dimensional hydrodynamic-biochemical coupling models were developed. For example, Skogen (1993) established NORWECOM (NORWegianECOlogical Model system) on the basis of the three-dimensional barometric hydrodynamic mode POM (Princeton Ocean Model), Delhez *et al.* (1994) established GHER (GeoHydrodynamics and Environmental Research) according to the ocean circulation model, Moll (1995), by coupling the one-dimensional vertical ecological model and three-dimensional barometric pressure hydrodynamic model HAMSOM (Hamburg Shelf Ocean Model), established the ECOHAM (Ecological North Sea Model Hamburg) model, and Baretta *et al.* (1995) established the ERSEM box model (European Regional Seas Ecosystem Model), which could be used to simultaneously analyze all chemical and biological components. Currently, marine ecosystem hydrodynamics models have been widely used in studies on the function and structure of marine ecosystems (Chen *et al.*, 1999), and have become important tools to simulate and predict the health of marine ecosystems and global changes (Friedrichs *et al.*, 2007).

In this chapter, a Finite-Volume Coastal Ocean Model (FVCOM) was used to calculate the tidal field in the Bohai Sea where the tidal field in Laizhou Bay was selected as an example. The nutrient-phytoplankton-zooplankton-detritus (NPZD) model was coupled based on the tidal current field in Laizhou Bay, which was used to simulate the growth of plankton and nutrient cycling.

7.1 Simulation of Tidal Field Data in the Bohai Sea

7.1.1 *Theoretical basis of the hydrodynamic model*

An FVCOM was used to construct a three-dimensional barotropic hydrodynamic model of the Bohai Sea, from which the tidal field data of Laizhou Bay was obtained. The FVCOM is an ocean model with a triangle unstructured horizontal grid, and a three-dimensional primitive equation of free surface (Chen, 2003). The data from this model was calculated through the finite volume method, which combined the advantages of the finite difference method and the finite element method in existing ocean models. The former could efficiently calculate discrete original differential equations, while the latter was convenient for fitting complex bank boundaries and handling local encryption (Chen, 2003). The closed-form equation of this model was the Mellor–Yamada 2.5-order turbulent closed sub-model. The vertical coordinate was transformed from σ to reflect the irregular bottom boundary. Data calculation was done with the integral form of the equation and a more accurate calculation format, which effectively solved the problem of mass conservation. In order to save calculation time, the calculation scheme adopted the method with split internal and external modes. The dry and wet judgment method was used to deal with the tidal moving boundary. This model effectively solved the problems of fitting shallow coasts, mass conservation, and calculating validity in data calculation, and was suitable for the simulation of the coastal tidal current field (Chen *et al.*, 1999; 2003; Chen, 2003).

(1) *The governing equations of ocean motion*

The set of governing equations for ocean motion in the Cartesian coordinate system includes the momentum equation, continuous equation, temperature equation, salinity equation, and density equation, as shown in Eqs. (7.1) to (7.7).

$$\frac{\partial u}{\partial t} + u \cdot \frac{\partial u}{\partial x} + v \cdot \frac{\partial u}{\partial y} + w \cdot \frac{\partial u}{\partial z} - f \cdot v$$
$$= -\frac{1}{\rho_0} \cdot \frac{\partial P}{\partial x} + K_m \cdot \frac{\partial^2 u}{\partial z^2} + F_u, \tag{7.1}$$
$$\frac{\partial v}{\partial t} + u \cdot \frac{\partial v}{\partial x} + v \cdot \frac{\partial v}{\partial y} + w \cdot \frac{\partial v}{\partial z} + f \cdot u$$
$$= -\frac{1}{\rho_0} \cdot \frac{\partial P}{\partial y} + K_m \cdot \frac{\partial^2 v}{\partial z^2} + F_v, \tag{7.2}$$
$$\frac{\partial P}{\partial z} = -\rho \cdot g, \tag{7.3}$$
$$\frac{\partial u}{\partial x} + \frac{\partial v}{\partial y} + \frac{\partial w}{\partial z} = 0, \tag{7.4}$$
$$\frac{\partial \theta}{\partial t} + u \cdot \frac{\partial \theta}{\partial x} + v \cdot \frac{\partial \theta}{\partial y} + w \cdot \frac{\partial \theta}{\partial z} = K_h \cdot \frac{\partial^2 \theta}{\partial z^2} + F_\theta, \tag{7.5}$$
$$\frac{\partial s}{\partial t} + u \cdot \frac{\partial s}{\partial x} + v \cdot \frac{\partial s}{\partial y} + w \cdot \frac{\partial s}{\partial z} = K_h \cdot \frac{\partial^2 s}{\partial z^2} + F_s, \tag{7.6}$$
$$\rho = \rho(\theta, s), \tag{7.7}$$

where, x, y, and z are East. North, and Vertical coordinates of the Cartesian coordinate system,

u, v, and w are velocity components in directions of x, y, and z,
Θ is the level temperature of seawater,
S is seawater salinity,
P is the density of seawater.
P is pressure,
F is the Cartesian parameter,
G is the gravitational acceleration,
K_m is the vertical vortex viscosity coefficient,
K_h is the vertical thermal diffusion coefficient,

F_u, F_v, F_θ, and F_s are diffusion terms of horizontal momentum, thermal, and salinity.

(2) Calculation of viscosity and thermal diffusivity of a vertical vortex K_m and K_h were calculated from the modified Mellor–Yamada 2.5-order turbulence closed-submodel, which are shown in Eqs. (7.8) and (7.9).

$$\begin{aligned}
&\frac{\partial q^2}{\partial t} + u \cdot \frac{\partial q^2}{\partial x} + v \cdot \frac{\partial q^2}{\partial y} + w \cdot \frac{\partial q^2}{\partial z} \\
&\quad = 2 \cdot (P_s + P_b - \varepsilon) + K_q \cdot \frac{\partial^2 q^2}{\partial z^2} + F_q, \qquad (7.8) \\
&\frac{\partial q^2 \cdot l}{\partial t} + u \cdot \frac{\partial q^2 \cdot l}{\partial x} + v \cdot \frac{\partial q^2 \cdot l}{\partial y} + w \cdot \frac{\partial q^2 \cdot l}{\partial z} \\
&\quad = l \cdot E_1 \cdot \left(P_s + P_b - \frac{\tilde{W}}{E_1} \cdot \varepsilon \right) + K_q \cdot \frac{\partial^2 q^2 \cdot l}{\partial z^2} + F_l, \qquad (7.9)
\end{aligned}$$

where

$q^2 = (u'^2 + v'^2)/2$ is turbulent energy,
l is the mixing length of turbulent,
K_q is the vertical diffusion coefficient of turbulent kinetic energy,
F_q and F_l are horizontal diffusion terms of turbulent kinetic energy and its mixing length;

Shear-generating term of turbulent kinetic energy

$$P_s = K_m \cdot (u_z^2 + v_z^2).$$

Buoyancy term:

$$P_b = (g \cdot K_h \cdot \rho_z)/\rho_0.$$

$\varepsilon = q^3/B_1 \cdot l$ is the turbulent energy consumption rate,

$\tilde{W} = 1 + E_2 \cdot l^2/(\kappa \cdot L)^2$ is an approximate function,

where $L^{-1} = (\zeta - z)^{-1} + (H + z)^{-1}$,
$\kappa = 0.4$ is the Von Carman constant,
H is the average water depth at rest,
Z is the height of free surface,
E_1 and B_1 are parameters to be determined.

The turbulent kinetic energy and its mixing length equations were closed according to Eq. (7.10).

$$K_m = l \cdot q \cdot S_m, \quad K_h = l \cdot q \cdot S_h, \quad K_q = 0.2 \cdot l \cdot q, \tag{7.10}$$

where S_m and S_h are stability functions as shown in Eq. (7.11),

$$S_m = \frac{0.4275 - 3.354 \cdot G_h}{(1 - 34.676 \cdot G_h) \cdot (1 - 6.127 \cdot G_h)}, \quad S_h = \frac{0.494}{1 - 34.676 \cdot G_h},$$
$$G_h = \frac{l^2 \cdot g}{q^2 \cdot \rho_0} \cdot \rho_z \tag{7.11}$$

(3) σ coordinate transformation in a vertical direction
The σ coordinate transformation in a vertical direction was defined as Eq. (7.12),

$$\sigma = \frac{z - \zeta}{H + \zeta} = \frac{z - \zeta}{D}, \tag{7.12}$$

where σ is -1 at the bottom and 0 at the surface. The governing equations and equations of the turbulent kinetic energy and mixing length could be converted accordingly in this coordinate system (Chen, 2003).

(4) Boundary conditions
In these boundary conditions, the surface ($\sigma = 0$) has the boundary condition as shown in Eq. (7.13):

$$\left(\frac{\partial u}{\partial \sigma}, \frac{\partial v}{\partial \sigma}\right) = \frac{D}{\rho_0 \cdot K_m} \cdot (\tau_{sx}, \tau_{sy}), \quad \omega = 0,$$
$$\frac{\partial \theta}{\partial \sigma} = \frac{D}{\rho \cdot c_p \cdot K_h} \cdot [Q_n(x, y, t) - SW(x, y, 0, t)],$$
$$\frac{\partial s}{\partial \sigma} = \frac{s \cdot (P - E) \cdot D}{K_h}, \quad q^2 \cdot l = 0, \quad q^2 = B_1^{2/3} \cdot u_{\tau s}^2. \tag{7.13}$$

At the bottom ($\sigma = -1$), the boundary conditions are given by Eq. (7.14).

$$\left(\frac{\partial u}{\partial \sigma}, \ \frac{\partial v}{\partial \sigma}\right) = \frac{D}{\rho_0 \cdot K_m} \cdot (\tau_{bx}, \ \tau_{by}), \quad \omega = 0, \quad \frac{\partial \theta}{\partial \sigma} = \frac{\partial s}{\partial \sigma} = 0,$$

$$q^2 \cdot l = 0, \quad q^2 = B_1^{2/3} \cdot u_{zb}^2,$$

$$(\tau_{sx}, \tau_{sy}) = C_d \cdot \sqrt{u_s^2 + v_s^2} \cdot (u_s, v_s) \quad \text{and}$$

$$(\tau_{bx}, \tau_{by}) = C_d \cdot \sqrt{u_b^2 + v_b^2} \cdot (u_b, v_b), \tag{7.14}$$

where u_s, u_b, v_s, and v_b are components along the x and y directions of the surface and bottom velocity, respectively, and C_d is the resistance coefficient, $Q_n(x, y, t)$ is the net surface heat flux, $SW(x, y, 0, t)$ is the short-wave radiation flux on the sea surface, c_p is the specific heat capacity of seawater, and $u_{\tau s}$ and $u_{\tau b}$ are friction speeds of the surface and bottom to the boundary layer, respectively.

Land boundary conditions were shown in Eq. (7.15).

$$v_n = 0, \quad \frac{\partial \theta}{\partial n} = 0, \quad \frac{\partial s}{\partial n} = 0, \tag{7.15}$$

where, n is boundary normal axis,

v_n is boundary normal velocity component.

The border was opened with the tide-driven approach, and tide wave ζ was represented by Eq. (7.16):

$$\zeta = \sum_{i=1} [f_i \cdot H_i \cdot \cos(\sigma_i \cdot t + V_i + \theta_i - G_i)], \tag{7.16}$$

where f_i, σ_i, V_i, and θ_i are modifications of the i-th intersection factor, angular velocity, hour angle, and retard angle.

H_i and G_i are harmonic constants as amplitude and the retard angle of the i-th tide, respectively.

7.1.2 *Construction of hydrodynamic model of Bohai*

The studied area of the model was divided by an unstructured triangular mesh to better fit the land boundary. The Bohai computational domain consisted of 33,757 nodes and 62,816 triangular grids, with a horizontal resolution of 200–800 m, and was divided into 8 σ layers vertically. The calculation area of the Bohai hydrodynamic model was shown in Fig. 7.1, and the open boundary was the eastern boundary. According to simulation results of the Bohai hydrodynamic model and reported literature (Yu, 2012) about the tidal current field in Laizhou Bay (as shown in Fig. 7.1), the North and East boundaries were chosen as open boundaries. The open boundary was driven by the M_2 sub-tide. The roughness of the sea floor was the software default value of 0.001, and the bottom friction coefficient was 0.0015, according to the simulation results and its verification. The external and internal time steps of the model were set to 2 and 12 seconds, respectively, and the calculation time was from January 1, 2010 to December 31, 2012. In the Laizhou Bay model (as shown in Fig. 7.1), the river runoff data was obtained from the *China River Sediment Bulletin* and its related historical data (Fig. 7.2).

7.1.3 *Simulation results of tide field in Bohai*

Through the simulation, the tide field in Bohai from January 1, 2010 to December 31, 2012, was obtained. The surface-water level,

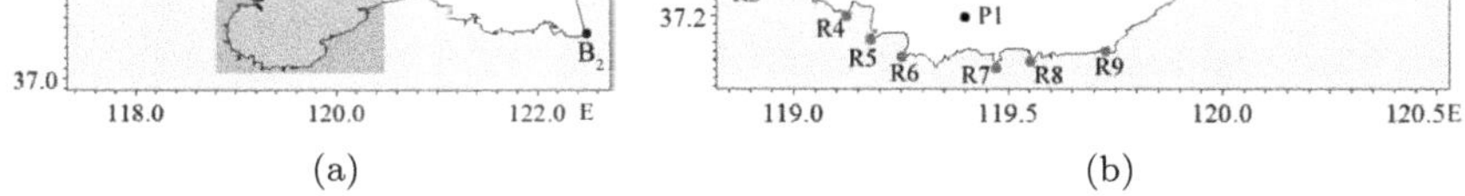

Fig. 7.1. Calculation area of the seawater hydrodynamic model of Bohai and Laizhou Bay: (a) Bohai, and (b) Laizhou Bay.

Notes: B_1B_2 is the open boundary of the calculation area of Bohai. L_1L_2 and L_2L_3 are the north and east opening boundaries of the calculation area of Laizhou Bay, respectively. V1 to V6 are the six measured points for the tide verification of Bohai. P1 to P5 are five representative points at different locations within Laizhou Bay. R1 to R11 refer to the ten major rivers entering the sea around the Yellow River and Laizhou Bay, corresponding to the Yellow River, Zhimai River, Xiaoqing River, Mi River, Bailang River, Yu River, Wei River, Jiaolai River, Sha River, Wang River, and Jie River.

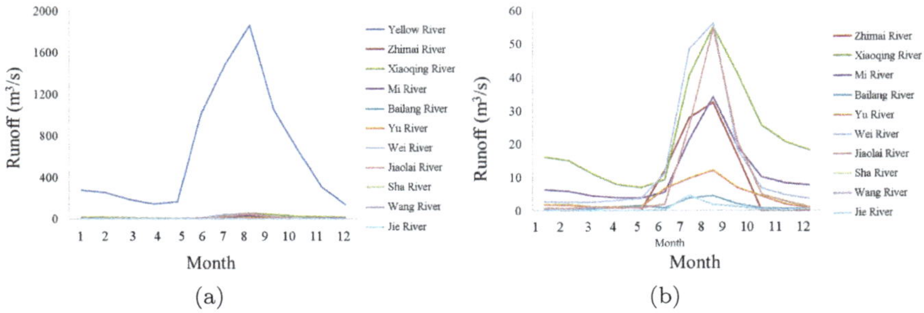

Fig. 7.2. Runoff of sea-entering rivers around the Yellow River and Laizhou Bay: (a) Yellow River, and (b) Laizhou Bay.

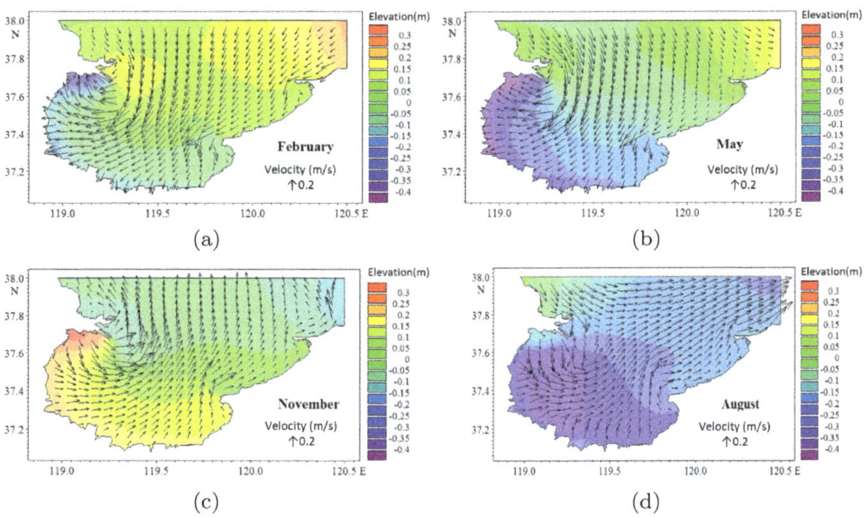

Fig. 7.3. Simulation results of tidal field in Laizhou Bay: (a) at 00:00 on February 15, (b) at 00:00 on May 15, (c) at 00:00 on August 15, and (d) at 00:00 on November 15.

tidal current velocity, and flow direction space-time distribution in Laizhou Bay are shown in Fig. 7.3.

7.1.4 *Verification of simulation results of Bohai seawater hydrodynamic model*

(1) Verification of the current velocity and direction of tidal field in the Bohai area

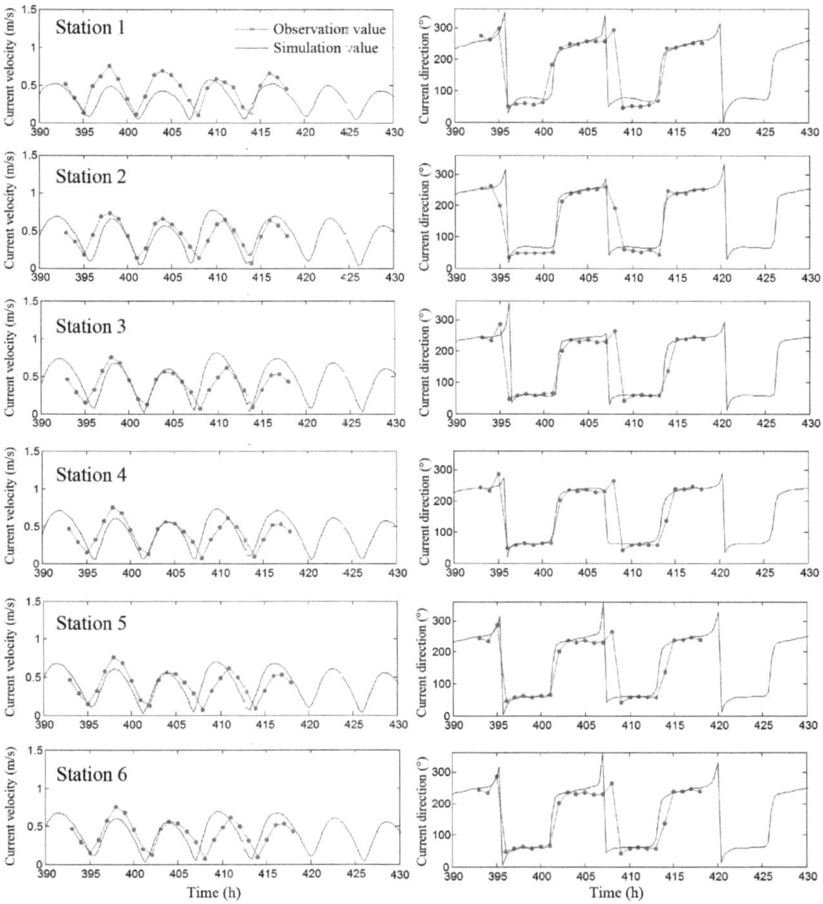

Fig. 7.4. Verification of the velocity and direction of the tidal current in the Bohai seawater hydrodynamic model.

Notes: The points of the verification location were mainly located in the sea near Xianren Island in Liaodong Bay (as shown in Fig. 7.1).

Data from six stations (Fig. 7.1) near Xianren Island in Liaodong Bay that was collected every 1 hour from 9 a.m. on April 16, 2010, to 10 a.m. on April 17, 2010, was used to verify the velocity and direction of the tidal current in the Bohai seawater hydrodynamic model, and the results are shown in Fig. 7.4. Since the tidal field obtained in this chapter is mainly used to provide hydrodynamic forcing for the ecological hydrodynamics model, it can be assumed that our verification effect is better than that previously reported.

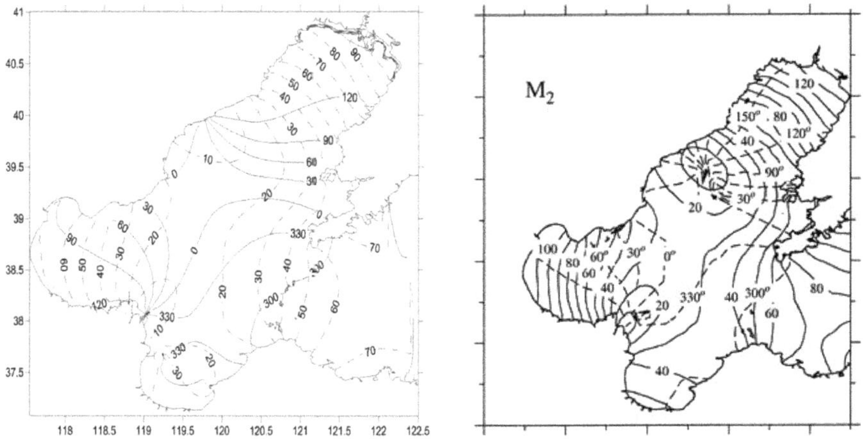

(a) (b)

Fig. 7.5. Simulation results of tide splitting and merging mapping in the Bohai area M_2 and the results from Chen *et al.* (2003): (a) simulation results, and (b) research results.

(2) Tide splitting and merging of the M2 tide field in Bohai

The Fourier analysis on M2 tide splitting was done on the simulation results of the tidal field of Bohai, and the M2 tide splitting and merging diagram is shown in Fig. 7.5, which are consistent with reported results (Dou *et al.*, 1993, Chen and Beardsley, 2003). The M2 tide splitting had an amphidromic point near the old Yellow River estuary and along the coast of Qinhuangdao. Its isochronous angle line rotated counterclockwise; its isoamplitude lines were parallel to the shoreline, and the amplitude of Laizhou Bay was about 0.2–0.4 m. Compared with previously reported results, the M2 tide splitting in this chapter was turned toward the land at the amphidromic point near Qinhuangdao. Because the purpose of this chapter is to provide hydrodynamic forcing for the marine ecological hydrodynamic model and mainly focuses on the waters in Laizhou Bay, it can be assumed that the above-mentioned simulation deviation has little effect on the simulation results of Laizhou Bay's ecological hydrodynamic model. It can also be assumed that the simulation results of the tidal field in the Bohai area are good.

7.2 Construction of Biological-physical Coupling Model in Laizhou Bay

7.2.1 *Theoretical basis of marine ecological hydrodynamic model*

Phytoplankton, zooplankton, detritus, and nutrients are the basic components of marine ecosystems. In this chapter, river source terms, such as tidal current drive, solar radiation, sea temperature, and nutrient salinity, are considered simultaneously to simulate the growth and nutrient salt cycling of phytoplankton in Laizhou Bay. The marine ecological hydrodynamic model in this chapter adopted a simple NPZD model, which had five state variables, including phytoplankton carbon content, zooplankton carbon content, detrital carbon content, and inorganic nitrogen and inorganic phosphorus concentrations in seawater, and were related to 50 parameters in total (Table 7.1). The conceptual model of the model in this chapter is shown in Fig. 7.6 (Shi *et al.*, 2014), and the specific variables and related formulas in the model can be found in literature from Shi *et al.* (2012).

Generally, the simulation of biological variables is based on carbon to study the growth, death, sedimentation of phytoplankton, and the grazing of zooplankton. The growth process of phytoplankton is mainly determined by its maximum growth rate, and at the same time, it is also limited by physical conditions such as solar radiation, seawater temperature, and nutrients. The death and sedimentation process of phytoplankton is limited by its own death, sedimentation rate, nitrogen, phosphorus, and nutrient salt. During the phytoplankton grazing process of phytoplankton, some physical and chemical conditions, such as the maximum grazing rate, seawater temperature, dissolved oxygen, and phytoplankton biomass, should be considered. Zooplankton is mainly affected by its own growth and death. Detritus mainly comes from the unmineralized part of dead phytoplankton and the excretion of zooplankton. The sink of detritus is mainly affected by its mineralization and sedimentation process. The nutrient cycle of the model in this chapter is mainly the inorganic nitrogen and inorganic phosphorus from rivers that enter the sea in Laizhou Bay. In addition to the release of nutrient salt into

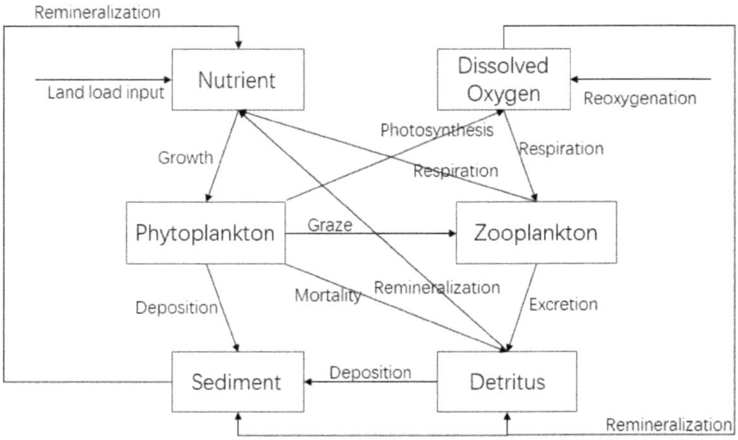

Fig. 7.6. Conceptual model of the ecological hydrodynamic model of Laizhou Bay (Shi *et al.*, 2014).

the sea from the ten seasonal rivers around Yellow River and Laizhou Bay, nutrient salt is also released from biochemical processes, mainly from zooplankton respiration and the mineralization of debris and sediment. The enrichment of nutrient salt is mainly from the phytoplankton uptake.

7.2.2 *Construction of the eco-hydrodynamic model of Laizhou Bay*

In the constructed Laizhou Bay marine ecological hydrodynamic model that was coupled with the Bohai hydrodynamic model in this chapter, the part encompassing Laizhou Bay was the hydrodynamic forcing. The effective light intensity of the sea surface in Laizhou Bay was obtained from existing literature. Short-wave radiation in the literature was analyzed according to selected National Centers for Environmental Prediction (NCEP). Seawater temperature was determined according to Bohai literature (Yang and Leng, 2013) and the fourth-quarter field survey data of Laizhou Bay (Fig. 7.7). Seawater salinity was set to a constant of 29.85, according to the field survey data. The initial conditions of the model were set to May, and the simulation time was from May 1, 2011 to December 31, 2012.

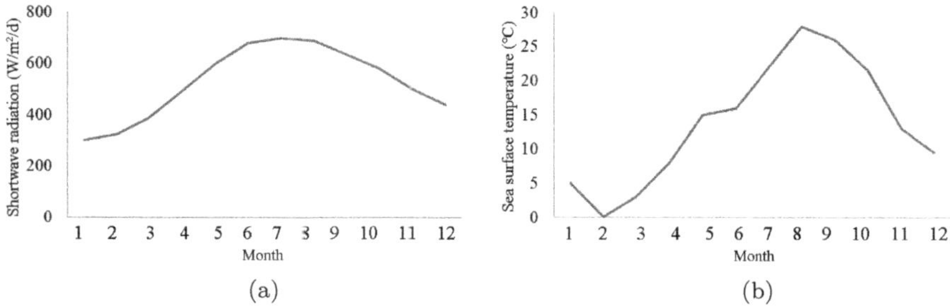

Fig. 7.7. Changes in sea surface short-wave radiation and sea surface temperature during the year in Laizhou Bay: (a) short-wave radiation from the sea surface of Laizhou Bay, and (b) sea surface temperature.

Notes: The short-wave radiation data of the sea surface in Laizhou Bay was from the reanalysis data of NCEP. Sea surface temperature data was from data reported by Yang *et al.* (2013) and the fourth-quarter field survey data of Laizhou Bay.

The initial field spatial interpolation data was mainly the field survey data near Laizhou Bay and the Yellow River Estuary in May 2007 and the Long Island sea area in May 2013. The simulation result was obtained with a constant of 0.108 mg/L. Open boundary conditions were obtained from the survey data of the Bohai area in the summer and winter of 2006, the spring and autumn of 2007, and reported literature (Yang *et al.*, 2013). The simulation results were obtained through time order interpolation data. Due to data availability, the terrestrial input was set to the concentrations of inorganic nitrogen and inorganic phosphorus entering the sea during the high water season of the Yellow River and ten seasonal rivers around Laizhou Bay (Fig. 7.8). The values of 50 parameters involved in the model and their related literature are shown in Table 7.1.

7.3 Simulation Results of the Laizhou Bay Ecological Hydrodynamic Model

Temporal and spatial variation of phytoplankton, zooplankton, detrital carbon content, and concentration of inorganic nitrogen and inorganic phosphorus in the seawater of Laizhou Bay from May 1, 2011

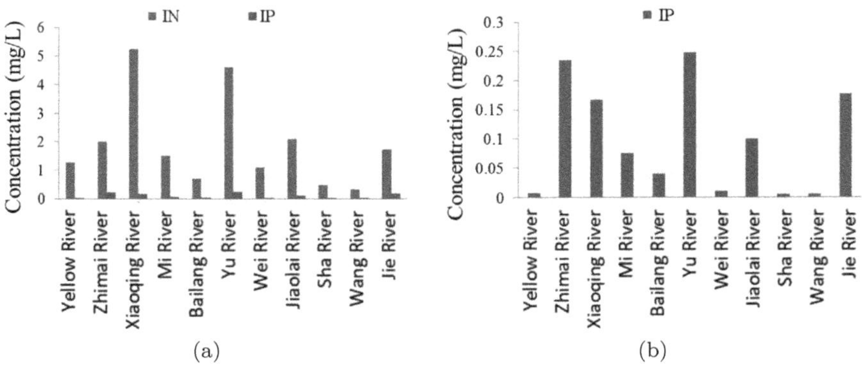

Fig. 7.8. Variations in the pollution concentration of seasonal rivers around the Laizhou Bay during the flood season (Zhang *et al.*, 2015).

to December 31, 2012, were obtained through simulation. With the whole year from January 1, 2012, to December 31, 2012, as an example, the characteristics of the simulated temporal and spatial variation were shown in Figs. 7.9 to 7.14.

Among those biological state variables, (1) the spatial distribution of phytoplankton carbon content in Laizhou Bay showed that the concentration in the near-shore sea area was greater than that in the bay mouth, basically, a peak in summer in the bay and more obvious when it was nearer to the shore, and the distribution characteristics were affected more significantly by the water depth, mainly due to the effect of water photoextinction on phytoplankton (Figs. 7.9 to 7.10). Generally, the near-shore shallow waters in the southwest showed a peak in summer and a sub-peak in autumn, the central waters in the middle of the bay had a peak in summer, and this peak became smaller as it approached the estuary until there was no significant peak in the middle of the estuary, and some regions showed small fluctuation throughout the year, (2) the concentration of zooplankton carbon in the bay mouth was greater than that in the coastal waters in spring, autumn, and winter. In summer, the concentration in the western and southern coastal waters was higher than that in the coastal waters. Most of them showed a summer peak in the bay, and the peak near the southwest was obvious. The sea area near the bay mouth showed a certain fluctuation due to the influence of the central Bohai Sea (Figs. 7.9 and 7.11), and (3) detrital carbon content at the bay mouth was greater in the coastal waters. Likewise, detrital carbon content in the western and southern waters were larger than

Table 7.1. Fifty parameters involved in the model and their values.

Number	Parameter symbol	Parameter meaning	Value	Unit	Reference
1	μ_{PRPC}	Growth rate of phytoplankton	1.0	d^{-1}	(Feng *et al.*, 2007)
2	μ_{DEPC}	Mortality of phytoplankton	0.05	d^{-1}	(Gao *et al.*, 1999)
3	μ_{PRZC}	Growth rate of phytoplankton	0.75	d^{-1}	
4	μ_{DEZC1}	Primary mortality of phytoplankton	0.05	d^{-1}	
5	μ_{DEZC2}	Secondary mortality of phytoplankton	6	d^{-1}	*
6	μ_{GRPC}	The maximum grazing rate	1.5	d^{-1}	*
7	K_{PC0}	Biomass coefficient of Phytoplankton with Level 0 grazing rate	3	no	(Shi *et al.*, 2012)
8	K_{PC1}	Biomass coefficient of Phytoplankton with Level 1 grazing rate	40	no	*
9	K_{CHmin}	The minimum production factor of chlorophyll	0.224	$W/(m^2 \cdot d)$	(Shi *et al.*, 2012)
10	K_{CHmax}	The maximum production factor of chlorophyll	6.16	$W/(m^2d)$	(Shi *et al.*, 2012)
11	PN_{min}	The minimum nitrogen concentration in phytoplankton cells	0.07	gN/gC	(Shi *et al.*, 2012)
12	PN_{max}	The maximum nitrogen concentration in phytoplankton cells	0.176	gN/gC	(Redfield *et al.*, 1963)
13	PP_{min}	The minimum phosphorus concentration in phytoplankton cells	0.002	gP/gC	(Shi *et al.*, 2012)
14	PP_{max}	The maximum phosphorus concentration in phytoplankton cells	0.024	gP/gC	(Redfield *et al.*, 1963)
15	F_{AC}	Dark response correction factor	1.3	no	(Shi *et al.*, 2012)
16	K_{ZN}	Nitrogen-carbon ratio of zooplankton	0.17	gN/gC	(Zhang *et al.*, 2002)
17	K_{ZP}	Phosphorus-carbon ratio of zooplankton	0.013	gP/gC	(Zhang *et al.*, 2002)
18	μ_{REZC}	Respiration constant of zooplankton	0.3	no	(Shi *et al.*, 2012)
19	bla	Extinction coefficient of water body	0.1	m^{-2}	(Shi *et al.*, 2012)
20	pla	Extinction coefficient of chlorophyll	20.0	$m^{-2}g^{-1}$	(Shi *et al.*, 2012)
21	dla	Extinction coefficient of organic debris	0.1	$m^{-2}g^{-1}$	(Shi *et al.*, 2012)
22	μ_{SEPC}	Sedimentation rate of phytoplankton	0.2	m/d	(Feng *et al.*, 2007)
23	μ_{SEDC}	Sedimentation rate of organic debris	0.05	m/d	(Shi *et al.*, 2012)
24	h_{PP}	Half-saturation constant of phosphorus	0.004	gP/gC	(Shi *et al.*, 2012)
25	μ_{UPPN}	Nitrogen uptake rate of phytoplankton under restricted conditions	0.3	d^{-1}	(Shi *et al.*, 2012)

(*Continued*)

Table 7.1. (*Continued*)

Number	Parameter symbol	Parameter meaning	Value	Unit	Reference
26	μ_{UPPP}	Phosphorus uptake rate of phytoplankton under restricted conditions	0.05	d^{-1}	(Shi *et al.*, 2012)
27	h_{UPPN}	Half-saturation constant of phytoplankton nitrogen uptake	0.112	mg/L	(Wang *et al.*, 2006)
28	h_{UPPP}	Half-saturation constant for phytoplankton phosphorus uptake	0.02	mg/L	(Wang *et al.*, 2006)
29	μ_{REDC}	Mineralization rate of organic debris	0.05	d^{-1}	(Shi *et al.*, 2012)
30	K_{REDN}	Nitrogen releasing proportion of debris mineralization	1	no	(Shi *et al.*, 2012)
31	K_{REDP}	Phosphorus releasing ratio of debris mineralization	1	no	(Shi *et al.*, 2012)
32	K_{REDEPC}	Direct mineralization rate of dead phytoplankton	0.1	no	(Shi *et al.*, 2012)
33	μ_{RESC}	Sediment mineralization rate	1.0	d^{-1}	(Shi *et al.*, 2012)
34	N_{REL}	Nitrogen releasing ratio from sediments under hypoxia	1	no	(Shi *et al.*, 2012)
35	P_{REL}	Phosphorus releasing ratio from sediments under hypoxia	1	no	(Shi *et al.*, 2012)
36	K_{RESN}	Nitrogen releasing ratio from sediment mineralization	0.05	$g/(m^2d)$	(Shi *et al.*, 2012)
37	K_{RESP}	Phosphorus releasing ratio from sediment mineralization	0.01	$g/(m^2d)$	(Shi *et al.*, 2012)
38	h_{DO1}	Half-saturation constant of oxygen in water	2.0	mg/L	(Shi *et al.*, 2012)
39	h_{DO2}	Half-saturation constant of oxygen in sediment	2.0	mg/L	(Shi *et al.*, 2012)
40	K_{DO}	Oxygen dependence coefficient	1	no	(Shi *et al.*, 2012)
41	K_{OC}	Production/consumption ratio of oxygen compared to carbon	3.5	gDO/gC	(Shi *et al.*, 2012)
42	μ_{REAR}	Atmospheric reoxygenation rate	1.5	d^{-1}	(Shi *et al.*, 2012)
43	β_{PRPC}	Temperature coefficient of phytoplankton growth	0.06	$°C^{-1}$	*
44	β_{GRPC}	Temperature coefficient of the maximum grazing rate	0.05	$°C^{-1}$	(Shi *et al.*, 2012)
45	β_{REDC}	Temperature coefficient of debris mineralization	0.13	$°C^{-1}$	(Shi *et al.*, 2012)
46	β_{α}	Temperature coefficient of light saturation intensity	0.04	$°C^{-1}$	(Shi *et al.*, 2012)
47	α	Light saturation intensity	150	$W/(m^2d)$	(Feng *et al.*, 2007)
48	β_{RESN}	Temperature coefficient of nitrogen releasing of sediment	0.1		(Shi *et al.*, 2012)
49	β_{RESP}	Temperature coefficient of phosphorus releasing of sediment	0.1	$°C^{-1}$	(Shi *et al.*, 2012)
50	β_{RESC}	Temperature coefficient of sediment mineralization	0.07	$°C^{-1}$	(Shi *et al.*, 2012)

Note: "*" indicates that the value is corrected in this chapter.

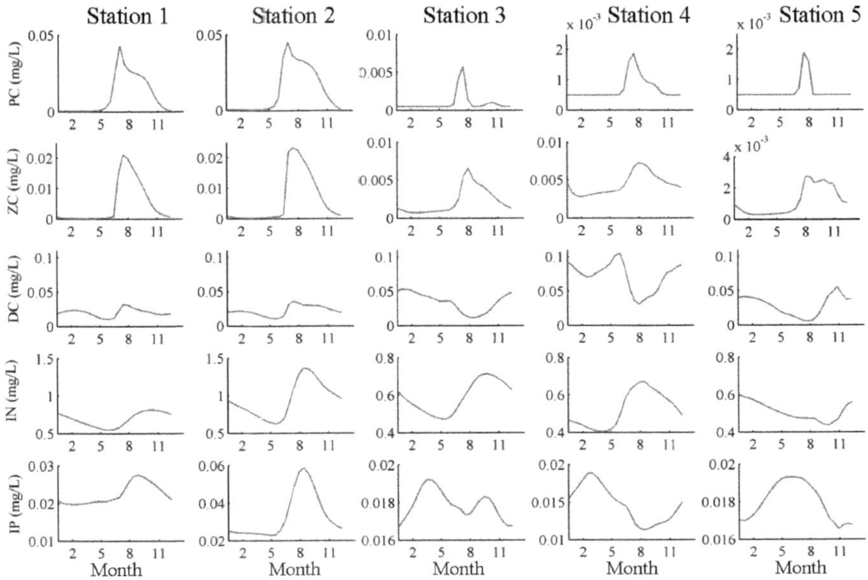

Fig. 7.9. Simulation results of state variables of the ecological model of Laizhou Bay with time.

Notes: With the period from January 1, 2012 to December 31, 2012, as an example, the representative location points are shown in Fig. 7.1, in which the representative Point 1 was located near the estuary in the south, Point 2 was near the estuary of the Xiaoqing River, Point 3 was located in the central waters of the bay, Point 4 was located near the estuary of the Yellow River, and Point 5 was located in the eastern waters of the bay mouth.

that in the eastern waters in summer. With the change of offshore distance, there were significantly different seasonal changes (Figs. 7.9 and 7.12). In the southwestern coastal shallow waters, there were summer peaks, sub-peaks in autumn and winter, and low peaks in spring. There were also low summer peaks in the middle of the bay and near the bay mouth, and the middle waters in the bay had winter peaks and spring-autumn sub-peaks.

According to the state variables of the nutrient salt, (1) the concentration of inorganic nitrogen in coastal waters was greater at the bay mouth, showing significantly different seasonal changes (Figs. 7.9 and 7.13). The sea area within the bay had summer peaks, sub-seasonal peaks in autumn and winter, and low peaks in spring, which were mainly affected by the discharge of nutrient salts from rivers into the sea. That of waters around the bay mouth showed small

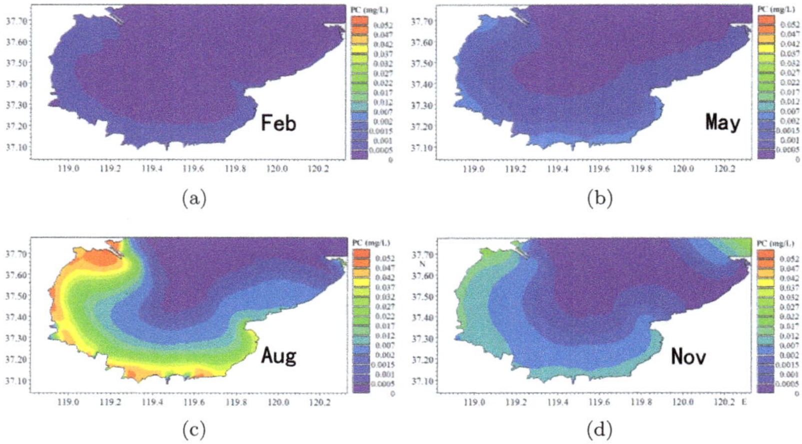

Fig. 7.10. Spatial distribution of phytoplankton carbon content in the Laizhou Bay: (a) at 00:00 on February 15, (b) at 00:00 on May 15, (c) at 00:00 on August 15, and (d) at 00:00 on November 15.

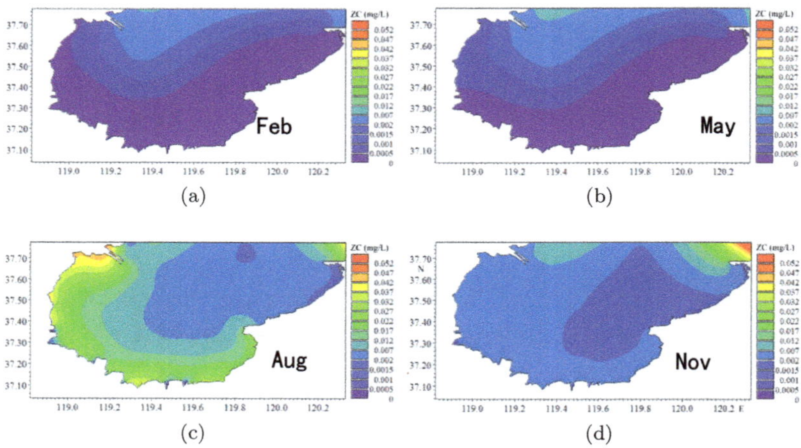

Fig. 7.11. Spatial distribution of simulated carbon content of zooplankton in the Laizhou Bay, (a) at 00:00 on February 15, (b) at 00:00 on May 15, (c) at 00:00 on August 15, and (d) at 00:00 on November 15.

fluctuations and was mainly affected by the central Bohai sea area, (2) inorganic phosphorus in seawater was mainly concentrated in the coastal waters of the southwest, especially in the vicinity of the Xiaoqing River Estuary; the concentration was relatively high in shallow coastal waters in summer, and that near the Yellow River Estuary

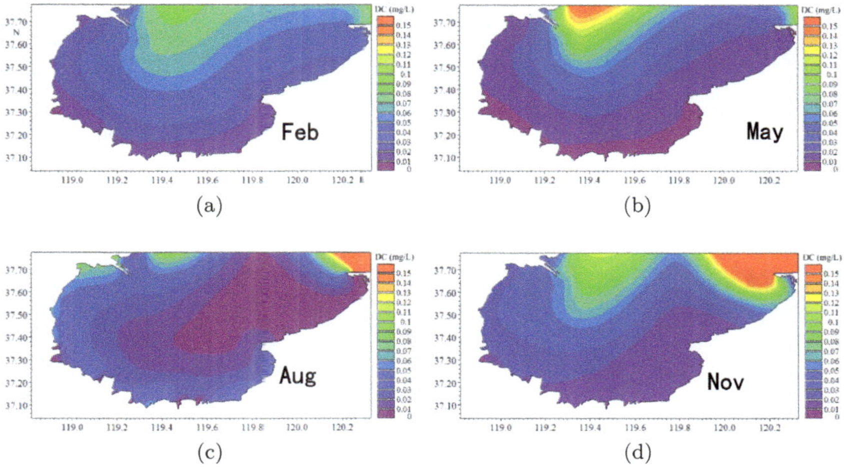

Fig. 7.12. Spatial distribution of simulated carbon content in the Laizhou Bay: (a) at 00:00 on February 15, (b) at 00:00 on May 15, (c) at 00:00 on August 15, and (d) at 00:00 on November 15.

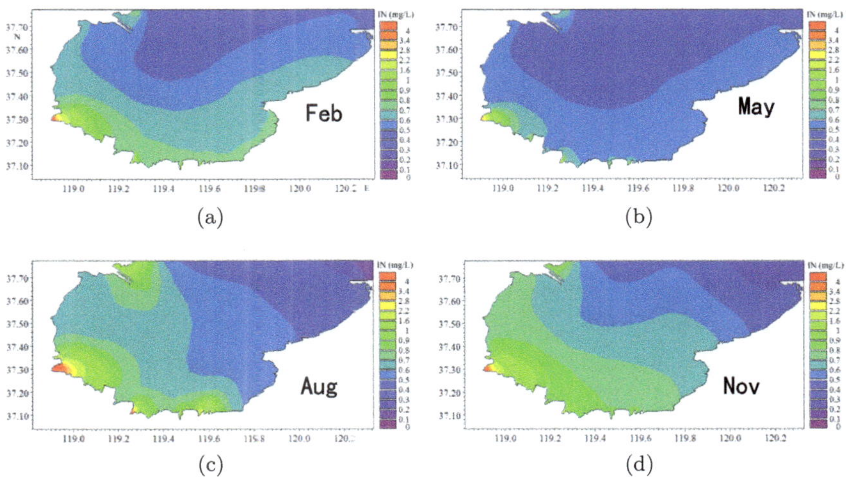

Fig. 7.13. Spatial distribution of simulated results of inorganic nitrogen concentration in the seawater of Laizhou Bay: (a) at 00:00 on February 15, (b) at 00:00 on May 15, (c) at 00:00 on August 15, and (d) at 00:00 on November 15.

was relatively high in winter and relatively low in summer. Seasonal changes in other sea areas were not significant relatively (Figs. 7.9 and 7.14).

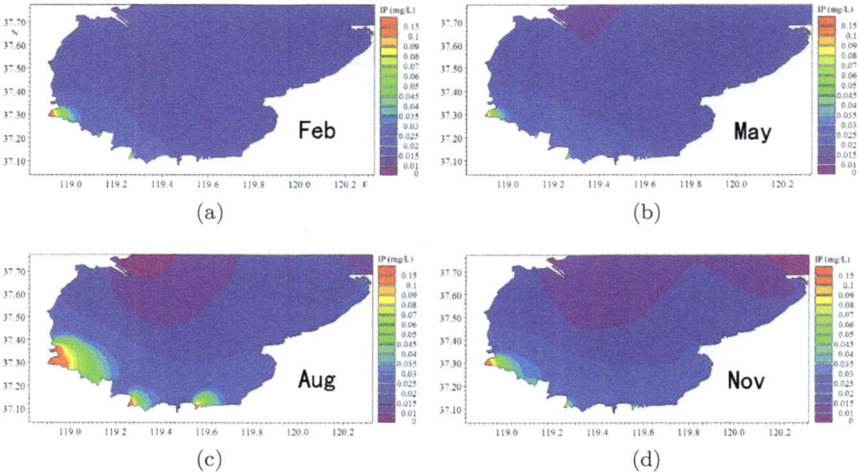

Fig. 7.14. Spatial distribution of simulated results of inorganic phosphorus concentration in the seawater of Laizhou Bay: (a) at 00:00 on February 15, (b) at 00:00 on May 15, (c) at 00:00 on August 15, and (d) at 00:00 on November 15.

7.4 Scenario Simulation of the Impact of River Inflow Pollution on Laizhou Bay Ecosystem

Reported simulations of the response of marine ecosystems to human activities are often based on marine ecological hydrodynamic models, in which different human activity impact schemes are designed through changing model parameter values, boundary conditions, and other model inputs. According to research needs, physical and chemical models and different nutrition levels models are used to run the multi-scenario simulation (Gibson and Spitz, 2011; Rombouts *et al.*, 2013; Shen *et al.*, 2016). Thus, the response characteristics of marine ecosystems to human disturbances have been studied, and socioeconomic activity control measures are designed with human disturbances as intermediate variables. In this chapter, mainly through changing the river input of nutrient salt, the response characteristics of near-shore marine ecosystems to typical human activities, such as river pollution into the sea, will be investigated. In order to simulate the response of the Laizhou Bay ecosystem to the nutrient salt coming from rivers, under scenarios with the half-cut overall concentration of inorganic nitrogen and inorganic phosphorus entering the sea from the river, water quality controlled within the second-class standards, and river discharge into the sea eliminated.

Characteristics of the state change of the NPZD model of Laizhou Bay under different scenarios were simulated. Under the current model settings, nutrient salt changes from rivers into the sea had little effect on the biomass of phytoplankton, zooplankton, and debris in Laizhou Bay. However, existing studies have shown that an inconsistent nutrient salt structure could cause changes in species types and their community succession (Sun *et al.*, 2001; Wei *et al.*, 2004; Peng *et al.*, 2012; Zheng *et al.*, 2016). For example, when the sea area of Laizhou Bay changed from nitrogen limitation to phosphorus limitation, the phytoplankton community changed from an absolute diatom domain to a codomain with diatoms and dinoflagellates (Zhang *et al.*, 2014). In the polluted Xiaoqing River Estuary, macrobenthos were dominated by polychaetes, while the central part of Laizhou Bay and the waters of the bay mouth were dominated by mollusks (Zhang, 2011).

According to the response simulation results of the inorganic nitrogen concentration in the seawater of Laizhou Bay to different concentrations of inorganic nitrogen entering the sea (Figs. 7.15 and 7.16), the inorganic nitrogen concentration in the seawater changed significantly, and the change in the coastal waters, in general, was greater than that in the vicinity of the bay mouth. The change in response degree with time showed that there was a large change in

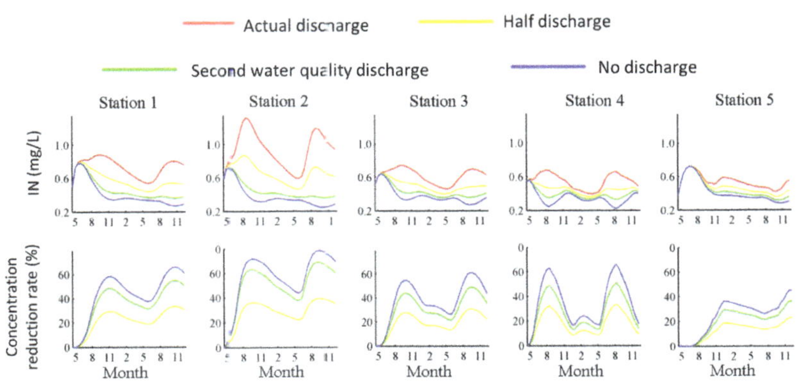

Fig. 7.15. The variation of inorganic nitrogen concentration in seawater with time under the scenarios with river nutrient salt entering the sea.

Notes: Using the period from May 1, 2011, to December 31, 2012, as an example, the representative location points are shown in Fig. 7.1. Point 1 was located near the mouth of the southern sea, Point 2 near the mouth of Xiaoqing River, Point 3 in the central waters of the bay, Point 4 near the Yellow River mouth, and Point 5 in the eastern waters of the bay mouth.

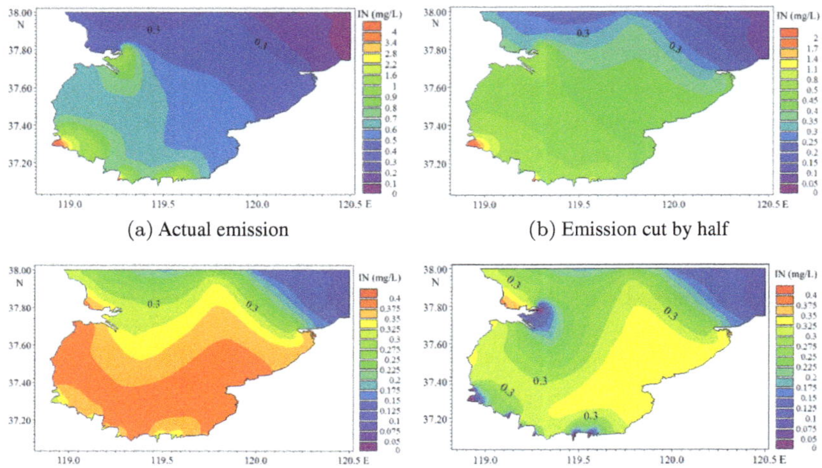

(a) Actual emission

(b) Emission cut by half

(c) Satisfied with Type 2 water quality

(d) Emissions eliminated

Fig. 7.16. Spatial distribution of inorganic nitrogen concentration in seawater under the scenarios with river nutrient salt entering the sea: (a) the actual concentration discharge, (b) concentration cut by half, (c) concentration control within the Type 2 water quality standard, and (d) pollutant discharge eliminated. All pictures showed the spatial distribution at 00:00 on August 15, 2012.

the waters in the bay during the high water period and that the waters near the bay mouth began to fluctuate slightly after a high water period and the change in the second high water period was relatively small. According to the actual concentration of inorganic nitrogen entering the sea, the inorganic nitrogen concentration in the seawater of Laizhou Bay was inferior to Type 4 water quality. When the inorganic nitrogen concentration of the river entering the sea was reduced by half, the waters in the middle and the vicinity of the bay mouth reached Type 4 water quality after a flood period. When the discharge concentration met Type 2 water quality standard or the discharge was eliminated, after a water-rich period, the concentration of inorganic nitrogen in the seawater of Laizhou Bay showed a small fluctuation and met Type 3 water quality standards (less than 0.4 mg/L). After a short period of time, the waters on the south side of the Yellow River Estuary reached Type 3 water quality before the first abundant period.

According to the simulation results of the response of the inorganic phosphorus concentration in the seawater of Laizhou Bay to the different concentrations of inorganic phosphorus in the river (Figs. 7.17 and 7.18), the inorganic phosphorus concentration in the

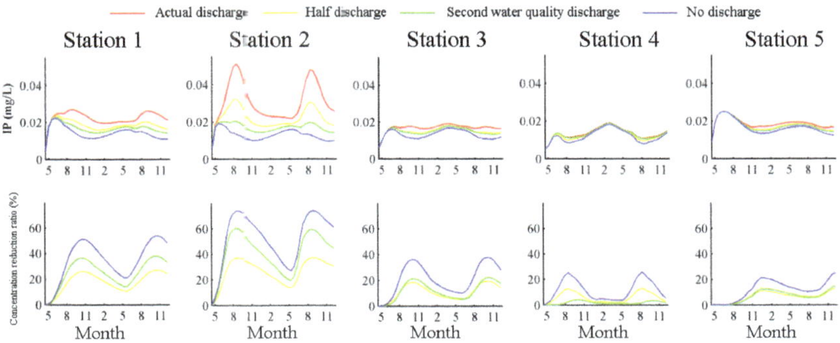

Fig. 7.17. Variation of inorganic phosphorus concentration in seawater over time under the scenarios with river nutrient salt entering into the sea.

Notes: Using the period from May 1, 2011 to December 31, 2012 as an example, the representative location points were shown in Fig. 7.1. Among them, the representative Point 1 was located near the southern estuary, Point 2 near the estuary of Xiaoqing River, Point 3 in the central waters of the bay, Point 4 near the Yellow River mouth, and Point 5 in the eastern waters of the bay mouth.

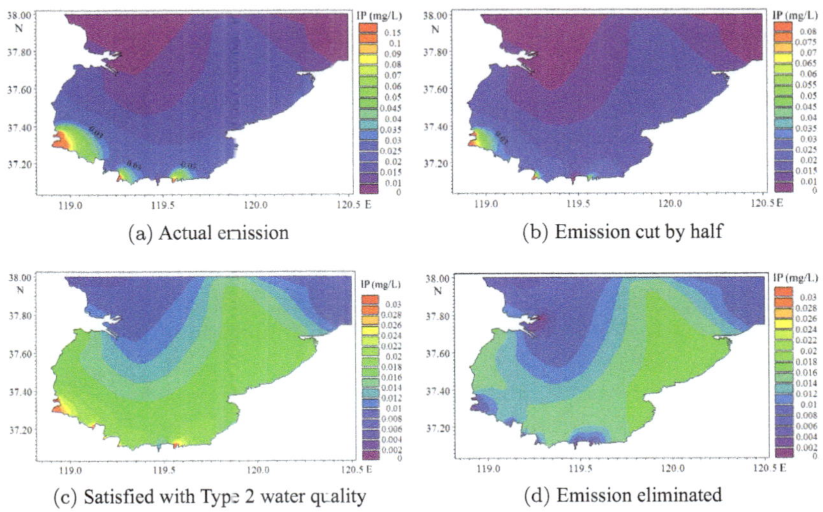

Fig. 7.18. Spatial distribution of inorganic phosphorus concentration in seawater under the scenarios with river nutrient salt entering the sea: (a) the actual concentration discharged, (b) emission concentration cut by half, (c) concentration control within the Type 2 water quality standard, and (d) eliminated pollutant discharging.

Note: All pictures showed the spatial distribution at 00:00 on August 15, 2012.

local waters near the estuary of the sea changed significantly, while its change in the central waters and the nearby sea area of the bay mouth was relatively small, and thus it had little effect on the overall spatial distribution pattern of inorganic phosphorus concentration. The change in response degree on the time scale was similar to that of inorganic nitrogen concentration in seawater.

References

Baretta J W, Ebenhoh W, Ruardij P. 1995. The European regional seas ecosystem model, a complex marine ecosystem model. *Netherlands Journal of Sea Research.* **33**(3): 233–246.

Chen C S. 2003. *Marine Ecosystem Dynamics and Models.* Beijing: Higher Education Press.

Chen C, Ji R, Zheng L, *et al.* 1999. Influences of physical processes on the ecosystem in Jiaozhou Bay: A coupled physical and biological model experiment. *Journal of Geophysical Research: Oceans.* **104**: 29925–29949.

Chen C, Liu H, Beardsley R C. 2003. An unstructured grid, finite-volume, three-dimensional, primitive equations ocean model: Application to coastal ocean and estuaries. *Journal of Atmospheric and Oceanic Technology.* **20**(1): 159–186.

Delhez E, Martin G. 1994. 3D Modelling of hydrodynamic and ecohydrodynamic processes on the north-western European continental shelf. *Bulletin De La Societe Royale Des Sciences De Liege* (Belgium). **63**: 1–2.

Dou Z X, Yang L W, Ozer J. 1993. Numerical simulation of Bohai 3D tide. *Haiyang Xuebao.* **15**(5): 1–15.

Feng S Z, Zhang J, Wei H. 2007. *Introduction to Environmental Dynamics of Bohai.* Beijing: Science Press.

Friedrichs M A M, Dusenberry J A, Anderson L A, *et al.* 2007. Assessment of skill and portability in regional marine biogeochemical models: Role of multiple planktonic groups. *Journal of Geophysical Research.* **112**(C8): C08001.

Gao H W, Sun W X, Zhai X M. 1999. Sensitive analysis of the parameters of a Pelagic ecosystem dynamic model. *Journal of Ocean University of Qingdao (Science Education).* **29**(3): 398–404.

Gibson G A, Spitz Y H. 2011. Impacts of biological parameterization, initial conditions, and environmental forcing on parameter sensitivity and uncertainty in a marine ecosystem model for the Bering Sea. *Journal of Marine Systems*. **88**(2): 214–231

Moll A. 1995. *Regionale Differenzierung der Primaerproduktion in der Nordsee: Untersuchung mit Einem drei-dimen-sionalen Model*. Hamburg: Zentrum fur meeresund Klimaforschung der Universitat Hamburg, 1–151.

Peng S, Qin X, Shi H, *et al.* 2012. Distribution and controlling factors of phytoplankton assemblages in a semi-enclosed bay during spring and summer. *Marine Pollution Bulletin*. **64**(5): 941–948.

Redfield A C, Ketchum B H, Richards F A. 1963. The influence of organisms on the composition of seawater. In: *The Sea*. New York: John Wiley, 26-27.

Rombouts I, Beaugrand G, Artigas L F, *et al.* 2013. Evaluating marine ecosystem health: Case studies of indicators using direct observations and modelling methods. *Ecological Indicators*. **24**: 353–365.

Shen C, Shi H, Liu Y, *et al.* 2016. Discussion of skill improvement in marine ecosystem hydrodynamic models based on parameter optimization and skill assessment. *Chinese Journal of Oceanology and Limnology*. DOI: 10.1007/s00343-016-5068-3.

Shi H H, Ding D W, Zheng W. 2012. *Key Technologies and Applications of Coastal Ecosystem Evaluation, Simulation and Regulation*. Beijing: Ocean Press.

Shi H H, Shen C C, Li F, *et al.* 2014. Parameter sensitivity analysis of a coupled biological-physical model in Jiaozhou Bay. *Acta Ecologica Sinica*. **34**(1): 41–49.

Skogen M D. 1993. *A User's Guide to NORVECOM: The Norwegian Ecological Model System*. Institute of Marine Research.

Sun J, Liu D, Qian S. 2001. Preliminary study on seasonal succession and development pathway of phytoplankton community in the Bohai Sea. *Acta Oceanologica Sinica*. **20**(2): 251–260.

Wang X L, Li K Q, Shi X Y. 2006. *Marine Environmental Capacity of Major Chemical Pollutants in Jiaozhou Bay*. Beijing: Science Press.

Wei H, Sun J, Moll A, *et al.* 2004. Phytoplankton hydrodynamics in the Bohai Sea — Observations and modelling. *Journal of Marine Systems*. **44**(3): 233–251.

Yang J Q, Leng Y. 2013. *Investigation and Research on the Bio-ecology of Bohai*. Beijing: Ocean Press.

Yu L L. 2012. *Studies on the Environmental Capacity of Nutrient Nitrogen and Phosphorus Salts in Laizhou Bay Based on Three-dimensional Marine Ecological Dynamic Model*. Qingdao: China Ocean University.

Zhang S S. 2011. *Establishment of Evaluation Index System of the Sedimentary Environment of Xiaoqing River Estuary based on Large Benthic Animals*. Qingdao: China Ocean University.

Zhang W C, Wang K, Xiao T. 2002. Some conversion factors in the studies on zooplankton biomass and energy. *Marine Sciences*. **26**(1): 33–36.

Zhang J F, Gao X L, Li P M, *et al.* 2015. Nutrient distribution characteristics and long-term trends in the southwest of the Laizhou Bay and its adjacent rivers. *Marine Science Bulletin*. **34**(2): 222–232.

Zhang J F, Gao X L, Zhuang W, *et al.* 2014. Analysis on the trend of fishery resources and environment change in Laizhou Bay. *Marine Lake Bulletin*. (3): 82–90.

Zheng W, Li F, Shi H, *et al.* 2016. Spatiotemporal heterogeneity of phytoplankton diversity and its relation to water environmental factors in the southern waters of Miaodao Archipelago, China. *Acta Oceanologica Sinica*. **35**(2): 46–55.

Chapter 8

Global Sensitivity Analysis and Parameter Optimization Methods: Application in Nearshore Water Quality-Physical Coupled Model

The water quality model is a type of quantitative mathematical model that describes water quality state variable variations of rivers, lakes, oceans, and other water bodies such as (DIN, DON, DO) with the spatiotemporal change under the action of physical, chemical, and biological processes. According to the solution characteristics and calculation dimension of the water quality model, the water quality model can be divided into a deterministic model and stochastic model, and a zero-dimensional model (Iriarte *et al.*, 2015), one-dimensional model (Pauer *et al.*, 2014), two-dimensional model (Bailey and Ahmadi 2014), and three-dimensional model (Zheng *et al.*, 2004; Li *et al.*, 2015; Laurent *et al.*, 2017; Lu *et al.*, 2017), respectively. The research into the water quality model is multidisciplinary and comprehensive, which involves the water environment and pollution control theory (Yu *et al.*, 1989).

The study of the water quality model dates back to the 1920s when Streeter and Phelps established the river oxygen balance model, namely the Streeter–Phelps model (S–P model). Since the S–P model is the first water quality model, many scholars have researched water quality models based on this model. In the 1970s and 1980s, research on water quality models developed rapidly and began to involve complex dynamic processes, such as LAKECO, WRMMS (Cleary,

1976), QUAL series models (Cole and Wells, 2006), and the three-dimensional model WASP model (Wool *et al.*, 2006). In recent years, the focus of water quality model research has expanded from point source pollution to non-point source pollution and from general water quality models to comprehensive marine dynamic water quality models. The most widely used models abroad include WASP, QUAL-II, CE-QUAL-RIV1, CE-QUAL-W2, and so on. These models all take rivers and lakes as research objects. At present, the three-dimensional hydrodynamic models used to simulate the spatiotemporal changes of ocean water quality are mainly the MIKE series models (Water, 2002) and FVCOM (Chen *et al.*, 2003). However, most of these models do not perform data assimilation, and the calculation accuracy of the models is significantly affected by the quality of the observed parameters. In addition, complex water quality models contain many parameters that are difficult to determine. The application scope of the model is also limited by the parameter adaptation area (Evans, 1999; Fan, 2009).

In response to this situation, some scholars not only studied the temporal and spatial distribution of parameters in ecological dynamics models but also optimized the parameters of the models (Hemmings *et al.*, 2004; Losa *et al.*, 2004, 2006; Fan and Lv, 2009; Li, 2013; Wang, 2013). These studies demonstrated that the commonly used four-dimensional variational method, namely the adjoint assimilation method, was very efficient. The method of assimilation can invert the parameters with spatiotemporal characteristics by assimilating the observation data and reducing the deviation between the simulation results and the observed values.

The essence of the parameter optimization method is to optimize the simulation results according to actual observation data and adjust the key influence parameters of the model, open boundary conditions, and the initial field of state variables to improve the model's simulation performance. This method is referred to as model assimilation in the marine and meteorological fields (Shen *et al.*, 2016). The data assimilation method was used for early meteorological forecasting but has been gradually applied to the forecasting of tides and waves. In recent years, it has also been applied to the dynamic model of marine ecosystems (Gregg *et al.*, 2009). Li (2013) summarized the current assimilation methods of marine ecosystem dynamics models and put forward the theory that commonly used data assimilation methods were mainly variational methods. The variational method was first proposed and used by Sasaki (1970). This method was first applied to the dynamic model of marine ecosystems in 1995, which successfully inverted the initial field and parameters of the model. Among various variational methods, the adjoint method is the most commonly used (McGillicuddy *et al.*, 1998; Lu, 2000; Zhao and Lu, 2008; Fan and Lv, 2009; Lü and Fan, 2009). The adjoint method describes the simulation results of the model by constructing a cost function and uses the Lagrange multiplier method to solve the conditional extreme value problem, to achieve the purpose of optimizing model parameters and improving simulation accuracy. In recent years, the adjoint assimilation method has been continuously developed and improved (Guo *et al.*, 2017; Liu *et al.*, 2017).

Existing studies on water quality models focused on water quality parameters, while few studies have integrated marine ecosystem dynamics models. In this chapter, by establishing a three-dimensional hydrodynamic-water quality coupled model, the assimilation method is used to assimilate the model and invert the spatiotemporal changes of the parameters to improve the simulation results of the model and accurately simulate the water quality parameters.

8.1 Major Biochemical Processes Considered in the Model

The water quality model established in this paper is an improved version based on the NPZD model, and the water quality models of Fennel *et al.* (2006) and Lu *et al.* (2017). In this model, the phytoplankton, zooplankton, and detritus in the nutrient-phytoplankton-zooplankton-detritus (NPZD) model are retained, and the nutrients in the NPZD model are extended to more state variables. Integrated with the water quality model, a three-dimensional hydrodynamic-water quality coupled model is established for studying various nutrients and spatiotemporal characteristics of dissolved oxygen.

The conceptual diagram of the model is shown in Figs. 8.1 and 8.2. The water quality model is mainly composed of nine state variables: Phytoplankton (P), zooplankton (Z), detritus (D), dissolved organic nitrogen (DON), ammonium nitrogen (NH_4), nitrate nitrogen (NO_3), dissolved organic phosphorus (DOP), phosphate (PO_4),

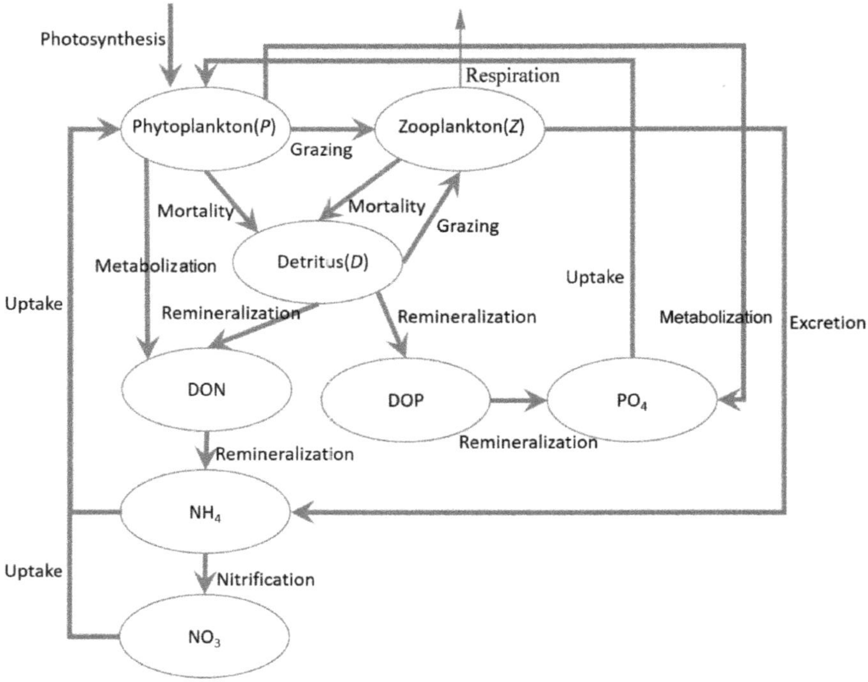

Fig. 8.1. Model concept model.

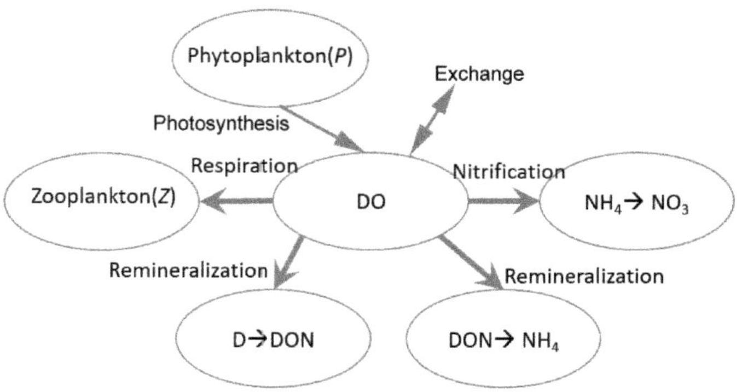

Fig. 8.2. Conceptual diagram of dissolved oxygen dynamics in the water quality model.

and dissolved oxygen (DO), which covers critical mechanisms such as photosynthesis, predation, mineralization, metabolism, absorption of nutrients, and redox reactions. Photosynthesis is the main energy input for this marine ecosystem. Equations of state variables can be expressed as (8.1) to (8.9), shown below. The main parameters of the model are shown in Table 8.1. In Eqs. (8.1) to (8.9), x, y, and z represent the east-west, north-south, and vertical directions in the Cartesian coordinate system, respectively, and u, v, and w correspond to the eastward velocity, northward velocity, and vertical velocity of the Cartesian coordinate system, respectively. A_h and A_v represent horizontal and vertical turbulent diffusion coefficients, respectively.

$$\begin{aligned}
\frac{\partial P}{\partial t} &= -u\frac{\partial P}{\partial x} - v\frac{\partial P}{\partial y} - w\frac{\partial P}{\partial z} + \frac{\partial}{\partial x}\left(A_h\frac{\partial P}{\partial x}\right) + \frac{\partial}{\partial y}\left(A_h\frac{\partial P}{\partial y}\right) \\
&\quad + \frac{\partial}{\partial z}\left(A_v\frac{\partial P}{\partial z}\right) + \left[\mu_{PPT_G}\left(1 - r_{PPT_E}\right) - \mu_{PPT_D}\right]P \\
&\quad - \mu_{PPT_z}Z, \qquad (8.1) \\
\frac{\partial Z}{\partial t} &= -u\frac{\partial Z}{\partial x} - v\frac{\partial Z}{\partial y} - w\frac{\partial Z}{\partial z} + \frac{\partial}{\partial x}\left(A_h\frac{\partial Z}{\partial x}\right) + \frac{\partial}{\partial y}\left(A_h\frac{\partial Z}{\partial y}\right) \\
&\quad + \frac{\partial}{\partial z}\left(A_v\frac{\partial Z}{\partial z}\right) + (\delta_{PPT_z}\mu_{PPT_z} + \delta_{DPT_z}\mu_{DPT_z} \\
&\quad - \mu_{ZPT_D} - \mu_{ZPT_N} - k_{ZPT_F})Z, \qquad (8.2) \\
\frac{\partial D}{\partial t} &= -u\frac{\partial D}{\partial x} - v\frac{\partial D}{\partial y} - w\frac{\partial D}{\partial z} + \frac{\partial}{\partial x}\left(A_h\frac{\partial D}{\partial x}\right) + \frac{\partial}{\partial y}\left(A_h\frac{\partial D}{\partial y}\right) \\
&\quad + \frac{\partial}{\partial z}\left(A_v\frac{\partial D}{\partial z}\right) + (\mu_{ZPT_D} + (1 - \delta_{PPT_z})\mu_{PPT_z} \\
&\quad - \delta_{DPT_z}\mu_{DPT_z})Z + \mu_{PPT_D}P - \mu_{DPT_B}D, \qquad (8.3) \\
\frac{\partial DON}{\partial t} &= -u\frac{\partial DON}{\partial x} - v\frac{\partial DON}{\partial y} - w\frac{\partial DON}{\partial z} + \frac{\partial}{\partial x}\left(A_h\frac{\partial DON}{\partial x}\right) \\
&\quad + \frac{\partial}{\partial y}\left(A_h\frac{\partial DON}{\partial y}\right) + \frac{\partial}{\partial z}\left(A_v\frac{\partial DON}{\partial z}\right) + \mu_{PPT_E}P \\
&\quad + (1 - r_{ZPT_N})\mu_{ZPT_N}Z + \mu_{DPT_DON}D \\
&\quad - \mu_{DON_NH_4} \cdot DON, \qquad (8.4)
\end{aligned}$$

Table 8.1. List of water quality model parameters.

Symbol	Parameter description	Value	Confidence interval	Unit
ρ	The ratio of effective light irradiance in photosynthesis	0.43	0.301–0.559	—
I_{opt}	Optimal light irradiance	72.5	50.750–94.250	W/m^2
k_{PPT_G}	Phytoplankton growth rate	0.8	0.560–1.040	1/day
k_{PPT_D}	Phytoplankton mortality rate	0.05	0.035–0.065	1/day
k_{PPT_z}	The ratio of phytoplankton predated by zooplankton	0.4	0.280–0.520	1/day
k_{ZPT_D}	Zooplankton mortality rate	0.05	0.035–0.065	1/day
k_{ZPT_N}	Zooplankton excretion rate	0.2	0.140–0.260	1/day
k_{ZPT_F}	The ratio of zooplankton predated by fish	0.1	0.070–0.130	1/day
k_{DPT_z}	The ratio of zooplankton predation on detritus	0.6	0.420–0.780	1/day
k_{DPT_B}	Demineralization rate of detritus	0.05	0.035–0.065	–
k_{DON_B}	DON remineralization rate	0.03	0.021–0.039	1/day
k_{DOP_B}	DOP remineralization rate	0.04	0.028–0.052	1/day
k_{NO3}	Half-saturated concentration of nitrate nitrogen	0.5	0.350–0.650	mmol N/m^3
k_{NH4}	The half-saturated concentration of ammonium nitrogen	0.5	0.350–0.650	mmol N/m^3
k_{PO4}	The half-saturated concentration of phosphate	0.03	0.021–0.039	mmol P/m^3
KS_{DPT}	The half-saturated concentration of zooplankton predation on detritus	0.7	0.490–0.910	mmol N/m^3
KS_{PPT}	The half-saturated concentration of zooplankton predation on phytoplankton	0.6	0.420–0.780	mmol N/m^3
P_{thre}	Threshold of phytoplankton overgrowth	0.12	0.084–0.156	mmol N/m^3
Γ_{PPT_G}	Temperature coefficient affecting phytoplankton growth Temperature coefficient affecting DOP remineralization	0.065	0.046–0.085	1/°C
Γ_{PPT_D}	Temperature coefficient affecting phytoplankton death	0.065	0.046–0.085	1/°C
Γ_{ZPT_N}	Temperature coefficient affecting zooplankton excretion	0.027	0.019–0.035	1/°C

Γ_{ZPT_D}	Temperature coefficient affecting zooplankton death	0.05	0.350–0.650	1/°C
Γ_{DOP_B}	Temperature coefficient affecting DOP remineralization	0.065	0.046–0.085	1/°C
Γ_{DPT_B}	Temperature coefficient affecting detritus remineralization	0.05	0.350–0.650	1/°C
δ_{PPT_z}	assimilation rate Assimilation rate of zooplankton predation on phytoplankton	0.8	0.560–1.040	–
δ_{DPT_z}	Assimilation rate of zooplankton predation on detritus	0.7	0.490–0.910	–
r_{ZPT_N}	The proportion of inorganic salts in zooplankton excrement	0.75	0.525–0.975	–
Q_0	Light coefficient	1368	–	W/m^2
R	Sea surface reflectance	0.378	–	–
lk_0	Optical attenuation coefficient	0.8	–	1/m
lk_1	Optical attenuation coefficient	0.0088	–	1/m (mg/Chla)
lk_2	Optical attenuation coefficient	0.054	–	1/m (mg/Chla)$^{2/3}$
$kext$	Optical attenuation coefficient	0.1	0.07–0.13	1/m
$r_{Chl/PN}$	The ratio of chlorophyll-a concentration to nitrogen	1.6	–	mg Chla/(mmol N)
r_{N_P}	Ratios of nitrogen to phosphorus in phytoplankton and zooplankton	16	–	mmol N/(mmol P)
k_{DON_NH4}	Remineralization rate of DON at 0°C	0.027	0.019–0.035	1/day
k_{DPT_DON}	Ratio of detritus decomposed to DON at 0°C	0.01	0.007–0.013	1/day
Γ_{DON_NH4}	Temperature coefficient affecting DON remineralization	0.056	0.039–0.073	°C
Γ_{DPT_DON}	Temperature coefficient affecting the decomposition of detritus into DON	0.063	0.044–0.082	°C
DO_{SDON}	Dissolved oxygen half-saturation coefficient of consumed dissolved oxygen during DON remineralization process	1.0	0.700–1.300	mg/l

(Continued)

Table 8.1. *(Continued)*

Symbol	Parameter description	Value	Confidence interval	Unit
DO_{SDPT}	Dissolved oxygen half-saturation coefficient of consumed dissolved oxygen during the detritus remineralization process	1.0	0.700–1.300	mg/l
μ_{SDO}	Oxygen exchange between dissolved oxygen and sediment	0.8	0.560–1.040	mg/m^2/day
k_{NH4_DON}	The oxidation rate of ammonium nitrogen at 0°C	0.053	0.037–0.069	1/day
Γ_{NH4_NO3}	Temperature coefficient of ammonium nitrogen oxidation process	0.062	0.043–0.081	°C
DO_{SNH4}	Dissolved oxygen half-saturation coefficient of consumed dissolved oxygen during ammonium nitrogen oxidation	0.5	0.350–0.650	mg/l
k_{ZPT_R}	Zooplankton respiration coefficient at 0°C	0.03	0.021–0.039	1/day
Γ_{ZPT_R}	Temperature coefficient of zooplankton respiration	0.061	0.043–0.079	°C

$$\begin{aligned}\frac{\partial NH_4}{\partial t} &= -u\frac{\partial NH_4}{\partial x} - v\frac{\partial NH_4}{\partial y} - w\frac{\partial NH_4}{\partial z} + \frac{\partial}{\partial x}\left(A_h\frac{\partial NH_4}{\partial x}\right) \\ &\quad + \frac{\partial}{\partial y}\left(A_h\frac{\partial NH_4}{\partial y}\right) + \frac{\partial}{\partial z}\left(A_v\frac{\partial NH_4}{\partial z}\right) \\ &\quad - \mu_{PPT\_G}\frac{L_{NH_4}}{L_{NH_4}+L_{NO_3}}P - \mu_{NH_4\_NO_3}NH_4 + \mu_{ZPT\_R}Z \\ &\quad + \mu_{DON\_NH_4}\cdot DON, \end{aligned} \tag{8.5}$$

$$\begin{aligned}\frac{\partial \mathrm{NO_3}}{\partial t} &= -u\frac{\partial \mathrm{NO_3}}{\partial x} - v\frac{\partial \mathrm{NO_3}}{\partial y} - w\frac{\partial \mathrm{NO_3}}{\partial z} + \frac{\partial}{\partial x}\left(A_h\frac{\partial \mathrm{NO_3}}{\partial x}\right) \\ &\quad + \frac{\partial}{\partial y}\left(A_h\frac{\partial \mathrm{NO_3}}{\partial y}\right) + \frac{\partial}{\partial z}\left(A_v\frac{\partial \mathrm{NO_3}}{\partial z}\right) \\ &\quad - \mu_{PPT\_G}\frac{L_{NO_3}}{L_{NH_4}+L_{NO_3}}P + \mu_{NH_4\_NO_3}NH_4, \end{aligned} \tag{8.6}$$

$$\begin{aligned}\frac{\partial DOP}{\partial t} &= -u\frac{\partial DOP}{\partial x} - v\frac{\partial DOP}{\partial y} - w\frac{\partial DOP}{\partial z} + \frac{\partial}{\partial x}\left(A_h\frac{\partial DOP}{\partial x}\right) \\ &\quad + \frac{\partial}{\partial y}\left(A_h\frac{\partial DOP}{\partial y}\right) + \frac{\partial}{\partial z}\left(A_v\frac{\partial DOP}{\partial z}\right) \\ &\quad + [\mu_{PPT\_E}P + (1-r_{ZPT\_N})\mu_{ZPT\_N}Z + \mu_{DPT\_B}D]/r_{N\_P} \\ &\quad - \mu_{DOP\_B}\cdot DOP, \end{aligned} \tag{8.7}$$

$$\begin{aligned}\frac{\partial PO_4}{\partial t} &= -u\frac{\partial PO_4}{\partial x} - v\frac{\partial PO_4}{\partial y} - w\frac{\partial PO_4}{\partial z} + \frac{\partial}{\partial x}\left(A_h\frac{\partial PO_4}{\partial x}\right) \\ &\quad + \frac{\partial}{\partial y}\left(A_h\frac{\partial PO_4}{\partial y}\right) + \frac{\partial}{\partial z}\left(A_v\frac{\partial PO_4}{\partial z}\right) \\ &\quad + (-\mu_{PPT\_G}P + r_{ZPT\_N}\mu_{ZPT\_N}Z)/r_{N\_P} + \mu_{DOP\_B}\cdot DOP, \end{aligned} \tag{8.8}$$

$$\begin{aligned}\frac{\partial DO}{\partial t} &= -u\frac{\partial DO}{\partial x} - v\frac{\partial DO}{\partial y} - w\frac{\partial DO}{\partial z} + \frac{\partial}{\partial x}\left(A_h\frac{\partial DO}{\partial x}\right) \\ &\quad + \frac{\partial}{\partial y}\left(A_h\frac{\partial DO}{\partial y}\right) + \frac{\partial}{\partial z}\left(A_v\frac{\partial DO}{\partial z}\right) + \mu_{PPT\_G}P \\ &\quad - \mu_{ZPT\_R}Z - \mu_{DPT\_DON}D - \mu_{DON\_NH_4}\cdot DON \\ &\quad - \mu_{NH_4\_NO_3}NH_4 - \frac{\mu_{SDO}\cdot DO}{h}. \end{aligned} \tag{8.9}$$

8.1.1 *Phytoplankton growth process*

Phytoplankton growth is affected by water temperature, light irradiance, and nutrient concentration. In Eq. (8.1), the growth process of phytoplankton is described by Eq. (8.10). Specifically, $\mu_{\mathrm{PPT\_G}}$ represents the limiting function of phytoplankton growth, and $r_{\mathrm{PPT\_E}}$ represents the metabolic action during phytoplankton growth.

$$\mu_{PPT\_G}(1 - r_{PPT\_E})P, \tag{8.10}$$

$$r_{PPT\_E} = \frac{0.12}{10 - I_{\mathrm{opt}}} I_H + \frac{0.24 I_{\mathrm{opt}} - 1.2}{I_{\mathrm{opt}} - 10}, \tag{8.11}$$

$$\mu_{PPT\_G} = k_{PPT\_G} \cdot f_{PPT\_G}(T) \cdot f_{PPT\_G}(I) \cdot f_{PPT\_G}(N), \tag{8.12}$$

wherein I_{opt} is the optimal light irradiance, and I_H is a function of the light irradiance as the variation of water depth. $f_{PPT\_G}(T)$, $f_{PPT\_G}(I)$, and I_H refer to the temperature limit function, light limit function, and nutrient salt limit function, respectively. The selection of each limit function is explained below.

(1) **Temperature restriction**. The growth process of phytoplankton is restricted by water temperature, and the temperature limit function is set to an exponential limit function (Eppley, 1972), as shown in Eq. (8.13):

$$f(T)_{PPT\_G} = e^{\Gamma_{PPT\_G} \cdot T}. \tag{8.13}$$

(2) **Light restriction**. Phytoplankton growth is affected by light irradiance. The effective light radiation intensity in photosynthesis is the main energy of the ecosystem. This part of the energy is mainly generated by light radiation reaching the sea surface with a wavelength between 400 nm and 700 nm (Li *et al.*, 2015). The intensity of light irradiance is affected by the rotation of the Earth, the absorption of atmospheric clouds, and the height angle of the Sun. Dobson and Smith (1988) proposed an empirical formula for calculating the effective light irradiance:

$$I = Q_0 S_h (A + B S_h)(1 - R), \tag{8.14}$$

wherein Q_0 represents the light irradiance constant. According to the study by Sellers (1965), the value of Q_0 is shown in Eq. (8.15);

R represents the sea surface reflectance (8.16).

$$Q_0 = 1368\ \mathrm{W/m^2}, \tag{8.15}$$

$$R = 0.378. \tag{8.16}$$

In Eq. (8.14), A and B are cloud parameters (Dobson and Smith, 1988), and S_h is a sine function of the Sun and local normal inclination angle. In Eq. (8.17), φ is the latitude; the northern hemisphere is positive and the southern hemisphere is negative. θ is the declination with a range of $-23.50°$ to $23.50°$. Finally, τ represents either day or night, which is expressed in Eq. (8.19).

$$S_h = \sin h = \sin\varphi \sin\theta + \cos\varphi \cos\theta \cos\tau, \tag{8.17}$$

$$\theta = \arcsin[\sin(23.5\pi/180)\sin(2\pi(t_d - 81)/365.25)], \tag{8.18}$$

$$\tau = \pi(t_h - 12)/12. \tag{8.19}$$

In the above equations, t_d is the number of days from January 1 in a year and t_h represents the hourly time of the corresponding day. The range of τ is $[-\pi, \pi]$, in which 0 means noon.

The effective light irradiance on the sea surface is calculated by Eq. (8.20):

$$\begin{cases} I_{PA} = I\rho = Q_0\rho S_h(A + BS_h)(1 - R) & \text{if} \quad S_h > 0 \\ I_{PA} = 0 & \text{if} \quad S_h \leq 0 \end{cases}. \tag{8.20}$$

The light irradiance intensity decreases with increasing water depth. According to the exponential function based on the light attenuation coefficient given by Ebenhöh *et al.* (1997), the function of light irradiance variation with water depth is shown in Eq. (8.21):

$$I_H = I_{PA}\frac{1}{H}\int_0^H e^{-\kappa z}dz, \tag{8.21}$$

wherein H and κ represent the water depth and light attenuation coefficient, respectively, which is mainly affected by the mutual occlusion of phytoplankton (Riley, 1956). The expression for κ is shown in Eqs. (8.22) to (8.25):

$$\kappa = \kappa_0 + \kappa_1(r_{chl/PN}P) + \kappa_2(r_{chl/PN}P)^{\frac{2}{3}}, \tag{8.22}$$

$$\kappa_0 = 0.8 \text{ /m}, \tag{8.23}$$

$$\kappa_1 = 0.0088 \text{ (mg Chla/m)}, \tag{8.24}$$

$$\kappa_2 = 0.054 \text{ (mg Chla)}^{\frac{2}{3}}\text{/m}. \tag{8.25}$$

According to the Steele equation (Steele, 1962) and the effective light irradiance calculated based on the above function, the light limit function is defined as Eq. (8.26):

$$f(I)_{PPT\_G} = \frac{I_H}{I_{\text{opt}}} e^{1-\frac{I_H}{I_{\text{opt}}}}. \tag{8.26}$$

(3) **Nutrient restriction**. Phytoplankton growth is restricted by the concentration of nutrients. Based on Liebig's law of minimum factors, we derived the nutrient limit function (8.27):

$$f(N)_{PPT\_G} = \min(L_N, L_P), \tag{8.27}$$

wherein L_N is the nitrogen limit function. According to the research results by Fennel *et al.* (2006), we derived Eq. (8.28):

$$L_N = L_{\text{NO}_3} + L_{\text{NH}_4}. \tag{8.28}$$

Based on the Michaelis–Menten formula (Schnoor, 1996; Fennel *et al.*, 2006), which is widely used in marine ecosystem dynamic models, we derived the corresponding limit functions of ammonium nitrogen, nitrate nitrogen, and phosphate.

$$L_{\text{NO}_3} = \frac{\text{NO}_3}{k_{\text{NO}_3} + \text{NO}_3} \cdot \frac{1}{1 + \text{NH}_4/k_{\text{NH}_4}}, \tag{8.29}$$

$$L_{\text{NH}_4} = \frac{\text{NH}_4}{k_{\text{NH}_4} + \text{NH}_4}, \tag{8.30}$$

$$L_P = \frac{\text{PO}_4}{k_{\text{PO}_4} + \text{PO}_4}. \tag{8.31}$$

8.1.2 The mortality of phytoplankton

The mortality of phytoplankton is mainly affected by water temperature. We use Eq. (8.32) to describe the mortality of phytoplankton. Mortality rates vary with water temperature (8.33) (Eppley, 1972):

$$\mu_{PPT\_D} P, \tag{8.32}$$

$$\mu_{PPT\_D} = k_{PPT\_D} \cdot e^{\Gamma_{PPT\_D} \cdot T}. \tag{8.33}$$

8.1.3 The predatory process of zooplankton

The predatory process of zooplankton mainly consists of the predation of phytoplankton and detritus, which is influenced by the biomass and detritus content of phytoplankton. The predatory process of zooplankton on phytoplankton and detritus is described by Eqs. (8.34) to (8.36). The relationships between the zooplankton predation rate and phytoplankton biomass, zooplankton predation rate, and detritus content, are described with Eqs. (8.35) to (8.37), which are similar to the Michaelis–Menten equation:

$$\mu_{PPT\_z} Z, \tag{8.34}$$

$$\mu_{PPT\_z} = \begin{cases} k_{PPT\_z} \dfrac{P - P_{\text{thre}}}{P - P_{\text{thre}} + KS_{PPT}} & \text{if} \quad P \geq P_{\text{thre}} \\ 0 & \text{else} \end{cases}, \tag{8.35}$$

$$\mu_{DPT\_z} Z, \tag{8.36}$$

$$\mu_{DPT\_z} = k_{DPT\_z} \frac{D}{D + KS_{DPT}}. \tag{8.37}$$

8.1.4 The mortality of zooplankton

Zooplankton is mainly affected by temperature. Similar to Eqs. (8.32) and (8.33), we derived zooplankton mortality in Eqs. (8.38) and (8.39) (Eppley, 1972):

$$\mu_{ZPT\_D} Z, \tag{8.38}$$

$$\mu_{ZPT\_D} = k_{ZPT\_D} e^{\Gamma_{ZPT\_D} \cdot T}. \tag{8.39}$$

8.1.5 The excretion and predation on zooplankton

The processes of excretion and predation on zooplankton are described in Eqs. (8.40) to (8.42). The excretion rate is mainly affected by temperature, as shown in Formula (8.41).

$$\mu_{ZPT\_N} Z, \tag{8.40}$$

$$\mu_{ZPT\_N} = k_{ZPT\_N} e^{\Gamma_{ZPT\_N} \cdot T}, \tag{8.41}$$

$$k_{ZPT\_F} Z. \tag{8.42}$$

8.1.6 The dynamic process of dissolved organic nitrogen

The dynamic process of dissolved organic nitrogen mainly includes the metabolic process of phytoplankton (P), the mortality of zooplankton (Z), the remineralization process of detritus (D), and the remineralization process of dissolved organic nitrogen (DON), as shown in Eq. (8.43) (Chapelle *et al.*, 1994; Li *et al.*, 2015). The remineralization rates of D and DON are shown in Eqs. (8.46) and (8.47), respectively.

$$\mu_{PPT\_E} P + (1 - r_{ZPT\_N}) \mu_{ZPT\_N} Z + \mu_{DPT\_DON} D - \mu_{DON\_NH_4} \cdot DON, \tag{8.43}$$

$$\mu_{PPT\_E} = r_{PPT\_E} \cdot \mu_{PPT\_G}, \tag{8.44}$$

$$\mu_{DPT\_DON} = k_{DPT\_DON} e^{\Gamma_{DPT\_DON} \cdot T}, \tag{8.45}$$

$$\mu_{DPT\_DON} = k_{DPT\_DON} e^{\Gamma_{DPT\_DON} \cdot T} \frac{DO}{DO_{S_{DPT}} + DO}, \tag{8.46}$$

$$\mu_{DON\_NH_4} = k_{DON\_NH_4} e^{\Gamma_{DON\_NH_4} \cdot T} \frac{DO}{DO_{S_{DON}} + DO}. \tag{8.47}$$

8.1.7 The dynamic process of ammonium nitrogen

The dynamic process of ammonium nitrogen (NH_4) mainly includes the uptake of phytoplankton (P), oxidation of NH_4, excretion of zooplankton (Z), and remineralization of dissolved organic nitrogen (DON), as shown in Eq. (8.48). Among them, the oxidation rate of

NH_4 is shown in Eq. (8.49) (Dueri *et al.*, 2009; Lu *et al.*, 2017).

$$-\mu_{PPT\_G}\frac{L_{\mathrm{NH_4}}}{L_{\mathrm{NH_4}}+L_{\mathrm{NO_3}}}P-\mu_{\mathrm{NH_4\_NO_3}}\mathrm{NH_4}$$
$$+\mu_{ZPT\_R}Z+\mu_{DON\_\mathrm{NH_4}}\cdot DON, \tag{8.48}$$

$$\mu_{\mathrm{NH_4\_NO_3}}=k_{\mathrm{NH_4\_NO_3}}e^{\Gamma_{\mathrm{NH_4\_NO_3}}\cdot T}\frac{DO}{DO_{S_{\mathrm{NH_4}}}+DO}. \tag{8.49}$$

8.1.8 *The dynamic process of nitrate nitrogen*

The dynamic process of nitrate nitrogen (NO_3) mainly includes the uptake of phytoplankton (P) and the oxidation process of ammonium nitrogen (NH_4), as shown in Formula (8.50):

$$-\mu_{PPT\_G}\frac{L_{\mathrm{NO_3}}}{L_{\mathrm{NH_4}}+L_{\mathrm{NO_3}}}P+\mu_{\mathrm{NH_4\_NO_3}}\mathrm{NH_4}. \tag{8.50}$$

8.1.9 *The dynamic process of dissolved oxygen*

The dynamic process of dissolved oxygen (DO) mainly includes oxygen produced by the photosynthesis of phytoplankton (P), oxygen absorption by zooplankton (Z) respiration, oxygen consumption by detritus (D) remineralization, oxygen consumption by dissolved organic nitrogen (DON) remineralization, oxygen consumption by ammonium nitrogen (NH_4) oxidation, and DO exchange with the bottom layer deposit, as shown in Eq. (8.51):

$$\mu_{PPT\_G}P-\mu_{ZPT\_R}Z-\mu_{DPT\_DON}D-\mu_{DON\_\mathrm{NH_4}}\cdot DON$$
$$-\mu_{\mathrm{NH_4\_NO_3}}\mathrm{NH_4}-\frac{\mu_{SDO}\cdot DO}{h}. \tag{8.51}$$

8.1.10 *Dynamics of phosphates and dissolved organic phosphorus*

Phosphate budgets include phytoplankton (P) uptake, zooplankton (Z) excretion, and remineralization of dissolved organic phosphorus (DOP), as shown in Eq. (8.52). The budgets of DOP include P metabolism, Z excretion, remineralization of detritus (D), and remineralization of DOP. The remineralization rate of DOP is mainly affected by temperature, which is expressed by the Eq. (8.54).

$$(-\mu_{PPT\_G}P + r_{ZPT\_N}\mu_{ZPT\_N}Z)/r_{N\_P} + \mu_{DOP\_B} \cdot DOP, \tag{8.52}$$

$$[\mu_{PPT\_E}P + (1 - r_{ZPT\_N})\mu_{ZPT\_N}Z + \mu_{DPT\_B}D]/r_{N\_P} - \mu_{DOP\_B} \cdot DOP, \tag{8.53}$$

$$\mu_{DOP\_B} = k_{DOP\_B}e^{\Gamma_{DOP\_B} \cdot T}. \tag{8.54}$$

8.2 The Finite Difference Scheme of the Model

Compared with the classic marine ecosystem dynamics model and water quality model, the structure of the water quality model is more complex and requires high-computational capability. Therefore, to improve the computational efficiency of this model, we have improved the finite difference scheme in numerical methods and introduced a second-order conservative characteristic finite difference scheme (Xu *et al.*, 2017).

The generalization of Eqs. (8.1) to (8.9) can be expressed as Eq. (8.55):

$$\frac{\partial C}{\partial t} = -u\frac{\partial C}{\partial x} - v\frac{\partial C}{\partial y} - w\frac{\partial C}{\partial z} + \frac{\partial}{\partial x}\left(A_h\frac{\partial C}{\partial x}\right) + \frac{\partial}{\partial y}\left(A_h\frac{\partial C}{\partial y}\right) + \frac{\partial}{\partial z}\left(A_v\frac{\partial C}{\partial z}\right) + S. \tag{8.55}$$

The land boundary condition of the model is set as a solid boundary condition, and the open boundary condition is set as a radiative boundary condition. In other words, the open boundary condition has the steady inflow and non-divergent outflow, as shown in Eq. (8.56), wherein V_n and n represent the flow velocity perpendicular to the direction of the open boundary normal and the normal direction, respectively.

$$\begin{cases} \dfrac{\partial C}{\partial t} = 0, & V_n \leq 0, \\ \dfrac{\partial C}{\partial n} = 0, & V_n > 0. \end{cases} \tag{8.56}$$

$C^n_{i,j,k}$ represents the value of the state variable on the feature vector $\vec{x} = (i\Delta x, j\Delta y, k\Delta z)$ for the corresponding moment $t^n = n\Delta t$. Δx and Δy indicate horizontal grid resolution in the east and north directions, respectively, while Δz and Δt represent vertical grid resolution and the time step, respectively. In the finite difference scheme, the $C^n_{i,j,k}$ value of the state variable t^n is known. To solve the state variable value $C^{n+1}_{i,j,k}$ at time t^{n+1}, the characteristic direction is $\vec{\tau}$, and $X(\vec{\tau}; \vec{x}, t^{n+1})$ is noted as the feature line, as shown in Eqs. (8.57) and (8.58):

$$\frac{dX(\vec{\tau}\ \vec{x}, t^{n+1})}{d\vec{\tau}} = U(X(\vec{\tau}; \vec{x}, t^{n+1}),\ \vec{\tau}), \tag{8.57}$$

$$X(\vec{\tau}; \vec{x}, t^{n+1}) = \vec{x}. \tag{8.58}$$

According to Eqs. (8.57) and (8.58), we solve the feature vector.

$$\vec{x}^* = X(t^n; \vec{x}, t^{n+1}) = \vec{x} - U(\vec{x}, t^{n+1})\Delta t. \tag{8.59}$$

The value of the state variables $\bar{C}^n_{i,j,k}$ on the feature vector $\vec{x}^*$ can be obtained by the interpolation of the value of the surrounding eight grid points: $C^n_{p,q,l}$, $C^n_{p+1,q,\,l}$, $C^n_{p,q+1,\,l}$, $C^n_{p+1,q+1,k}$, $C^n_{p,q,\,l+1}$, $C^n_{p+1,q,\,l+1}$, $C^n_{p,q+1,\,l+1}$, and $C^n_{p+1,q+1,\,l+1}$, which is described in Eq. (8.60) (Xu *et al.*, 2017):

$$\begin{aligned} \bar{C}^n_{i,j,k} = {} & a_{000}C^n_{p,q,l} + a_{100}C^n_{p+1,q+1,l} + a_{010}C^n_{p,q+1,l} + a_{011}C^n_{p+1,q+1,l} \\ & + a_{001}C^n_{p,q,\,l+1} + a_{101}C^n_{p+1,q,\,l+1} + a_{011}C^n_{p,q+1,\,l+1} \\ & + a_{111}C^n_{p+1,q+1,l+1}, \\ & \text{where} \quad x_p < \bar{x}_i < x_{p+1}, \quad y_q < \bar{y}_j < y_{q+1}, \quad z_l < \bar{z}_k < z_{l+1} \end{aligned} \tag{8.60}$$

and
$$\begin{cases} a_{000} = \dfrac{x_{p+1} - \bar{x}_i}{\Delta x}\dfrac{y_{q+1} - \bar{y}_j}{\Delta y}\dfrac{z_{l+1} - \bar{z}_k}{\Delta z} \\ a_{100} = \dfrac{\bar{x}_i - x_p}{\Delta x}\dfrac{y_{q+1} - \bar{y}_j}{\Delta y}\dfrac{z_{l+1} - \bar{z}_k}{\Delta z} \\ a_{010} = \dfrac{x_{p+1} - \bar{x}_i}{\Delta x}\dfrac{\bar{y}_j - y_q}{\Delta y}\dfrac{z_{l+1} - \bar{z}_k}{\Delta z} \\ a_{110} = \dfrac{\bar{x}_i - x_p}{\Delta x}\dfrac{\bar{y}_j - y_q}{\Delta y}\dfrac{z_{l+1} - \bar{z}_k}{\Delta z} \\ a_{001} = \dfrac{x_{p+1} - \bar{x}_i}{\Delta x}\dfrac{y_{q+1} - \bar{y}_j}{\Delta y}\dfrac{\bar{z}_k - z_l}{\Delta z} \\ a_{101} = \dfrac{\bar{x}_i - x_p}{\Delta x}\dfrac{y_{q+1} - \bar{y}_j}{\Delta y}\dfrac{\bar{z}_k - z_l}{\Delta z} \\ a_{011} = \dfrac{x_{p+1} - \bar{x}_i}{\Delta x}\dfrac{\bar{y}_j - y_q}{\Delta y}\dfrac{\bar{z}_k - z_l}{\Delta z} \\ a_{111} = \dfrac{\bar{x}_i - x_p}{\Delta x}\dfrac{\bar{y}_j - y_q}{\Delta y}\dfrac{\bar{z}_k - z_l}{\Delta z} \end{cases}. \tag{8.61}$$

Then, we use the tridiagonal matrix method to solve the implicit scheme according to Eq. (8.62) to solve for $C_{i,j,k}^{n+1}$.

$$\begin{aligned} \frac{C_{i,j,k}^{n+1} - \bar{C}_{i,j,k}^{n}}{\Delta t} &= \frac{A_h}{\Delta x}\left(\frac{C_{i+1,j,k}^{n+1} - C_{i,j,k}^{n+1}}{\Delta x} - \frac{C_{i,j,k}^{n+1} - C_{i-1,j,k}^{n+1}}{\Delta x}\right) \\ &\quad + \frac{A_h}{\Delta y}\left(\frac{C_{i,j+1,k}^{n+1} - C_{i,j,k}^{n+1}}{\Delta y} - \frac{C_{i,j,k}^{n+1} - C_{i,j-1,k}^{n+1}}{\Delta y}\right) \\ &\quad + \frac{A_v}{\Delta z}\left(\frac{C_{i,j,k+1}^{n+1} - C_{i,j,k}^{n+1}}{\Delta z} - \frac{C_{i,j,k}^{n+1} - C_{i,j,k-1}^{n+1}}{\Delta z}\right) + S. \end{aligned} \tag{8.62}$$

Based on the second-order conservation feature finite difference scheme to solve the water quality model, we can enlarge the time step to ten times the original, which considerably improves the calculation efficiency of the model.

The adjoint assimilation method based on the Lagrange multiplier method has been widely used in parameter optimization of marine ecosystem dynamics models (Zhao and Lu, 2008; Fan and Lv, 2009; Gregg *et al.*, 2009; Li, 2013; Shen, 2015; Liu *et al.*, 2017). In the adjoint assimilation method, we use the difference between the simulation results of the model and the observations as the driving force of the adjoint assimilation method and use the Lagrange multiplier method to solve the conditional extreme value of the error. As a result, the inversion is achieved to optimize the model parameters and improve the accuracy of the model simulation.

The process of the adjoint assimilation method can be summarized in Fig. 8.3. In the adjoint assimilation method, we first give the initial guess value of the parameters. According to these parameters, we calculate the water quality pattern and obtain the model simulation value.

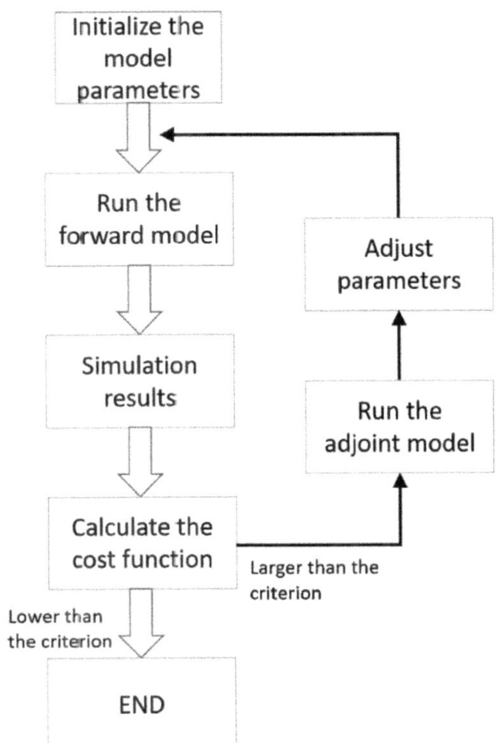

Fig. 8.3. Concept diagram of the adjoint assimilation method.

The model is evaluated according to the absolute average error between the observed data and simulation value. If the error is sufficiently less than the specified error, the model is stopped, and the simulation results are output. Otherwise, the reverse model is calculated, and the cost function is calculated to observe the gradient of each parameter. Subsequently, the value of the model parameters is adjusted according to the gradient algorithm. The adjusted parameters are then recalculated into the water quality model, which is repeated in this pattern. The mean absolute error of model simulation and observation data drives the calculation of the adjoint assimilation pattern.

In the adjoint assimilation pattern, we first define the cost function of the model as shown by Eq. (8.63):

$$\begin{aligned} J = \frac{1}{2}\sum[&K_1(P-\overline{P})^2 + K_2(Z-\overline{Z})^2 + K_3(D-\overline{D})^2 \\ &+K_4(DON-\overline{DON})^2 + K_5(\mathrm{NH}_4-\overline{\mathrm{NH}_4})^2 \\ &+K_6(\mathrm{NO}_3-\overline{\mathrm{NO}_3})^2 + K_7(DOP-\overline{DOP})^2 \\ &+K_8(\mathrm{PO}_4-\overline{\mathrm{PO}_4})^2 + K_9(DO-\overline{DO})^2], \end{aligned} \tag{8.63}$$

wherein $\overline{P}$, $\overline{Z}$, $\overline{D}$, $\overline{DON}$, $\overline{\mathrm{NH}_4}$, $\overline{\mathrm{NO}_3}$, $\overline{DO}$, $\overline{\mathrm{PO}_4}$, and $\overline{DOP}$ represent the observed values of phytoplankton (P), zooplankton (Z), detritus (D), dissolved organic nitrogen (DON), ammonium nitrogen (NH_4), nitrate nitrogen (NO_3), dissolved organic phosphorus (DOP), phosphate (PO_4), and dissolved oxygen (DO) at the corresponding positions and time, respectively. If the errors of the simulation results of the model are uncorrelated and their weights are the same, the weight coefficients K_1-K_9 can be simplified as follows. If there are observed values at the corresponding time and space, the weight is assumed as 1. Otherwise, the weight is assumed as 0.

Based on the cost function, we define the Lagrange function:

$$\begin{aligned} L(C) = J + \int_{\Sigma} [&P^*G(P) + Z^*G(Z) + D^*G(D) + C^*_{DON}G(DON) \\ &+ C^*_{\mathrm{NH}_4}G(\mathrm{NH}_4) + C^*_{\mathrm{NO}_3}G(NO_3) + C^*_{DOP}G(DOP) \\ &+ C^*_{\mathrm{PO}_4}G(\mathrm{PO}_4) + C^*_{DO}G(DO)], \end{aligned} \tag{8.64}$$

wherein $G(P)$, $G(Z)$, $G(D)$, $G(DON)$, $G(\mathrm{NH_4})$, $G(NO_3)$, $G(DOP)$, $G(\mathrm{PO_4})$, and $G(DO)$ represent the shift of all the items at the right side of the equation to the left side of the equation in Formulas (8.1) to (8.9). P^*, Z^*, D^*, C^*_{DON}, $C^*_{\mathrm{NH_4}}$, $C^*_{\mathrm{NO_3}}$, C^*_{DOP}, $C^*_{\mathrm{PO_4}}$, and C^*_{DO} represent the adjoint variables of phytoplankton (P), zooplankton (Z), detritus (D), dissolved organic nitrogen (DON), ammonium nitrogen (NH_4), nitrate nitrogen (NO_3), dissolved organic phosphorus (DOP), phosphate (PO_4), and dissolved oxygen (DO), respectively (corresponding to Lagrange operator).

The conditional extreme value is solved according to the Lagrange multiplier method. Assuming:

$$\frac{\partial L}{\partial P}=0,\quad \frac{\partial L}{\partial Z}=0,\quad \frac{\partial L}{\partial D}=0,\quad \frac{\partial L}{\partial DON}=0,\quad \frac{\partial L}{\partial \mathrm{NH_4}}=0$$
$$\frac{\partial L}{\partial \mathrm{NO_3}}=0,\quad \frac{\partial L}{\partial DOP}=0,\quad \frac{\partial L}{\partial \mathrm{PO_4}}=0,\quad \frac{\partial L}{\partial DO}=0. \tag{8.65}$$

All partial differentials in (8.65) are solved to get the adjoint equations for each adjoint variable:

(1) Adjoint equations corresponding to the adjoint variables of phytoplankton

$$\begin{aligned}
-\frac{\partial P^*}{\partial t}-\frac{\partial}{\partial z}\left(A_V\frac{\partial P^*}{\partial z}\right) &= \frac{\partial(uP^*)}{\partial x}+\frac{\partial(vP^*)}{\partial y}+\frac{\partial(wP^*)}{\partial z}\\
&\quad+\frac{\partial}{\partial x}\left(A_h\frac{\partial P^*}{\partial x}\right)+\frac{\partial}{\partial y}\left(A_h\frac{\partial P^*}{\partial y}\right)\\
&\quad+\left[\mu_{PPT_G}(1-r_{PPT_E})-\mu_{PPT_D}\right]P^*+\mu_{PPT_D}D^*\\
&\quad+\mu_{PPT_E}C^*_{DON}-\mu_{PPT_G}\frac{L_{\mathrm{NH_4}}}{L_{\mathrm{NH_4}}+L_{\mathrm{NO_3}}}C^*_{\mathrm{NH_4}}\\
&\quad-\mu_{PPT_G}\frac{L_{\mathrm{NO_3}}}{L_{\mathrm{NH_4}}+L_{\mathrm{NO_3}}}C^*_{\mathrm{NO_3}}+\mu_{PPT_G}C^*_{DO}-\mu_{PPT_G}C^*_{\mathrm{PO_4}}\\
&\quad+\mu_{PPT_E}C^*_{DOP}-(P-\bar{P}).
\end{aligned} \tag{8.66}$$

(2) Adjoint equations corresponding to the adjoint variables of zooplankton

$$
\begin{aligned}
&-\frac{\partial Z^*}{\partial t}-\frac{\partial}{\partial z}\left(A_v\frac{\partial Z^*}{\partial z}\right)=\frac{\partial\left(uZ^*\right)}{\partial x}+\frac{\partial\left(vZ^*\right)}{\partial y}+\frac{\partial\left(wZ^*\right)}{\partial z}\\
&+\frac{\partial}{\partial x}\left(A_h\frac{\partial Z^*}{\partial x}\right)+\frac{\partial}{\partial y}\left(A_h\frac{\partial Z^*}{\partial y}\right)-\mu_{PPT\_z}P^*+(\delta_{PPT\_z}\mu_{PPT\_z}\\
&+\delta_{DPT\_Z}\mu_{DPT\_z}-\mu_{ZPT\_D}-\mu_{ZPT\_N}-k_{ZPT\_F})\,Z^*\\
&+\left(\mu_{ZPT\_D}+(1-\delta_{PPT\_z})\,\mu_{PPT\_z}-\delta_{DPT\_z}\mu_{DPT\_z}\right)D^*\\
&+\mu_{ZPT\_R}C^*_{\mathrm{NH_4}}+(1-r_{ZPT\_N})\,\mu_{ZPT\_N}C^*_{DON}-\mu_{ZPT\_R}C^*_{DO}\\
&+r_{ZPT\_N}\mu_{ZPT\_N}C^*_{\mathrm{PO_4}}/r_{N\_P}+(1-r_{ZPT\_N})\\
&\times\mu_{ZPT\_N}C^*_{DOP}/r_{N\_P}-(Z-\bar{Z}).
\end{aligned}
\tag{8.67}
$$

(3) Adjoint equations corresponding to the adjoint variables of detritus

$$
\begin{aligned}
&-\frac{\partial D^*}{\partial t}-\frac{\partial}{\partial z}\left(A_v\frac{\partial D^*}{\partial z}\right)=\frac{\partial\left(uD^*\right)}{\partial x}+\frac{\partial\left(vD^*\right)}{\partial y}+\frac{\partial\left(wD^*\right)}{\partial z}\\
&+\frac{\partial}{\partial x}\left(A_h\frac{\partial D^*}{\partial x}\right)+\frac{\partial}{\partial y}\left(A_h\frac{\partial D^*}{\partial y}\right)+\delta_{DPT\_z}k_{DPT\_z}\\
&\times\frac{KS_{DPT}ZZ^*}{(D+KS_{DPT})^2}-\mu_{DPT\_B}D^*+\mu_{DPT\_DON}C^*_{DON}\\
&-\mu_{DPT\_DON}C^*_{DO}+\mu_{DPT\_B}C^*_{DOP}/r_{N\_P}-(D-\bar{D}).
\end{aligned}
\tag{8.68}
$$

(4) Adjoint equations corresponding to the adjoint variables of dissolved organic nitrogen

$$
\begin{aligned}
&-\frac{\partial C^*_{DON}}{\partial t}-\frac{\partial}{\partial z}\left(A_v\frac{\partial C^*_{DON}}{\partial z}\right)=\frac{\partial(uC^*_{DON})}{\partial x}+\frac{\partial(vC^*_{DON})}{\partial y}\\
&+\frac{\partial(wC^*_{DON})}{\partial z}+\frac{\partial}{\partial x}\left(A_h\frac{\partial C^*_{DON}}{\partial x}\right)+\frac{\partial}{\partial y}\left(A_h\frac{\partial C^*_{DON}}{\partial y}\right)\\
&+\mu_{DON\_\mathrm{NH_4}}(-C^*_{DON}+C^*_{\mathrm{NH_4}}-C^*_{DO})-(DON-\overline{DON}).
\end{aligned}
\tag{8.69}
$$

(5) Adjoint equations corresponding to the adjoint variables of ammonium nitrogen

$$
\begin{aligned}
&-\frac{\partial C^*_{\mathrm{NH_4}}}{\partial t}-\frac{\partial}{\partial z}\left(A_v\frac{\partial C^*_{\mathrm{NH_4}}}{\partial z}\right)=\frac{\partial(uC^*_{\mathrm{NH_4}})}{\partial x}+\frac{\partial(\nu C^*_{\mathrm{NH_4}})}{\partial y}\\
&\quad+\frac{\partial(wC^*_{\mathrm{NH_4}})}{\partial z}+\frac{\partial}{\partial x}\left(A_h\frac{\partial C^*_{\mathrm{NH_4}}}{\partial x}\right)+\frac{\partial}{\partial y}\left(A_h\frac{\partial C^*_{\mathrm{NH_4}}}{\partial y}\right)\\
&\quad+(1-r_{PPT\_E})\cdot k_{PPT\_G}\cdot f_{PPT\_G}(T)\cdot f_{PPT\_G}(I)\\
&\quad\cdot\frac{f_{PPT\_G}(N)}{\partial \mathrm{NH_4}}\cdot P\cdot P^*-k_{PPT\_G}\cdot f_{PPT\_G}(T)\cdot f_{PPT\_G}(I)\\
&\quad\cdot\frac{\partial f_{PPT\_G}(N)}{\partial \mathrm{NH_4}}\frac{L_{\mathrm{NH_4}}}{L_{\mathrm{NH_4}}+L_{\mathrm{NO_3}}}\cdot P\cdot C^*_{\mathrm{NH_4}}-k_{PPT\_G}\cdot f_{PPT\_G}(T)'\\
&\quad\cdot f_{PPT\_G}(I)\cdot\frac{\partial f_{PPT\_G}(N)}{\partial \mathrm{NH_4}}\frac{L_{\mathrm{NO_3}}}{L_{\mathrm{NH_4}}+L_{\mathrm{NO_3}}}\cdot P\cdot C^*_{\mathrm{NO_3}}\\
&\quad-\mu_{PPT\_G}\frac{\partial\left(L_{\mathrm{NH_4}}/L_N\right)}{\partial \mathrm{NH_4}}\cdot P\cdot C^*_{\mathrm{NH_4}}-\mu_{PPT\_G}\frac{\partial\left(L_{\mathrm{NO_3}}/L_N\right)}{\partial \mathrm{NH_4}}\\
&\quad\cdot P\cdot C^*_{\mathrm{NO_3}}-\mu_{\mathrm{NH_4\_NO_3}}C^*_{\mathrm{NH_4}}+\mu_{\mathrm{NH_4\_NO_3}}C^*_{\mathrm{NO_3}}\\
&\quad-\mu_{\mathrm{NH_4\_NO_3}}C^*_{DO}-(\mathrm{NH_4}-\overline{\mathrm{NH_4}}).
\end{aligned}
\tag{8.70}
$$

$$
\frac{\partial f_{PPT\_G}(N)}{\partial \mathrm{NH_4}}=\begin{cases}\dfrac{k_{\mathrm{NO_3}}}{k_{\mathrm{NO_3}}+\mathrm{NO_3}}\cdot\dfrac{k_{\mathrm{NH_4}}}{(k_{\mathrm{NH_4}}+\mathrm{NH_4})^2}, & \text{if}\quad L_N<L_P\\ 0, & \text{if}\quad L_N\geq L_P,\end{cases}
\tag{8.71}
$$

$$
\frac{\partial f_{PPT\_G}(N)}{\partial \mathrm{NO_3}}=\begin{cases}\dfrac{k_{\mathrm{NO_3}}}{(k_{\mathrm{NO_3}}+\mathrm{NO_3})^2}\cdot\dfrac{1}{1+\mathrm{NH_4}/k_{\mathrm{NH_4}}}, & \text{if}\quad L_N<L_P\\ 0, & \text{if}\quad L_N\geq L_P,\end{cases}
\tag{8.72}
$$

$$\frac{\partial f_{PPT\_G}(N)}{\partial \mathrm{PO_4}} = \begin{cases} 0, & \text{if} \quad L_N < L_P \\ \dfrac{k_{\mathrm{P}}}{(k_{\mathrm{P}} + \mathrm{PO_4})}, & \text{if} \quad L_N \geq L_P, \end{cases} \tag{8.73}$$

$$\frac{\partial \left(L_{\mathrm{NH_4}}/L_{\mathrm{N}}\right)}{\partial \mathrm{NH_4}} = \frac{k_{\mathrm{NH_4}} \cdot \mathrm{NO_3} \cdot (k_{\mathrm{NO_3}} + \mathrm{NO_3})}{(k_{\mathrm{NH_4}} \cdot \mathrm{NO_3} + k_{\mathrm{NO_3}} \cdot \mathrm{NH_4} + \mathrm{NH_4} \cdot \mathrm{NO_3})^2}, \tag{8.74}$$

$$\frac{\partial \left(L_{\mathrm{NH_4}}/L_N\right)}{\partial \mathrm{NO_3}} = \frac{-k_{\mathrm{NH_4}} \cdot k_{\mathrm{NO_3}} \cdot \mathrm{NH_4}}{(k_{\mathrm{NH_4}} \cdot \mathrm{NO_3} + k_{\mathrm{NO_3}} \cdot \mathrm{NH_4} + \mathrm{NH_4} \cdot \mathrm{NO_3})^2}, \tag{8.75}$$

$$\frac{\partial \left(L_{\mathrm{NO_3}}/L_N\right)}{\partial \mathrm{NH_4}} = \frac{-k_{\mathrm{NH_4}} \cdot \mathrm{NO_3} \cdot (k_{\mathrm{NO_3}} + \mathrm{NO_3})}{(k_{\mathrm{NH_4}} \cdot \mathrm{NO_3} + k_{\mathrm{NO_3}} \cdot \mathrm{NH_4} + \mathrm{NH_4} \cdot \mathrm{NO_3})^2}, \tag{8.76}$$

$$\frac{\partial \left(L_{\mathrm{NO_3}}/L_{\mathrm{N}}\right)}{\partial \mathrm{NO_3}} = \frac{k_{\mathrm{NH_4}} \cdot k_{\mathrm{NO_3}} \cdot \mathrm{NH_4}}{(k_{\mathrm{NH_4}} \cdot \mathrm{NO_3} + k_{\mathrm{NO_3}} \cdot \mathrm{NH_4} + \mathrm{NH_4} \cdot \mathrm{NO_3})^2}. \tag{8.77}$$

(6) Adjoint equations corresponding to the adjoint variables of nitrate nitrogen

$$\begin{aligned} &-\frac{\partial C^*_{\mathrm{NO_3}}}{\partial t} - \frac{\partial}{\partial z}\left(A_v \frac{\partial C^*_{\mathrm{NO_3}}}{\partial z}\right) = \frac{\partial \left(u C^*_{\mathrm{NO_3}}\right)}{\partial x} + \frac{\partial \left(\nu C^*_{\mathrm{NO_3}}\right)}{\partial y} \\ &+ \frac{\partial \left(w C^*_{\mathrm{NO_3}}\right)}{\partial z} + \frac{\partial}{\partial x}\left(A_h \frac{\partial C^*_{\mathrm{NO_3}}}{\partial x}\right) + \frac{\partial}{\partial y}\left(A_h \frac{\partial C^*_{\mathrm{NO_3}}}{\partial y}\right) \\ &+ (1 - r_{PPT\_E}) \cdot k_{PPT\_G} \cdot f_{PPT\_G}(T) \cdot f_{PPT\_G}(I) \\ &\cdot \frac{\partial f_{PPT\_G}(N)}{\partial \mathrm{NO_3}} \cdot P \cdot P^* - k_{PPT\_G} \cdot f_{PPT\_G}(T) \cdot f_{PPT\_G}(I) \\ &\cdot \frac{\partial f_{PPT\_G}(N)}{\partial \mathrm{NO_3}} \frac{L_{\mathrm{NH_4}}}{L_{\mathrm{NH_4}} + L_{NO_4}} \cdot P \cdot C^*_{\mathrm{NH_4}} - k_{PPT\_G} \cdot f_{PPT\_G}(T) \\ &\cdot f_{PPT\_G}(I) \cdot \frac{\partial f_{PPT\_G}(N)}{\partial \mathrm{NO_3}} \frac{L_{\mathrm{NO_3}}}{L_{\mathrm{NH_4}} + L_{\mathrm{NO_3}}} \cdot P \cdot C^*_{\mathrm{NO_3}} \end{aligned}$$

$$-\mu_{PPT\_G}\frac{\partial\left(L_{\mathrm{NH_4}}/L_N\right)}{\partial \mathrm{NO_3}}\cdot P\cdot C^*_{\mathrm{NH_4}}-\mu_{PPT\_G}\frac{\partial\left(L_{\mathrm{NO_3}}/L_N\right)}{\partial \mathrm{NO_3}}$$
$$\cdot P\cdot C^*_{\mathrm{NO_3}}-\left(\mathrm{NO_3}-\overline{\mathrm{NO_3}}\right). \tag{8.78}$$

(7) Adjoint equations corresponding to the adjoint variables of dissolved organic phosphorus

$$-\frac{\partial C^*_{DOP}}{\partial t}-\frac{\partial}{\partial z}\left(A_v\frac{\partial C^*_{DOP}}{\partial z}\right)=\frac{\partial\left(uC^*_{DOP}\right)}{\partial x}+\frac{\partial\left(\nu C^*_{DOP}\right)}{\partial y}$$
$$+\frac{\partial\left(wC^*_{DOP}\right)}{\partial z}+\frac{\partial}{\partial x}\left(A_h\frac{\partial C^*_{DOP}}{\partial x}\right)+\frac{\partial}{\partial y}\left(A_h\frac{\partial C^*_{DOP}}{\partial y}\right)$$
$$+\mu_{DOP\_B}\left(C^*_{\mathrm{PO_4}}-C^*_{DOP}\right)-\left(DOP-\overline{DOP}\right). \tag{8.79}$$

(8) Adjoint equations corresponding to the adjoint variables of phosphates

$$-\frac{\partial C^*_{\mathrm{PO_4}}}{\partial t}-\frac{\partial}{\partial z}\left(A_v\frac{\partial C^*_{\mathrm{PO_4}}}{\partial z}\right)=\frac{\partial\left(uC^*_{\mathrm{PO_4}}\right)}{\partial x}+\frac{\partial\left(\nu C^*_{\mathrm{PO_4}}\right)}{\partial y}$$
$$+\frac{\partial\left(wC^*_{\mathrm{PO_4}}\right)}{\partial z}+\frac{\partial}{\partial x}\left(A_h\frac{\partial C^*_{\mathrm{PO_4}}}{\partial x}\right)+\frac{\partial}{\partial y}\left(A_h\frac{\partial C^*_{\mathrm{PO_4}}}{\partial y}\right)$$
$$+\left(1-r_{PPT\_E}\right)\cdot k_{PPT\_G}\cdot f_{PPT\_G}(T)\cdot f_{PPT\_G}(I)$$
$$\cdot\frac{\partial f_{PPT\_G}(N)}{\partial \mathrm{PO_4}}\cdot P\cdot P^*-\left(\mathrm{PO_4}-\overline{\mathrm{PO_4}}\right). \tag{8.80}$$

(9) Adjoint equations corresponding to the adjoint variables of dissolved oxygen

$$-\frac{\partial C^*_{DO}}{\partial t}-\frac{\partial}{\partial z}\left(A_v\frac{\partial C^*_{DO}}{\partial z}\right)=\frac{\partial\left(uC^*_{DO}\right)}{\partial x}+\frac{\partial\left(vC^*_{DO}\right)}{\partial y}$$
$$+\frac{\partial\left(wC^*_{DO}\right)}{\partial z}+\frac{\partial}{\partial x}\left(A_h\frac{\partial C^*_{DO}}{\partial x}\right)+\frac{\partial}{\partial y}\left(A_h\frac{\partial C^*_{DO}}{\partial y}\right)$$
$$+k_{DPT\_DON}e^{r_{DPT\_DON}\cdot T}\frac{DO_{S_{DPT}}}{\left(DO_{S_{DPT}}+DO\right)^2}D\cdot C^*_{DON}$$

$$- k_{\mathrm{NH_4\_NO_3}} e^{\Gamma_{\mathrm{NH_4\_NO_3}} \cdot T} \frac{DO_{S_{\mathrm{NH_4}}}}{(DO_{S_{\mathrm{NH_4}}} + DO)^2} \mathrm{NH_4}$$

$$\cdot (C^*_{\mathrm{NH_4}} - C^*_{\mathrm{NO_3}}) + k_{DON\_\mathrm{NH_4}} e^{\Gamma_{DON\_\mathrm{NH_4}} \cdot T}$$

$$\times \frac{DO_{S_{DON}}}{(DO_{S_{DON}} + DO)^2} DON (C^*_{\mathrm{NH_4}} - C^*_{DON})$$

$$- \frac{\mu_{SDO} \cdot C^*_{DO}}{h} - (DO - \overline{DO}). \tag{8.81}$$

8.3 The Setting of Water Quality Model

In the water quality model, the open-source datasets used include dissolved oxygen data, nitrate nitrogen data, and phosphate data from CARS2009, and chlorophyll-a concentration from SeaWIFS. We interpolated the above data based on the modified Cressman interpolation method (Liu *et al.*, 2016) to obtain the initial field of the water quality model (Figs. 8.4 to 8.8).

Fig. 8.4. Interpolation result of chlorophyll-a data from the SeaWIFS dataset. Unit: mmol/l.

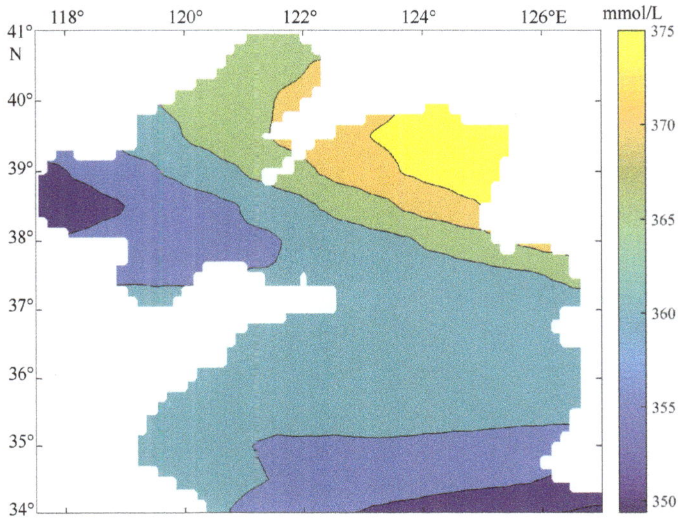

Fig. 8.5. Interpolation result of dissolved oxygen data from the CARS2009 dataset.

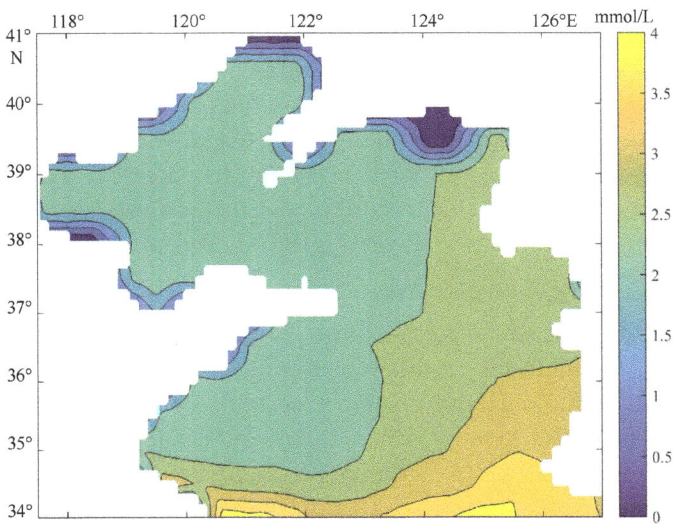

Fig. 8.6. Interpolation result of nitrate nitrogen data from the CARS2009 dataset.

(1) CARS2009 (CSIRO Atlas of Regional Seas)

CARS is a digital data set with a climatic state or seasonal average water mass characteristics. It consists of gridded observation data

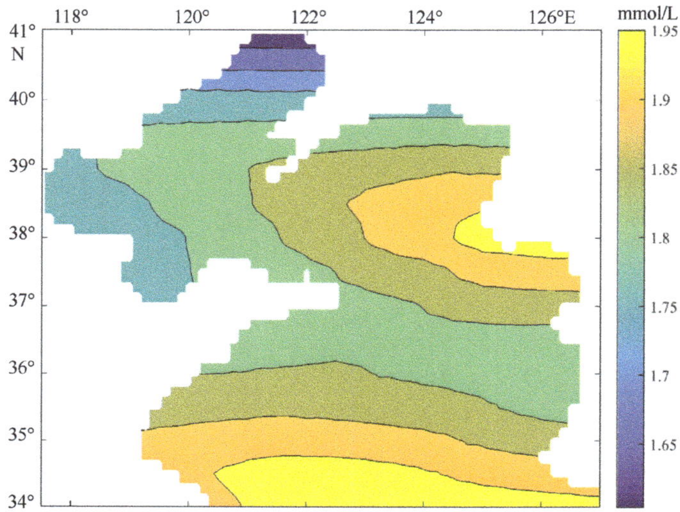

Fig. 8.7. Interpolation result of phosphate data from the CARS2009 dataset.

over a specific period of time and these grid data are time-averaged. The lattice method is a weighted least square smoothing method (LOESS smoother; Ridgway *et al.*, 2002).

The CARS2009 data covers marine areas around the world, with a horizontal grid resolution of $0.5° \times 0.5°$ and a vertical z-coordinate. It is divided into 79 layers with a range of 0–5500 m. The data set contains temperature (°C), salinity (psu), dissolved oxygen (ml/l), nitrate nitrogen (μmol/l), silicate (μmol/l), and phosphate (μmol/l).

(2) WOA09 (World Ocean Atlas, 2009)

The WOA09 data set is a grid data set based on field observations. The time scale includes monthly, seasonal, and annual averages. The spatial scale is the national marine area with the z coordinate, which is divided into 102 layers with a range of 0–5500 m. The data set includes temperature (°C), salinity (psu), density (kg/m^3), dissolved oxygen (ml/l), nitrate nitrogen (μmol/l), silicate (μmol/l), and phosphate (μmol/l). The horizontal grid resolution of temperature and salinity is 0.25°, and the horizontal grid resolution of the remaining data is 1°.

(3) SeaWIFS (Sea-Viewing Wide Field-of-View Sensor)

SeaWIFS is a satellite detector used to collect global marine organism data. It is mainly used to observe the chlorophyll-a concentration of

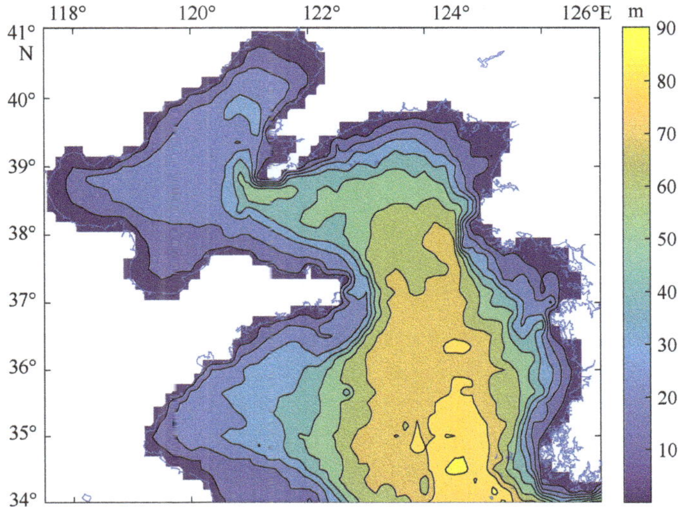

Fig. 8.8. Depth maps of the Bohai and Yellow Sea areas in the main study area of the model.

phytoplankton with a horizontal grid resolution of 5′ and a time scale of seasonal average, which only collect surface data.

To improve the calculation efficiency of the model, the internal and external mode decomposition calculation method was adopted in the POM (Princeton Ocean Model). Considering the characteristics of climatic state circulation in the calculation process, we used NCEP/NCAR reanalyzed wind field data and WOA09 (World Ocean Atlas, 2009) climatic state temperature and salinity data. The open boundary of the model is set at 34°N. The open boundary condition is driven by the tidal level, which includes four constituents: M_2, S_2, K_1, and O_1. The bottom friction is set to second-order bottom friction with a bottom friction coefficient of 1.5×10^{-3} (Liu *et al.*, 2017). The tidal constituent data are mainly from TOPEX/PoSeidon (TPXO; Egbert *et al.*, 2002). The reliability of the tidal current field simulated by the POM model has been demonstrated in studies by Wang (2013).

Based on the background flow field calculated by the POM and the above model settings, we performed a three-dimensional marine hydrodynamic-water quality coupled model calculation. The starting date of the water quality model is February 1, 2016, and the ending date is September 1, 2016. The selection of the date is mainly related to the time of existing ocean observation data. In this calculation, the units of all state variables except dissolved oxygen are converted into mmol N/L, and the unit of dissolved oxygen is mg/L. Based on existing observation data, we define the cost function of the water quality model as (8.82):

$$J = \frac{1}{2}[K_1(P - \overline{P})^2 + K_6(\mathrm{NO_3} - \overline{\mathrm{NO_3}})^2 + K_8(\mathrm{PO_4} - \overline{\mathrm{PO_4}})^2 + K_9(DO - \overline{DO})^2], \tag{8.82}$$

wherein the definitions of K_1, K_6, K_8, and K_9 are similar to those of Formula (8.63). The upper bar represents the average value.

The cost function is the driving force of the assimilation process of the water quality model, which drives the model parameters to continuously optimize and adjust the process. In Eq. (8.80), it is difficult to evaluate the simulation accuracy of each state variable because multiple state variables are involved. Therefore, we still need to calculate the mean absolute error (MAE) of each parameter as a criterion to evaluate the accuracy of the model.

8.4 Global Sensitivity Analysis

Based on the model settings in Section 8.5, we performed numerical experiments on the model. According to the MAE and cost function, the simulation results of the model follow: The MAEs of phytoplankton, nitrate nitrogen, phosphate, and dissolved oxygen were 0.016 mmol N/L, 0.026 mmol N/L, 0.001 mmol N/L, and 0.04 mg/L, respectively. These results demonstrate that the simulation accuracy of the model is high, and the simulation results are of good authenticity.

Owing to the large number of parameters, we need to calculate and rank the sensitivity of the model before optimizing the parameters. Using the Morris method as the global sensitivity analysis in Section 5.3, we calculated the mean value of the basic effects of each parameter by repeated sampling experiments, thereby obtaining the global sensitivity of each parameter. In Chapter 5, we validated the reliability of the global sensitivity analysis method and found that there was a strong correlation between sensitive parameters and other parameters. Therefore, in this section, we mainly calculate the mean primary effects of each parameter, namely, the global sensitivity of the parameter. To facilitate the analysis, we normalized the results of the globalized sensitivity to [0, 1].

According to the sensitivity analysis results in Figs. 8.9 to 8.17, it can be seen that among the 33 main parameters of the water quality model, there was a large gap between the sensitivity of different state variables with respect to different parameters. Since the Morris method is a qualitative global sensitivity analysis, it would be reasonable to assume the following: Variables with a sensitivity of greater than 0.8 are very sensitive parameters, parameters with a sensitivity of greater than 0.6 but less than 0.8 are more sensitive, and the remaining parameters are insensitive. According to these criteria, we can classify the model parameters. The global sensitivity analysis results of the model are shown in Table 8.2. From Table 8.2, the following conclusions can be drawn.

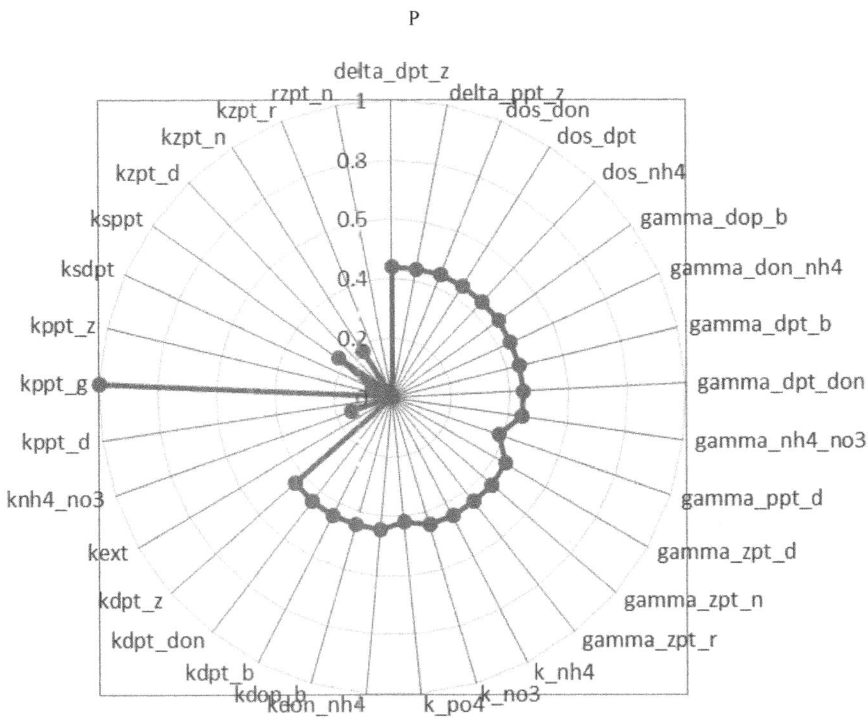

Fig. 8.9. Sensitivity ranking of each parameter with respect to phytoplankton biomass.

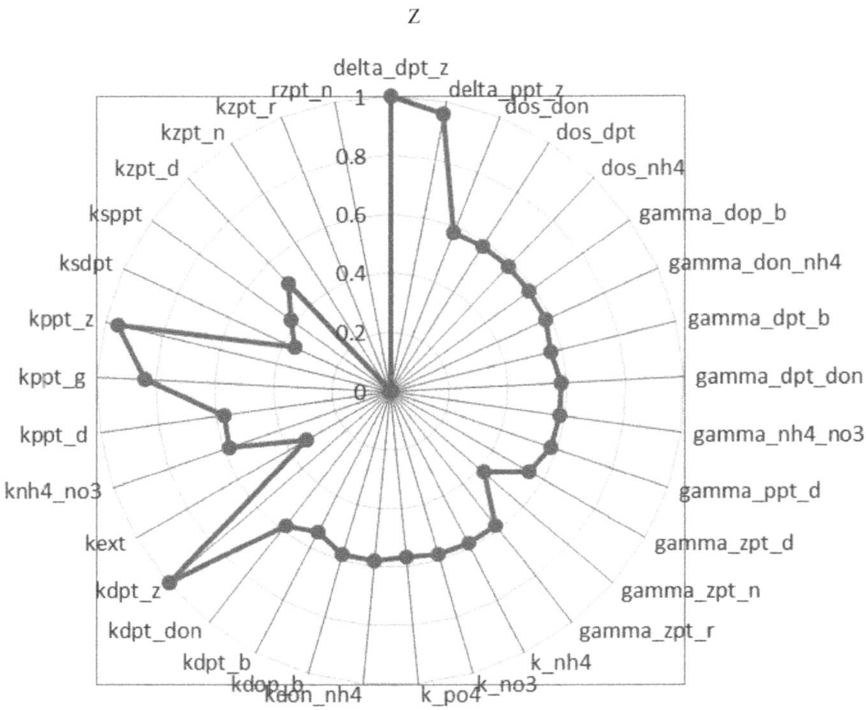

Fig. 8.10. Sensitivity ranking of each parameter with respect to zooplankton biomass.

(1) For phytoplankton biomass, the most sensitive parameter is the phytoplankton maximum growth rate (k_{PPT_G}), and the other parameters are not sensitive. This result indicates that the maximum growth rate parameter significantly affects the phytoplankton biomass simulation process.
(2) Regarding zooplankton biomass, very sensitive parameters mainly include the assimilation rate of zooplankton predation on detritus (δ_{DPT_z}), the assimilation rate of zooplankton predation on phytoplankton (δ_{PPT_z}) and the proportion of zooplankton predation on detritus (k_{DPT_z}), phytoplankton-predated rate by zooplankton (k_{PPT_z}), and the maximum phytoplankton growth rate (k_{PPT_G}). All other parameters are not sensitive. This result shows that the predation process of zooplankton and the growth process of phytoplankton significantly affect the simulation process of zooplankton biomass.

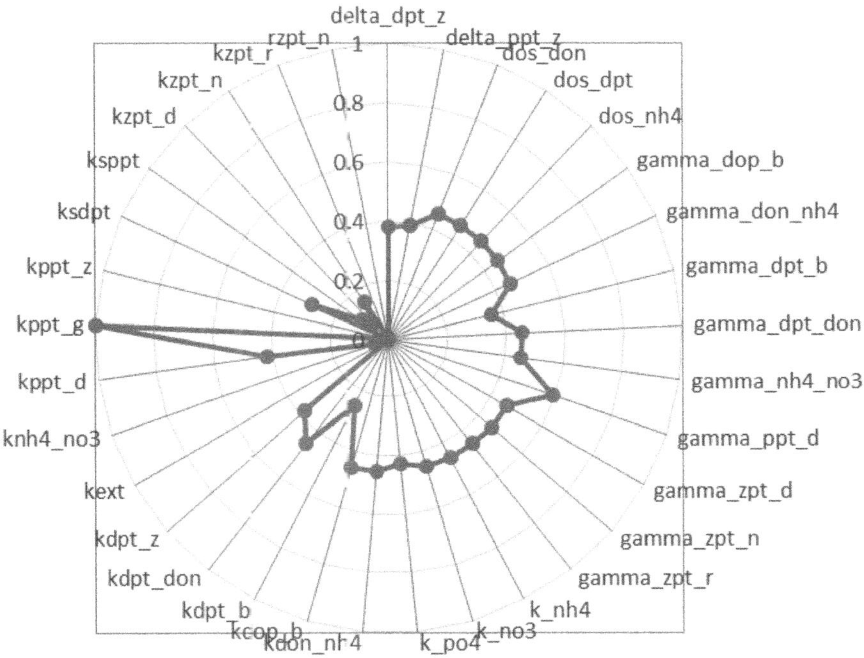

Fig. 8.11. Sensitivity ranking of each parameter with respect to detritus concentration.

(3) For detritus content, the most sensitive parameter is the maximum growth rate of phytoplankton (k_{PPT_G}). The more sensitive parameter is the temperature coefficient (Γ_{PPT_D}) that affects phytoplankton death. This result indicates that the detritus concentration is mainly related to the growth and mortality of phytoplankton.

(4) For the dissolved organic nitrogen concentration, the very sensitive parameters include the assimilation rate of the zooplankton predation on detritus (δ_{DPT_z}), assimilation rate of the zooplankton predation on phytoplankton (δ_{PPT_z}), dissolved oxygen half-saturation coefficient (DO_{SDON}) of consumed dissolved oxygen during the DON remineralization process, dissolved oxygen half-saturation coefficient (DO_{SDPT}) of consumed dissolved oxygen during demineralization process, dissolved oxygen half-saturation coefficient (DO_{SNH}) of consumed dissolved

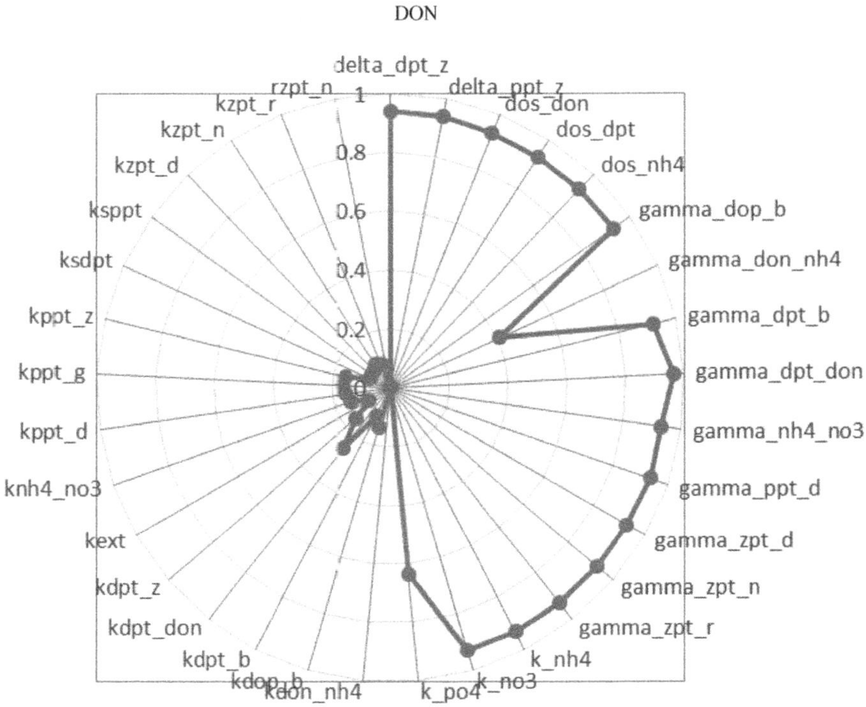

Fig. 8.12. Sensitivity ranking of each parameter with respect to dissolved organic nitrogen.

oxygen during ammonium nitrogen oxidation process, temperature coefficient (Γ_{DOP_B}) affecting DOP remineralization, temperature coefficient (Γ_{DPT_B}) affecting detritus remineralization, temperature coefficient (Γ_{DPT_DON}) affecting detritus decomposition to DON, temperature coefficient (Γ_{NH4_NO3}) of the ammonium nitrogen oxidation process, temperature coefficient (Γ_{PPT_D}) affecting phytoplankton death, temperature coefficient (Γ_{ZPT_D}) affecting zooplankton death, temperature coefficient (Γ_{ZPT_N}) affecting zooplankton excretion, temperature coefficient of zooplankton respiration (Γ_{ZPT_R}), half-saturated concentration of ammonium nitrogen (k_{NH4}), and half-saturated concentration of nitrate nitrogen (k_{NO3}). All others are insensitive parameters. Changes in the concentration of dissolved organic nitrogen are sensitive to many parameters.

(5) For ammonium nitrogen concentration, the most sensitive parameters are phytoplankton mortality (k_{PPT_D}) and light

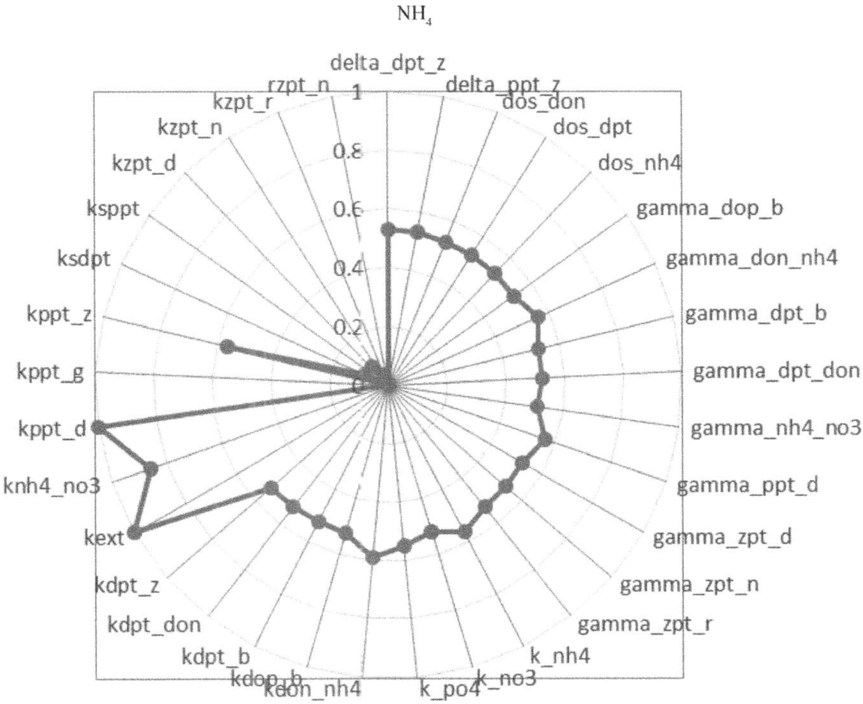

Fig. 8.13. Sensitivity ranking of each parameter with respect to ammonium nitrogen.

attenuation coefficient (*kext*). The more sensitive parameter is the remineralization rate of DON (k_{DON_NH4}) at 0°C.

(6) Very sensitive parameters for nitrate nitrogen concentration include phytoplankton mortality (k_{PPT_D}), ammonium nitrogen oxidation rate (k_{NH4_DON}), and light attenuation coefficient (*kext*) at 0°C. The more sensitive parameters include the assimilation rate of zooplankton predation on detritus (δ_{DPT_z}), the assimilation rate of zooplankton predation on phytoplankton (δ_{PPT_z}), the dissolved oxygen half-saturation coefficient (DO_{SDON}) of dissolved oxygen consumed during DON remineralization, dissolved oxygen half-saturation coefficient (DO_{SDPT}) of consumed dissolved oxygen during detritus remineralization process, dissolved oxygen half-saturation coefficient (DO_{SNH4}) of consumed dissolved oxygen during the oxidation of ammonium nitrogen, temperature coefficient (Γ_{DOP_B})

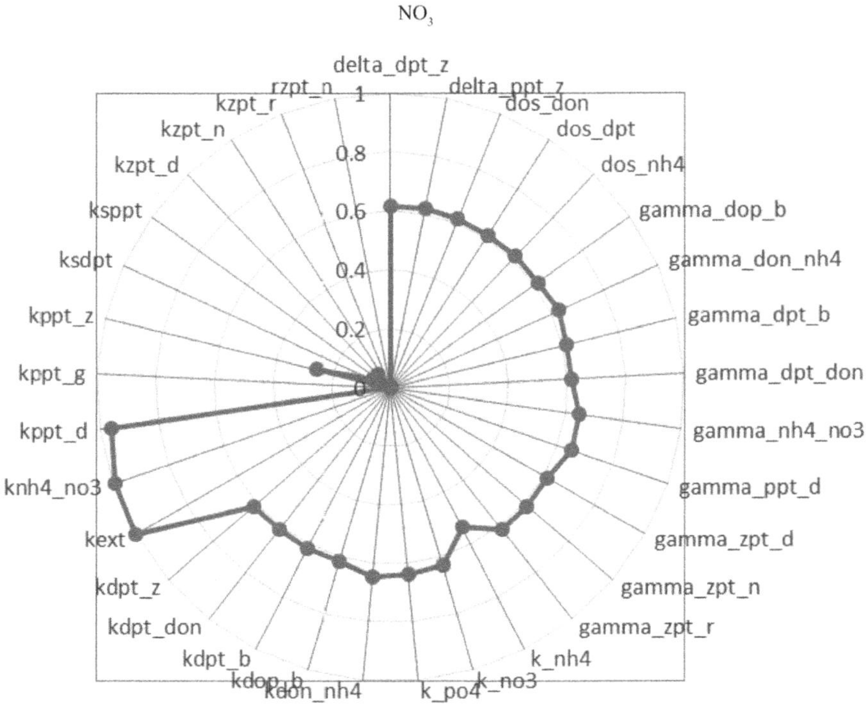

Fig. 8.14. Sensitivity ranking of each parameter with respect to nitrate nitrogen.

affecting DOP remineralization, temperature coefficient affecting remineralization of detritus (Γ_{DPT_B}), temperature coefficient affecting DON remineralization (Γ_{DON_NH4}), temperature coefficient affecting fragmentation to DON (Γ_{DPT_DON}), temperature coefficient of ammonium nitrogen oxidation process (Γ_{NH4_NO3}), temperature coefficient affecting phytoplankton mortality (Γ_{PPT_D}), temperature coefficient affecting zooplankton of mortality (Γ_{ZPT_D}), temperature coefficient of zooplankton excretion (Γ_{ZPT_N}), temperature coefficient of zooplankton respiration (Γ_{ZPT_R}), half-saturated concentration of nitrate nitrogen (k_{NO3}), half-saturated concentration of phosphate (k_{PO4}), the remineralization rate of DON (k_{DON_NH4}) at 0°C, the remineralization rate of detritus (k_{DPT_B}), the proportion of detritus decomposed to DON at 0°C (k_{DPT_DON}), and the ratio of zooplankton predation on detritus (k_{DPT_z}). Changes

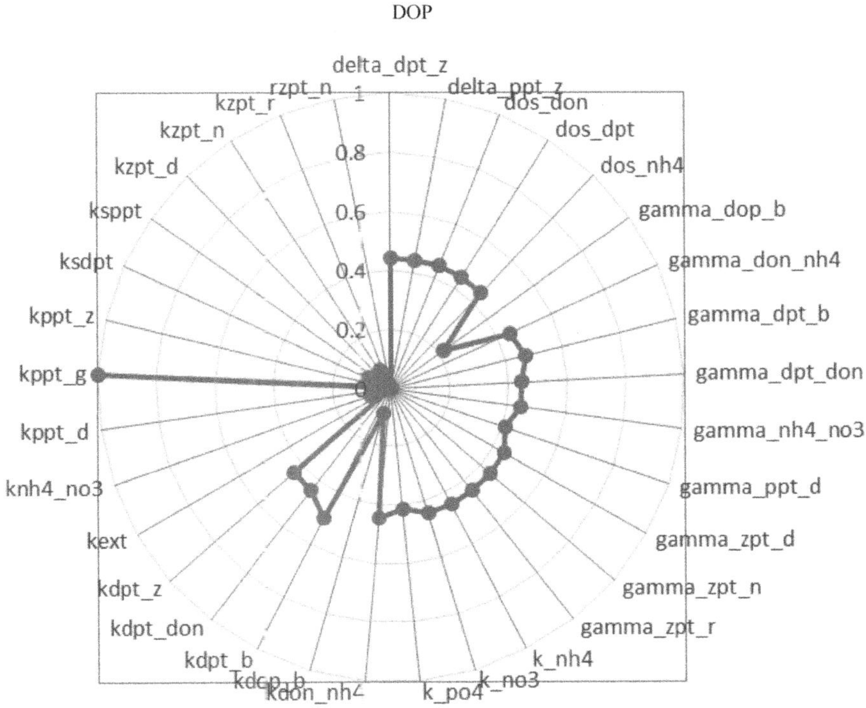

Fig. 8.15. Sensitivity ranking of each parameter with respect to dissolved organic phosphorous.

in the concentration of nitrate nitrogen are sensitive to many parameters.

(7) For the dissolved organic phosphorus concentration, the very sensitive parameter is the maximum growth rate of phytoplankton (k_{PPT_G}). The dissolved organic phosphorus concentration is not sensitive to other parameters.

(8) For phosphate concentration, the most sensitive parameters include phytoplankton mortality (k_{PPT_D}) and light attenuation coefficient ($kext$). The more sensitive parameters are the remineralization rate of DON (k_{DON_NH4}) at 0°C and the DOP remineralization rate (k_{DOP_B}).

(9) For dissolved oxygen concentration, the most sensitive parameter is the maximum growth rate of phytoplankton (k_{PPT_G}). The dissolved oxygen concentration is not sensitive to other parameters.

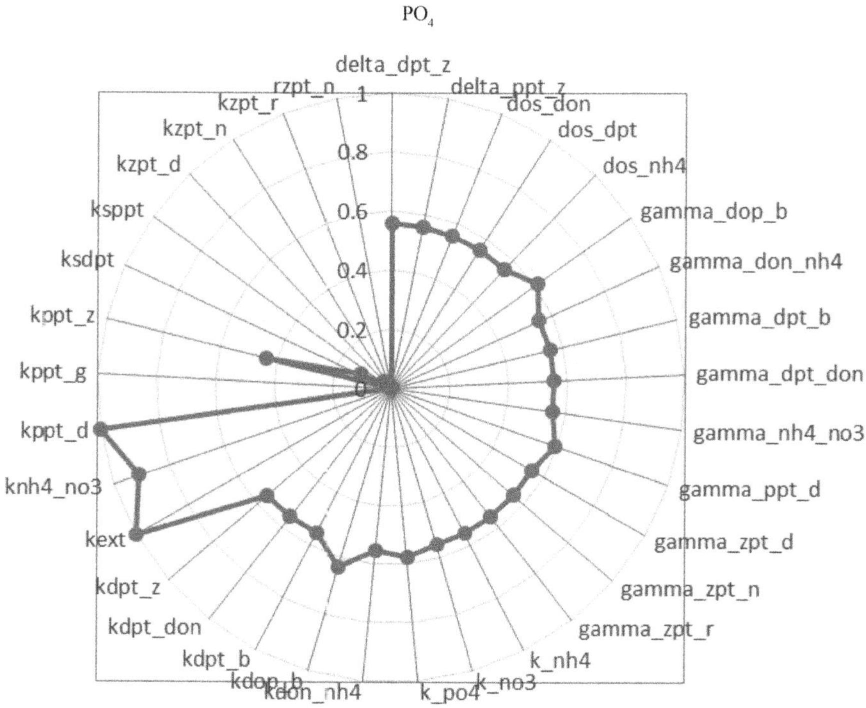

Fig. 8.16. Sensitivity ranking of each parameter with respect to phosphate.

In summary, we discovered that four state variables were very sensitive to the maximum phytoplankton growth rate (k_{PPT_G}), and three state variables were very sensitive to the phytoplankton mortality (k_{PPT_D}) and light attenuation coefficient ($kext$). Phytoplankton biomass is critical to the dynamic balance of the entire water quality model.

8.5 Parameter Optimizations

According to the conclusion of the global sensitivity analysis in Section 8.6, we performed a parameter optimization experiment on the maximum growth rate (k_{PPT_G}) of phytoplankton. The optimization method was the adjoint assimilation method in Section 8.4.

To optimize the model parameters, we also need to calculate the gradient function of the cost function with respect to the parameters

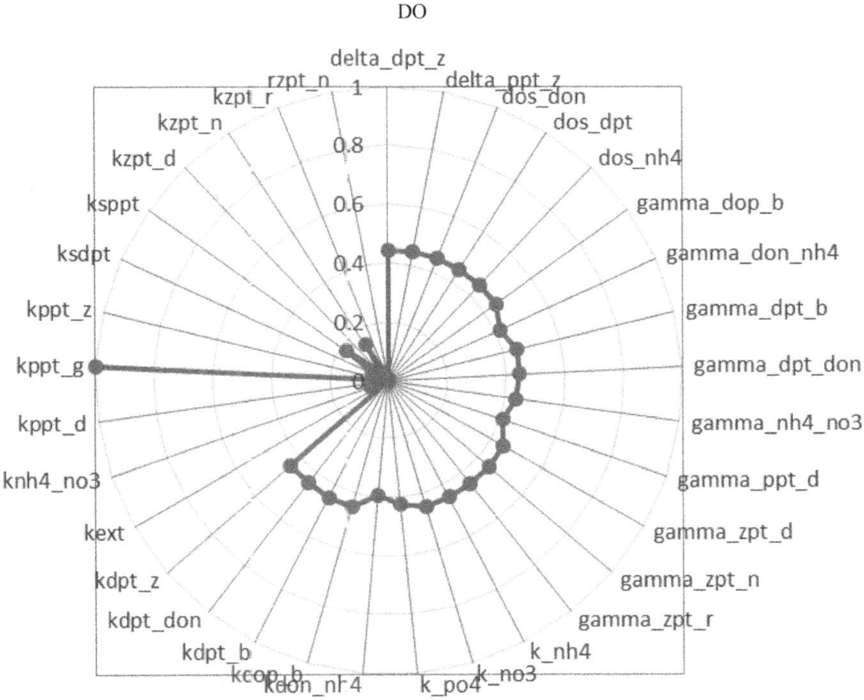

Fig. 8.17. Sensitivity ranking of each parameter with respect to dissolved oxygen.

to be optimized. According to the Lagrange multiplier method, the Lagrange function and the cost function are Eqs. (8.64) and (8.82), respectively. Equation (8.80) is obtained by assuming Lagrange's partial derivative about the maximum growth rate (k_{PPT_G}) of phytoplankton is 0.

$$\frac{\partial L}{\partial k_{PPT\_G}} = 0. \tag{8.83}$$

Formula (8.85) is obtained by solving the above equation.

$$\begin{aligned}\frac{\partial J}{\partial k_{PPT\_G}} &= -(1 - r_{PPT\_E}) f_{PPT\_G}(T) \cdot f_{PPT\_G}(I) \cdot f_{PPT\_G}(N) P \cdot P^* \\ &\quad + f_{PPT\_G}(T) \cdot f_{PPT\_G}(I) \cdot f_{PPT\_G}(N) \frac{L_{\mathrm{NH_4}}}{L_{\mathrm{NH_4}} + L_{\mathrm{NO_3}}} P \cdot C^*_{\mathrm{NH_4}}\end{aligned}$$

Table 8.2. Global sensitivity analysis results.

State variables	Very sensitive parameters	More sensitive parameters
Phytoplankton (P)	K_{PPT_G}	—
Zooplankton (Z)	δ_{DPT_z}, δ_{PPT_z}, k_{DPT_z}, k_{PPT_z}	—
Detritus (D)	k_{PPT_G}	Γ_{PPT_D}
Dissolved Organic Nitrogen (DON)	δ_{DPT_z}, δ_{PPT_z}, DO_{SDON}, DO_{SDPT}, DO_{SNH4}, Γ_{DOP_B}, Γ_{DPT_DON}, Γ_{DPT_DON}, Γ_{NH4_NO3}, Γ_{PPT_D}, Γ_{ZPT_D}, Γ_{ZPT_N}, Γ_{ZPT_R}, k_{NH4}, k_{NO3},	—
Ammonium (NH_4)	k_{PPT_D}, $kext$	k_{DON_NH4}
Nitrate Nitrogen (NO_3)	k_{PPT_D}, $kext$, k_{NH4_DON}	δ_{DPT_z}, δ_{PPT_z}, DO_{SDON}, DO_{SDPT}, DO_{SNH4}, Γ_{DOP_B}, Γ_{DPT_B}, Γ_{DON_NH4}, Γ_{DPT_DON}, Γ_{NH4_NO3}, Γ_{PPT_D}, Γ_{ZPT_D}, Γ_{ZPT_N}, Γ_{ZPT_R}, k_{NO3}, k_{PO4}, k_{DON_NH4}, k_{DPT_B}, k_{DPT_DON}, k_{DPT_z}
Dissolved Organic Phosphorus (DOP)	k_{PPT_G}	—
Phosphate (PO_4)	k_{PPT_D}, $kext$, k_{NH4_DON}	k_{DOP_B}
Dissolved Oxygen (DO)	k_{PPT_G}	—

$$
\begin{aligned}
&+ f_{PPT\_G}(T)\cdot f_{PPT\_G}(I)\cdot f_{PPT\_G}(N)\frac{L_{\mathrm{NO_3}}}{L_{\mathrm{NH_4}}+L_{\mathrm{NO_3}}}P\cdot C^*_{\mathrm{NO_3}}\\
&- f_{PPT\_G}(T)\cdot f_{PPT\_G}(I)\cdot f_{PPT\_G}(N)P\cdot C^*_{DO}\\
&+ f_{PPT\_G}(T)\cdot f_{PPT\_G}(I)\cdot f_{PPT\_G}(N)P\cdot C^*_{\mathrm{PO_4}}/r_{N\_P}. \quad (8.84)
\end{aligned}
$$

From the above formulas, the gradient expression of the maximum growth rate (k_{PPT_G}) of the phytoplankton with respect to the cost function is obtained.

Based on the above settings, we designed a set of parameter optimization experiments and set the number of assimilation times of the model to 50. In this set of experiments, since we know the maximum growth rate of the phytoplankton (k_{PPT_G}), we can calculate whether the inversion parameters are accurate. According to the parameter optimization results given in Figs. 8.18 and 8.19, the cost function and MAE of the water quality model both decreased significantly when we optimized the maximum growth rate (k_{PPT_G}) of phytoplankton, and both reached stability around step 40. The MAE of the parameter also decreased with the progress of the assimilation process. The results of this group of experiments show that the assimilation method exhibited a better optimization effect on the inversion of water quality model parameters and the optimization process for sensitive parameters could significantly improve the model's simulation accuracy.

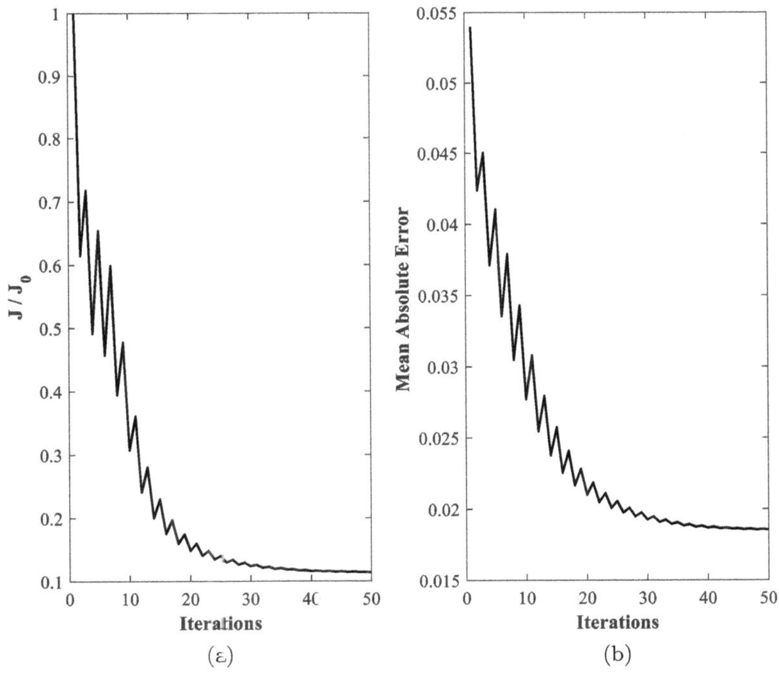

Fig. 8.18. The cost function and the mean absolute error (MAE) of the water quality model simulation results are a function of the number of assimilation curves. The horizontal axis is the number of the water quality model's assimilation times. The vertical axis (a) represents the cost function, and the vertical axis (b) represents the MAE of the simulation results.

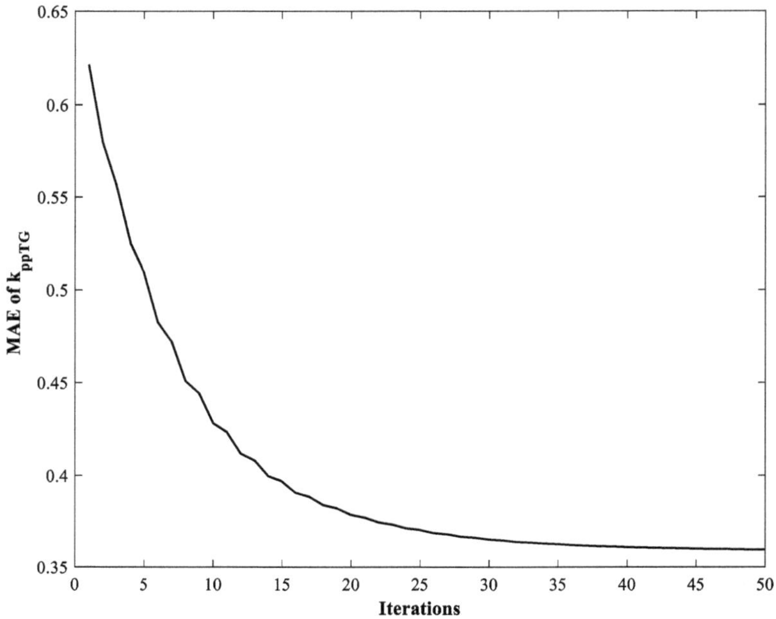

Fig. 8.19. When the maximum phytoplankton growth rate (k_{PPT_G}) is inverted, the mean absolute error (MAE) curve of the maximum phytoplankton growth rate obtained from the inversion is shown. The horizontal axis represents the number of assimilations, and the vertical axis represents the MAE. The unit is 1/day.

8.6 Summary

In this chapter, we established a three-dimensional hydrodynamic-water quality model and conduct research on model optimization. First, we applied the improved Cressman interpolation method to the interpolation process of open-source data and obtained the initial field of the water quality model. Based on the established model, we carried out experimental studies on the global sensitivity of 33 parameters of the water quality model. It was found that four state variables were very sensitive to the maximum phytoplankton growth rate (k_{PPT_G}) and phytoplankton mortality (k_{PPT_D}), and three state variables were very sensitive to the light attenuation coefficient ($kext$). Considering phytoplankton biomass is critical to the dynamic balance of the entire water quality model, we carried out parameter optimization experiments for the maximum phytoplankton growth rate (k_{PPT_G}). It was found that the assimilation method exhibited a better optimization effect in inverting the water quality model parameters, and the optimization process for sensitive parameters could significantly improve the simulation accuracy of the model.

Acknowledgement

This chapter partially reused the content of the following reference. Liu, Y.; Jiang, F.; Zhao, Z.; Tana; Lv, X. Advancing Three-Dimensional Coupled Water Quality Model of Marine Ranches: Model Development, Global Sensitivity Analysis, and Optimization Based on Observation System. *Journal of Marine Science and Engineering*. 2022, 10, 1028. https://doi.org/10.3390/jmse10081028 (changes were made). This article is an open access article distributed under the terms and conditions of the Creative Commons Attribution (CC BY) license (https://creativecommons.org/licenses/by/4.0/).

References

Bailey R T, Ahmadi M. 2014. Spatial and temporal variability of in-stream water quality parameter influence on dissolved oxygen and nitrate within a regional stream network. *Ecological Modelling*. **277**: 87–96.

Chapelle A, Lazure P, Ménesguen A. 1994. Modelling eutrophication events in a coastal ecosystem. Sensitivity analysis. *Estuarine, Coastal and Shelf Science*. **39**(6): 529–548.

Chen C S, Liu H D, Beardsley R C. 2003. An unstructured grid, finite-volume, three-dimensional, primitive equations ocean model: Application to coastal ocean and estuaries. *Journal of Atmospheric and Oceanic Technology*. **20**(1): 159–186.

Cleary R. 1976. *Mathematical Models. Boundaries of Analysis*. Cambridge, MA: Ballinger.

Cole T M, Wells S A. 2006. CE-QUAL-W2: A two-dimensional, laterally averaged, hydrodynamic and water quality model, version 3.5.

Dobson F W, Smith S D. 1988. Bulk models of solar radiation at sea. *Quarterly Journal of the Royal Meteorological Society*. **114**(479): 165–182.

Dueri S, Dahllöf I, Hjorth M, *et al.* 2009. Modeling the combined effect of nutrients and pyrene on the plankton population: Validation using mesocosm experiment data and scenario analysis. *Ecological Modelling.* **220**(17): 2060–2067.

Ebenhöh W, Baretta-Bekker J G, Baretta J W. 1997. The primary production module in the marine ecosystem model ERSEM II, with emphasis on the light forcing. *Journal of Sea Research.* **38**(3-4): 173–193.

Egbert G D, Erofeeva S Y. 2002. Efficient inverse modeling of barotropic ocean tides. *Journal of Atmospheric and Oceanic Technology.* **19**(2): 183–204.

Eppley R W. 1972. Temperature and phytoplankton growth in the sea. *Fishery Bulletin.* **70**(4): 1063–1085.

Evans G T. 1999. The role of local models and data sets in the joint global ocean flux study. *Deep Sea Research Part I: Oceanographic Research Papers.* **46**: 1369–1389.

Fan W. 2009. *Numerical Simulation and Assimilation of Marine Ecosystem Dynamics.* Qingdao: Ocean University of China.

Fan W, Lv X. 2009. Data assimilation in a simple marine ecosystem model based on spatial biological parameterizations. *Ecological Modelling.* **220**(17): 1997–2008.

Fennel K, Wilkin J, Levin J, *et al.* 2006. Nitrogen cycling in the Middle Atlantic Bight: Results from a three-dimensional model and implications for the North Atlantic nitrogen budget. *Global Biogeochemical Cycles.* **20**: 1–14.

Gregg W W, Friedrichs M A M, Robinson A R, *et al.* 2009. Skill assessment in ocean biological data assimilation. *Journal of Marine Systems.* **76**(1–2): 16–33.

Guo Z, Pan H, Fan W, Lv X. 2017. Application of surface spline interpolation in inversion of bottom friction coefficients. *Journal of Atmospheric and Oceanic Technology.* **34**(9), 2021–2028.

Hemmings J C, Srokosz M A, Challenor P, Fasham M J. 2004. Split-domain calibration of an ecosystem model using satellite ocean colour data. *Journal of Marine Systems.* **50**(3): 141–179.

Iriarte A, Villate F, Uriarte I, *et al.* 2015. Dissolved oxygen in a temperate estuary: The influence of hydro-climatic factors and eutrophication at seasonal and inter-annual time scales. *Estuaries and Coasts.* **38**(3): 1000–1015.

Laurent A, Fennel K, Cai W J, *et al.* 2017. Eutrophication-induced acidification of coastal waters in the northern Gulf of Mexico: Insights into origin and processes from a coupled physical-biogeochemical model. *Geophysical Research Letters.* **44**(2): 946–956.

Li K, Li Z, Li Y, *et al.* 2015. A three-dimensional water quality model to evaluate the environmental capacity of nitrogen and phosphorus in Jiaozhou Bay, China. *Marine Pollution Bulletin.* **91**(1): 306–316.

Li X Y. 2013. *Inversion of the Spatiotemporal Distribution of Control Parameters in Marine Ecosystem Dynamics Model.* Qingdao; Ocean University of China.

Liu, Y., Yu, J., Shen, Y., Lv, X., 2016. A modified interpolation method for surface total nitrogen in the Bohai Sea. *Journal of Atmospheric and Oceanic Technology*. **33**(7): 1509–1517.

Liu Y, Shen Y, Lv X, Liu Q. 2017. Numeric modelling and risk assessment of pollutions in the Chinese Bohai Sea. *Science China Earth Sciences.* **60**(8): 1546–1557.

Losa S N, Kivman G A, Ryabchenko V A. 2004. Weak constraint parameter estimation for a simple ocean ecosystem model: What can we learn about the model and data? *Journal of Marine Systems.* **45**(1): 1–20.

Losa S N, Vézina A, Wright D, *et al.* 2006. 3D ecosystem modelling in the North Atlantic: Relative impacts of physical and biological parameterizations. *Journal of Marine Systems.* **61**(3): 230–245.

Lu D, Li K, Liang S, *et al.* 2017. A coastal three-dimensional water quality model of nitrogen in Jiaozhou Bay linking field experiments with modelling. *Marine Pollution Bulletin.* **114**(1): 53–63.

Lü X Q, Fan W. 2009. Inversion study of spatial variation of parameters in marine ecosystem model using adjoint method. *Journal of Ocean University of China.* **39**(5): 846–854.

Lv X Q. 2000. *Research on Adjoint Methods in Numerical Modeling of Oceans.* Qingdao: Institute of Oceanology, Chinese Academy of Sciences.

McGillicuddy D, Lynch D, Moore A, *et al.* 1998. An adjoint data assimilation approach to diagnosis of physical and biological controls on Pseudocalanus spp. in the Gulf of Maine–Georges Bank region. *Fisheries Oceanography.* **7**(3–4): 205–218.

Pauer J J, DePetro P A, Anstead A M, Lehrter J C. 2014. Application of a one-dimensional model to explore the drivers and lability of carbon in the northern Gulf of Mexico. *Ecological Modelling.* **294**: 59–70.

Ridgway K R, Dunn J R, Wilkin J L. 2002. Ocean interpolation by four-dimensional least squares — Application to the waters around Australia. *Journal of Atmospheric and Oceanic Technology.* **19**(9): 1357–1375.

Riley G A. 1956. Oceanography of Long Island Sound 1952–1954 II. Physical oceanography. *Series I: Bulletin of the Bingham Oceanographic Collection, 1925–1959.* **15**: 15–46.

Sasaki Y. 1970. Some basic formalisms in numerical variational analysis. *Monthly Weather Review.* **98**: 875–883.

Schnoor J L. 1996. *Environmental Modelling.* New York: John Wiley & Sons Inc, 185–229.

Sellers W D. 1965. *Physical Climatology.* Chicago: University of Chicago Press, 272.

Shen Y L. 2015. *Inversion of Marine Ecosystem Dynamics Model Parameters and Its Application.* Qingdao: Ocean University of China.

Shen C, Shi H, Liu Y, *et al.* 2016. Discussion of skill improvement in marine ecosystem dynamic models based on parameter optimization and skill assessment. *Chinese Journal of Oceanology and Limnology.* **34**(4): 683–696.

Steele J H. 1962. Environmental control of photosynthesis in the sea. *Limnology Oceanography.* **7**(2): 137–150.

Wang C H. 2013. *Research and Application of Dynamic Assimilation Model of Marine Ecosystem.* Qingdao: Ocean University of China.

Water D H I. 2002. Environment: MIKE11: A modelling system for rivers and channels — User guide.

Wool T A, Ambrose R B, Martin J L, *et al.* 2006. Water quality analysis simulation program (WASP), 2006. User's manual, version 6.

Yu C Z, Markovsky, Li Y L. 1989. *Principles of Pollutant Diffusion and Transport in Water Environment and Water Quality Model.* Beijing: China Environmental Science Press.

Zhao Q, Lu X. 2008. Parameter estimation in a three-dimensional marine ecosystem model. *Journal of Marine Systems.* **74**(1–2): 443–452.

Zheng L, Chen C, Zhang F Y. 2004. Development of water quality model in the Satilla River Estuary, Georgia. *Ecological Modelling.* **178**(3): 457–482.

Zheng W, Shi H H, Fang G H, *et al.* 2012. Global sensitivity analysis of a marine ecosystem dynamic model of the Sanggou Bay. *Ecological Modelling.* **247**: 83–94.

Chapter 9

Construction of Marine Ecosystem Dynamics Model Based on an Artificial Neural Network and Cellular Automata

There are two main traditional methods to construct an ecological dynamics model. The first one is based on basic physical laws (i.e., conservation of matter), biological laws (i.e., the law of minimum limitation), biological processes (i.e., photosynthesis, respiration, and biological pump), and biological relationships (i.e., feeding and competition), etc., to construct the dynamic model with approximate methods (Franks and Chen, 1996; Chen *et al.*, 1999; Gao *et al.*, 2009). Another method is based on statistics and observational statistics. Both traditional dynamic modeling and statistical modeling inevitably involve scientists' subjective factors, such as approximations or artificial assumptions. On the other hand, the localization characteristics of these marine ecological dynamic models are not significant. Therefore, the accuracy of these models is usually not satisfactory in practice. In recent years, with the rapid development of computer technology, computer-based modeling methods such as artificial neural networks and cellular automata have become popular. The neural network method adopts a self-adaptive "black box" mode. When samples are sufficient, relatively high simulation accuracy can often be obtained (Olden, 2000; Li *et al.*, 2007). In the cellular automata method, based on the understanding of the dynamic mechanism of system change, the basic rules of cellular diffusion are set to reveal the change rules of complex systems (Ding *et al.*, 2009; Matsinos and Troundis, 2002).

9.1 Biomass Simulation of Coastal Phytoplankton Based on Artificial Neural Network

Because the current coastal eutrophication has attracted attention all over the world, the red tide caused by the rapid growth of phytoplankton has become one of the hotly studied issues in marine ecology (Halpern *et al.*, 2008; Graneli *et al.*, 2008; Gao *et al.*, 2009). Many complex factors affect the growth of phytoplankton, and these include not only the constraints of biological processes or nutrients, but also the limitations of basic environmental factors such as light, seawater temperature, and salinity. Because the system has numerous components related to each other, the change mechanism of the phytoplankton ecosystem is very complicated and presents significant nonlinear characteristics. Mathematical models have since become important tools to explore the response of phytoplankton growth to changes in environmental factors. For example, the models, which are based on artificial neural networks, have the ability to simulate complex nonlinear systems, and some scientists have used feed-forward neural networks to simulate the time and magnitude of algal blooms (Yabunaka *et al.*, 1997; Recknagel, 1997; Maier and Dandy, 2001; Hou *et al.*, 2004; Recknagel *et al.*, 2006; Li *et al.*, 2007). Compared with other methods, neural network models have higher accuracy to predict algae concentration and become a useful tool to simulate environmental systems (Wang *et al.*, 2006; Ding *et al.*, 2009). However, in the prediction and simulation of marine phytoplankton, this method has been rarely applied. In this section, an artificial neural network with Back Propagation (BP) will be used to construct the coastal ecosystem dynamics model, which will be used to simulate coastal phytoplankton biomass and its influencing factors (Zheng *et al.*, 2012).

9.1.1 *Construction of coastal ecosystem dynamics model based on BP artificial neural network*

The BP artificial neural network model is one of the most widely used neural networks in recent years, in which the essential characteristics of development are obtained through input data and then used to solve the problem of simulation and prediction. The topology of the BP artificial neural network is a three-layer network structure, including input layers, hidden layers, and output layers. In this section, based on the BP artificial neural network, we will predict changes in phytoplankton biomass through seawater environmental parameters. The model constructing steps are as follows:

Step 1: Data standardization method. The input data is mapped into the range 0–1 with the maximum-minimum normalization method to ensure that the network has good convergence and mapping capability.

Step 2: The quantity determination of hidden layer neurons. On the premise of ensuring a certain accuracy and sufficiently high generalization ability, the number of hidden layer neurons is determined according to three basic principles, which are: (1) the structure is as compact as possible and the number of hidden layer nodes is as small as possible, (2) the number of hidden layer neurons must be less than that of training samples, (3) the number of training samples must be greater than the connection weight of the network model (usually 2–10 times). A BP neural network is established separately for each number of hidden layer neurons that meets the requirements. Other parameter settings of the network are the same. After training the built network, the validation set is simulated for each network to obtain simulated values, which are then compared with the real values to obtain the mean square error, before finding the number of hidden layers corresponding to the smallest mean square error as the optimal number of hidden layers for the following test.

Step 3: Artificial neural network training. The obtained data is divided into two types: training set and validation set. Each set contains input and output subsets. Also, the test set samples should be independent of the training set and representative. Its sample selection should be as dispersed as possible, and the number of samples should not be less than 10% of that of the training set samples. The error goal, learning rate, and transfer function are chosen to normalize the data according to the previously mentioned method and determine the range of hidden neurons. Then, the network will be trained to choose the network with the optimal quantity of hidden neurons to be the simulation and prediction network.

Step 4: Test on BP artificial neural network. With the average absolute relative error between the network output and original data in the tested sample (the average value is obtained according to the absolute value of each relative error) as the main indicator, the effectiveness of the network is tested. Meanwhile, the correlation coefficient between the observed and predicted values of the test sample is used as the reference.

Step 5: Response of network output to input changes. After network training is successful, it can be used for prediction and simulation. In order to reveal the internal mechanism of the built artificial neural network model, especially to study the sensitivity of the network output to the input, the global sensitivity is used to analyze the input of all training samples to calculate the response of phytoplankton biomass to the change of various environmental factors under different environmental conditions (Cossarini and Solidoro, 2008). The distribution of these input samples is highly discrete and basically represents different initial positions.

9.1.2 *Case study of coastal phytoplankton biomass simulation based on artificial neural network*

Phytoplanktons are the foundations of the marine food chain and play a very important role in the marine nutrient cycle (Bricker *et al.*, 2003). The marine ecosystem is a complex system. Its change is affected by multiple factors at the same time and there is no mature method to predict its change. Phytoplankton biomass is the most important coastal ecosystem factor as an important indicator of the vitality and health of marine ecosystems. Meanwhile, the rapid growth of phytoplankton can easily lead to the blooming of marine harmful algae. But, if phytoplankton growth is too slow, the entire ecosystem will lack vitality. Therefore, studies on the changes of coastal phytoplankton biomass and its impact mechanism are of great significance for ecosystem-based marine management. Here, we first study chlorophyll-a, which is the representative factor of phytoplankton biomass, with Jiaozhou Bay, Quanzhou Bay, and the coast of Guangxi as examples (representing coastal ecosystems of the Bohai Sea, Yellow Sea, East China Sea, and the South China Sea), and is based on the BP coastal ecological model to construct an artificial neural network that simulates the dynamic changes of phytoplankton biomass in different sea areas. Secondly, the reliability of the model is verified by the observation data at different stations in different months. Finally, the parameter sensitivity analysis model of an artificial neural network model is established to analyze the contribution of the input parameters of each model to the prediction results. The main application cases include the following aspects:

(1) *Phytoplankton biomass simulation in Jiaozhou Bay*
In this study, we used monthly survey data from January to December 2003 in nine stations in Jiaozhou Bay. Water environmental indicators included dissolved oxygen (DO) concentration, pH value, chemical oxygen demand (COD), soluble dissolved inorganic nitrogen (DIN) concentration, soluble dissolved inorganic phosphorus (DIP) concentration, soluble silicate concentration (SiO_3-Si), petroleum concentration (Oil), sea surface temperature (SST), and surface chlorophyll-a (Chl-a) concentration that represented the phytoplankton biomass. After repeated experiments, according to the convergence and error of the network, a three-layer network with three hidden neurons was selected as the prediction network. The number of hidden neurons was set to 3, the number of network trainings was set to 50,000, the training error limit was set to 0.01, and the learning rate was set to 0.02. After the training was completed, the validation set was simulated in the human network to obtain the network simulation value. Then, these simulation values were normalized to finally obtain the simulated value of phytoplankton biomass (or Chl-a concentration) in Jiaozhou Bay. The average relative error of the network was 0.84%, the average absolute error was 11.33%, and the correlation coefficient between predicted and observed values of test samples was 0.8943. Therefore, the BP artificial neural network could meet the predicted accuracy requirements, and the network training was successful and could be used in prediction studies. Through examining the response of the network's output to input changes, it could be found that the average sensitivity of the phytoplankton biomass in Jiaozhou Bay to various environmental factors was: (in the order from large to small) DO, COD, DIP, SST, pH, Oil, DIN, and SiO_3-Si.

(2) *Simulation of primary productivity in Jiaozhou Bay*
All data was used to simulate the phytoplankton biomass in Jiaozhou Bay, and the output of the network was used to predict the primary productivity. After repeated experiments, the number of hidden neurons in the network was set to 5, the number of network training times was 50,000, the training error limit was set to 0.01, and the learning rate was set to 0.02. After the training was completed, the validation set was simulated in the network to obtain the network simulation value. These simulation values were normalized to finally obtain the simulation value of the primary productivity of Jiaozhou Bay. The average relative error of the network was 5.67%, the average absolute error was 22.86%, and the correlation coefficient between the predicted value and the observed value of the primary productivity test sample was 0.8146. The prediction accuracy of this BP artificial neural network model was slightly lower than that of the biomass (Chl-a concentration) model, but it could meet the prediction accuracy requirements to a certain extent. The high-to-low order of the average sensitivity of primary productivity to various environmental factors was pH, DO, SST, DIP, Oil, COD, DIN, and SiO_3-Si.

(3) *Phytoplankton biomass simulation in Quanzhou Bay*
According to the continuous observation data at eight stations in Quanzhou Bay from April to December 2008, the studied environmental indicators included water depth (H), sea surface temperature (SST), transparency (Tr), pH value, sea surface salinity (SSS), seawater chemistry oxygen demand (COD), soluble dissolved inorganic phosphorus (DIP was replaced with PO_4-P concentration), nitrite (NO_2-N), nitrate (NO_3-N), ammonium (NH_3-N), and surface chlorophyll-a (Chl-a) concentration that represented the phytoplankton biomass. After repeated experiments, according to the convergence and error of the network, a three-layer network (BP) with three hidden neurons was selected as the prediction network. The training times were 50,000, the training error goal was set to 0.01, and the learning rate was set to 0.02. The average relative error of the network was −3.46%, the average absolute error was 10.53%, and the correlation coefficient between the predicted value and the observed value of phytoplankton biomass BP network in Quanzhou Bay was 0.964, which met the general prediction accuracy requirements. The high-to-low order of the average sensitivity of phytoplankton biomass

to various environmental factors was pH, SST, SSS, COD, NH_3-N, NO_2-N, NO_3-N, H, DIP, and Tr.

(4) *Biomass simulation of coastal phytoplankton in Guangxi*
The biological and chemical survey data at 25 stations and 4 voyages in July 2006, October 2006, December 2006, and April 2007 in Guangxi Coast were used. The used environmental indicators included total nitrogen (TN), total phosphorus (TP), soluble phosphate (DIP), DOP, dissolved oxygen (DO), soluble silicate (SiO_3-Si), petroleum (Oil), transparency (Tr), and the surface chlorophyll-a (Chl-a) concentration that was used to represent phytoplankton biomass. After repeated experiments, according to the convergence and error of the network, a three-layer network (BP) containing two hidden neurons was selected as the prediction network. The number of trainings was 50,000, the training error limit was set to 0.01, and the learning rate was set to 0.02. The average relative error of the network was 2.29%, the average absolute error was 15.74%, and the correlation coefficient between the predicted value and observed value of the validation set was 0.935, which met the accuracy requirements of general prediction. The high-to-low order of the average sensitivity of phytoplankton biomass to various environmental factors was DO, SiO_3-Si, TN, TP, DOP, Oil, Tr, and DIP.

(5) *Biomass simulation of phytoplankton in Laizhou Bay*
According to the observation data from 30 stations and 2 voyages in Laizhou Bay in May and August 2004, the used environmental indicators included sea surface temperature (SST), transparency (Tr), water color, sea surface salinity (SSS), pH, DO, COD, DIP, DIN, suspended material concentration, and the surface chlorophyll-a (Chl-a) concentration that was used to represent the phytoplankton biomass. After repeated experiments, according to the convergence and error of the network, a three-layer network (BP) containing two hidden neurons was selected as the prediction network. The number of trainings was 50,000, the training error limit was set to 0.01, and the learning rate was set to 0.02. The average relative error of the network was −8%, and the average absolute error was 17%. The correlation coefficient between the predicted value and the observed value of phytoplankton biomass in the validation set from Laizhou Bay was 0.951, which met the general prediction accuracy requirements. The average sensitivity of phytoplankton biomass in Laizhou Bay to

various environmental factors was greater than 1, and the order (from large to small) was SST, DO, COD, pH, DIP, suspended material, Tr, DIN, SSS, and water color.

9.1.3 *Summary*

In this section, the established model, which is based on an artificial neural network, shows high accuracy in the simulation of change dynamics of coastal phytoplankton after data standardization, network optimization, network training, and model validity testing. In particular, through the sensitivity analysis, the model could reveal the response of phytoplankton biomass to changes in environmental factors, and then to "whiten" the "black box" network. It is also important to note that there is often a strong correlation between environmental factors in coastal ecosystems. Due to the high degree of adaptability and nonlinear simulation capabilities of artificial neural networks, the same method can be used to train the network and build a neural network model when the combination of environmental factors changes (or decreases in number). Compared with other models with fixed formulas, the network model is more applicable and generalizable. The sensitivity analysis also shows that the artificial neural network method can also be used to calculate the response of the network output (phytoplankton biomass) to the input (aquatic environmental factors) changes, which can reveal the dynamic mechanism of phytoplankton changes to a certain degree.

9.2 Cellular Automata-based Risk Assessment and Early Warning Model of Jellyfish Injury

Jellyfish continue to bloom in worldwide coastal waters, such as the Germany Gulf (Greve, 1994), Florida coast (Radwan *et al.*, 2001), and the Jiaozhou Bay. Relevant reports have shown that jellyfish blooms are increasing (Purcell *et al.*, 2007), which are related to climate change, eutrophication of water bodies, and over-fishing (Molinero *et al.*, 2009; Dong *et al.*, 2010). In addition, jellyfish blooms cause a series of socioeconomic problems. For example, they severely disrupt the fishing production by: (1) blocking and destroying fishing nets, (2) reducing fishing production, (3) hurting fishermen, (4) increasing the workload of cleaning fishing nets, and (5) causing fishing vessels to capsize (Quiñones *et al.*, 2012; Lucas *et al.*, 2014). In 2009, Japan's economic loss due to jellyfish blooms amounted to $125 million (Lucas *et al.* 2014). Jellyfish blooms also seriously threaten coastal tourism and the personal safety of tourists and cause inestimable economic loss. An Adriatic jellyfish bloom in the summer of 1978 caused nearly 250,000 tourists on the Yugoslav coast to be stung by jellyfish (Maretic *et al.*, 1979). According to hospital statistics, since 1983, there have been more than 2,000 cases with 13 deaths caused by jellyfish stings in coastal areas of China (Dong *et al.*, 2010). Based on statistical data, there are 2,050 deaths from jellyfish stings each year in the Philippines (Lippmann *et al.*, 2011), and at least 30,000 people have been stung by jellyfish along the coast of Spain (Ghermandi *et al.*, 2015). Therefore, in order to reduce the casualties caused by jellyfish, the according management measures should be formulated. The related numerical models and simulations could provide decision-making references for the early warning of jellyfish risks.

In order to build a jellyfish early warning simulation model, a reasonable discrete dynamic model framework is needed. However, it is difficult for the hydrodynamic model to simulate jellyfish movement because the jellyfish movement in water bodies has huge randomness and the calculation amount of the hydrodynamic model is relatively large. In addition, a jellyfish is usually large and is not suitable to be treated as a particle in coastal waters. However, conventional statistical methods, such as the Bayesian network method widely used in risk assessment and multivariate analysis, cannot simulate the dynamic model of spatiotemporal discreteness (Phan *et al.*, 2014). Methods such as Markov chains are not suitable to simulate the interaction between multiple targets that change with time and space. Although the random walk method and cellular automata (CA) method are similar to some extent, the CA method is more suitable for simulating the coupling between tourists and jellyfish movements. Because the CA method includes discrete space, discrete time, and discrete state space, it can simulate the interactions and complex relationships between various components, such as the simulation of rail transit (Li *et al.*, 2005), forest fire diffusion, and propagation (Alexandridis *et al.*, 2008), aircraft takeoff and landing time setting, the ecological invasion of a certain plant (Barona *et al.*, 2014), and the expansion process of cities (Liu *et al.*, 2014). Here, we mainly use the cellular automata theory to simulate the jellyfish hurting process in near-shore baths, to analyze the key influencing factors of jellyfish hurting, and to explore the early-warning mechanism of jellyfish in near-shore baths.

9.2.1 *Construction of jellyfish early warning model that is based on cellular automata*

Unlike other CA models, the early warning model of near-shore bath jellyfish contains two simulation targets with interactions, including tourists and jellyfish, and the time and space changes are completely different. Therefore, the CA model has been improved here to simulate two targets (tourists and jellyfish). Additionally, because the hurting process of jellyfish is affected by many factors, several parameters are added into the model to explore the influence of these factors, and then these parameters are significantly analyzed. The modeling process is described in detail below:

(1) *Introduction to cellular automata (CA)*
Different from other dynamic system models, CA does not require specific functions or equations, and only needs to describe the interaction between grids to operate normally. Therefore, CA has a relatively high variability without restriction from the governing equations. CA is just a general term for a series of models (Wolfram, 1983; Wainer, 2014).

CA can be considered as a functional in grid space as:

$$A = (L_{\mathrm{d}}, S, N, \mathrm{f}), \tag{9.1}$$

where A is the CA system,

L_{d} is grid space,
S is state space corresponding to grid space,
N is neighbor space,
F is the state transition function.

With a regular square grid as an example, its grid space L_{p} is set to the average diameter of a jellyfish. Figure 9.1 shows an example of observation space. The observation space contains the largest area

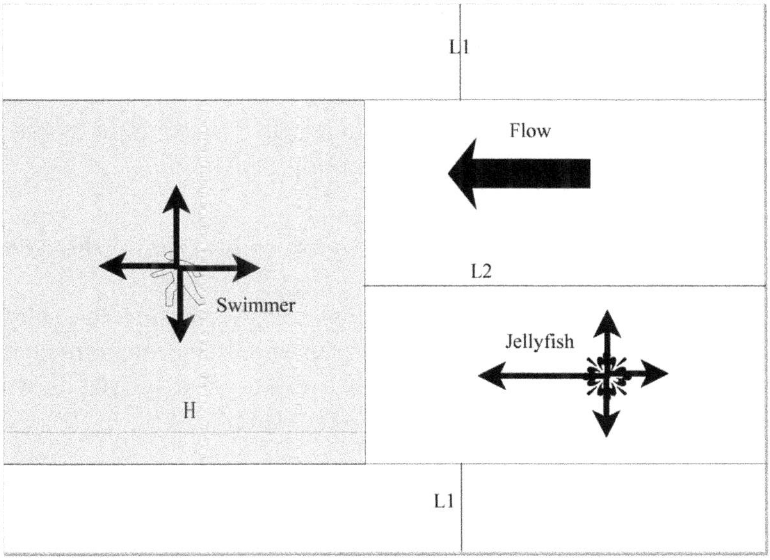

Fig. 9.1. An example of observation area settings in the model.

where tourists swim (the gray area in Fig. 9.1). Because the jellyfish outside the gray area may enter and hurt tourists, the observation space is enlarged to the entire space shown in Fig. 9.1.

The state space S is corresponding to the grid space. In the model, the sum of the sparse matrix is used to store the state space of tourists and jellyfish, respectively, which is exactly the improvement of CA to make the model more efficient.

In Fig. 9.1, H, $L1$, and $L2$ respectively represent the maximum swimming distance of the tourists in a specific direction, the maximum distance of movement of the jellyfish perpendicular to the flow direction, and the transport distance of the water flow within a specific time. The gray square area represents the largest area for tourists to move in four directions.

When the observation time length is t_{sum}, the maximum value of wind speed and velocity are v_{flow}, the jellyfish floating speed is v_{jelly}, and the tourist swimming speed is v_{peo}, the following can be obtained:

$$L2 = t_{\text{sum}} v_{\text{flow}},$$
$$L1 = t_{\text{sum}} v_{\text{jelly}},$$
$$H = t_{\text{sum}} v_{\text{peo}}.$$

The dimensions of matrices *hp* and *fp* are equal to the dimensions of grid space. Next, we introduce the two possible states for each cell.

(1) *State 0*: $fp^t(i, j) = 0$ *or* $hp^t(i, j) = 0$.
This state indicates that there are no jellyfish or tourists in the corresponding network in line i and column j at time t.

(2) *State 1*: $fp^t(i, j) = 1$ *or* $hp^t(i, j) = 1$.
This state indicates that there are jellyfish or tourists in the network in line i and column j at time t.

Neighbor space N is also called local rule. We define the neighbor radius in the model as 1 m, which captures jellyfish movement in an area that is approximately 1 m from tourists. If a tourist is within this 1 m radius, they will be considered "stung".

d is defined as:

$$\mathrm{d} = 1/\mathrm{diameter},$$
$$\mathrm{d} = \mathrm{int(d)},$$

where the diameter is the average radius of jellyfish.

The model is used to search the sparse matrix $hp^t(i, j)$ within each time step to find all grid positions corresponding to $hp^t(i, j) = 1$. Then, the model is used to search from the i–d line to $i + d$ line and from j–d column to $j+d$ column of the sparse matrix *fp* to determine whether $hp^t(i, j) = 1$ exists in this cycle. If so, the tourists are stung by jellyfish within this time step and the experiment is immediately terminated, and the next set of experiments is started. Otherwise, the model simulation continues within the next time step.

The state transition function f is also called a renewal rule. Before each renewal, the model will generate a random number within [0, 1]. The state space of the tourists is renewed as follows:

(1) $hp^{t+1}(i, j-1) = hp^t(i, j)$.
If the random number belongs to $[0, 0.25)$, $hp^t(i, j) = 0$,

(2) $hp^{t+1}(i, j+1) = hp^t(i, j)$.
If the random number belongs to $[0.25, 0.5)$,
$hp^t(i, j) = 0$,

(3) $hp^{t+1}(i-1, j) = hp^t(i, j)$.
If the random number belongs to $[0.5, 0.75)$,
$hp^t(i, j) = 0$,

(4) $hp^{t+1}(i+1,j) = hp^t(i,j)$.
If the random number belongs to [0.75, 1],
$hp^t(i,j) = 0$.

The direction of the jellyfish state space that is perpendicular to the flow direction is renewed as follows:

(1) $fp^{t+1}(i+1,j) = fp^t(i,j)$.
If the random number belongs to [0, 0.5],

$$fp^t(i,j) = 0,$$

(2) $fp^{t+1}(i-1,j) = fp^t(i,j)$.

If the random number belongs to [0.5, 1],

$$fp^t(i,j) = 0.$$

The direction of the jellyfish state space that is parallel to the flow direction is renewed as follows:

$$fp^{t+1}(i,j-1) = fp^t(i,j),$$
$$fp^t(i,j) = 0.$$

Before each renewal of the sparse matrix *fp*, the model will judge whether the tourists are stung or not according to local rules. According to the simulation above, the being-hurt frequency of tourists within n times in the bath can be calculated. (Without loss of generality, it can also be used as the probability of being stung when n tourists swim at the same time.) Since the state space of the visitor is only composed of 0 and 1, the event of each tourist playing in the bath can be regarded as a Bernoulli trial. According to Bernoulli's law of large numbers (Tome *et al.*, 2015), the probability can be approximated with frequency when n is large enough. Therefore, in this model, n is defined as 1,000, and the frequency of jellyfish injury is defined as the probability of jellyfish injury. In order to reduce the effect of random errors, each experiment is repeated five times.

(2) *Risk analysis of jellyfish injury*
With the jellyfish injury model, we can not only calculate the probability of jellyfish injury but also analyze the potential impacts of each parameter on the model to find the key influencing factors, thus,

to provide a theoretical basis for subsequent model correction and parameter optimization. Significant parameters have greater impacts on the simulation results of the model and are also the determinants of the risk of jellyfish injury. They should be optimized as the main parameters for parameter optimization. Meanwhile, insignificant parameters can be set to initial values without adjustment. Relatively significant parameters will be adjusted according to the specific conditions of the studied area. Parameters of the model include flow velocity (FV), jellyfish abundance (JA), the average diameter of jellyfish (DJ), simulation time (OT), tourist swimming speed (TV), and jellyfish swimming speed (JV). The initial values of these parameters are set to 0.7 m/s, 10 ind/ha, 0.035 m, 3 min, 0.1 m/s, and 0.075 m/s, respectively. These initial values will vary with the studied area. In order to explore the interaction between the parameter and the probability of jellyfish injury, we change the value of one parameter at a time to calculate the probability of jellyfish injury. The used significance analysis method is the analysis of variance (ANOVA).

In this chapter, through the adjustment of parameters, one by one, to calculate the probability of jellyfish injury, the main influencing factors of jellyfish injury probability have been identified, and a relationship between the jellyfish injury probability and its main influencing factors has been obtained quantitatively.

(3) *Early warning mechanism for jellyfish injury*

The main goal here is to establish an early warning mechanism to reduce casualties. According to the jellyfish injury probability calculated by the model, managers of the bath area can set different warning levels to estimate the jellyfish injury probability in advance. When there are strong winds and waves or the number of jellyfishes increases dramatically, the corresponding management measures can be implemented.

As shown in Fig. 9.2, you can enter various parameters in the GUI interface to run the program or generate animations. Each parameter shown in Fig. 9.2 is the default value. The parameters include the tourist swimming speed, jellyfish abundance, jellyfish floating speed, total observation time, maximum value of the velocity or wind speed, and average diameter of jellyfish. According to the input parameters, the model follows the process described above to randomly generate

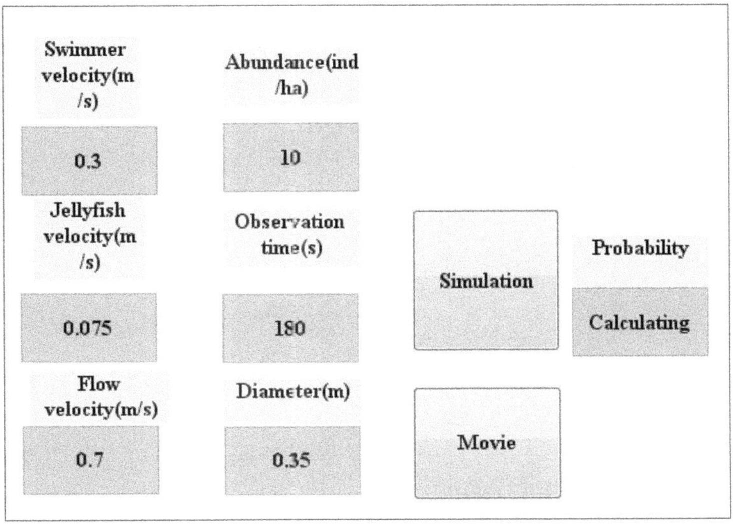

Fig. 9.2. GUI interface of the model.

a computing space containing jellyfish and tourists and to calculate the risk of jellyfish injury. The calculated injury probability is 0.308, as shown in Fig. 9.2.

According to the calculated probability of jellyfish injury, three early warning levels of jellyfish injury can be set: warning ($5\% >$ probability $\geq 3\%$), danger ($10\% >$ probability $\geq 5\%$) and closing the bath (probability $\geq 10\%$). Different baths can also set different warning levels according to their specific conditions.

9.2.2 *Risk assessment of jellyfish injury and analysis on early warning results*

The ANOVA results are shown in Table 9.1 and Fig. 9.3, in which p values of FV, DJ, JA, and OT are all 0.000 and very significant, while SV (p-value $= 0.366 > 0.05$) and JV (p-value $= 0.299 > 0.05$) do not affect the probability of jellyfish injury significantly; these two parameters can be set to the initial values of the model without adjustment. The significance of parameters can be used to analyze and evaluate the model and explain the correlation between the parameters and the probability of jellyfish injury. According to Fig. 9.3, FV, DJ,

Table 9.1. One-way ANOVA results of significance analysis through SPSS.

Parameter	FV/(m/s)	DJ/m	JA/(ind/ha)	OT/min	SV/(m/s)	JV/(m/s)
Values used in ANOVA analysis	0.3, 0.6, 0.9, 1.2, 1.5	0.2, 0.35, 0.5, 0.65, 0.8, 0.95	2, 5, 10, 15, 20, 25, 30, 35, 40, 45	5, 8, 10, 13, 15, 18, 20	0.1, 0.2, 0.3, 0.4, 0.5, 0.6	0.01, 0.02, 0.03, 0.04, 0.05, 0.06, 0.07
F-test	170.230	514.162	192.208	358.777	1.141	1.278
p-value	0.000	0.000	0.000	0.000	0.366	0.299
Significance	significant	significant	significant	significant	insignificant	insignificant

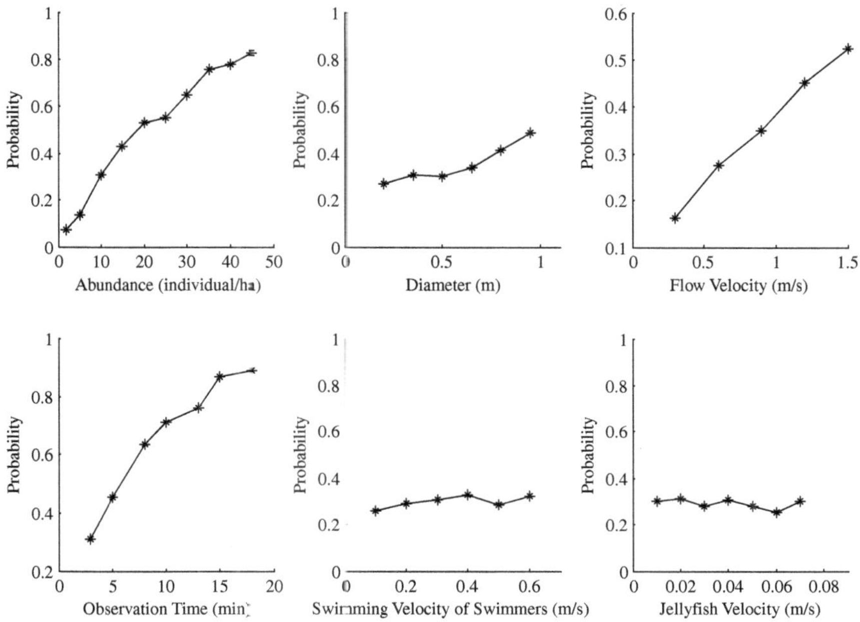

Fig. 9.3. Dot-line chart of various parameters and probability of jellyfish injury.

JA, and OT are positively correlated to the probability of jellyfish injury. When the abundance of jellyfish is increased from 2 ind/ha to 45 ind/ha, the probability of jellyfish injury is increased significantly from 0.74% to 82.6%; when the velocity is increased from 0.3 m/s to 1.5 m/s, which belongs to strong flow, the probability of jellyfish injury is increased from 16.4% to 52.4%. Therefore, jellyfish abundance, observation time, and water flow speed (or wind speed) all have significant impacts on the probability of jellyfish injury. In any follow-up jellyfish early warning work, we can focus on observing the water flow speed and jellyfish abundance near the shore. According to the observation, we can develop a reasonable early warning plan.

Since the traditional CA model is not completely suitable for simulating the probability of jellyfish injury, we have improved the CA model in this chapter to simulate the change process of two targets at the same time. Compared with methods such as the hydrodynamic model and Bayesian network, the CA model has the following advantages: (1) After the sparse matrix is introduced, the CA model can simulate the process of spatiotemporal change with a small amount of calculation, and (2) it can also simulate the interaction between different targets (van Vliet *et al.*, 2012). Because jellyfish bloom already has a huge negative impact on fisheries and tourism, according to the simulated probability of jellyfish injury, managers of bath areas can develop corresponding warning plans to reduce the number of casualties.

It should be noted that the jellyfish outbreak process is an important type of ecological dynamics process, but this chapter does not include the simulation of jellyfish growth, which is acceptable for the study on the short-term process of swimming. Regarding the long-term jellyfish growth and its diffusion mechanism on a sea area scale, special research is needed on the basis of accurately describing its key biological and dynamic processes.

9.3 Summary

Based on the improved CA model to observe multiple targets and introduced sparse matrix method to improve the model's computing efficiency and statistical theory, a jellyfish early warning model has been constructed. The risk analysis has been done with the significance analysis method. The results show that the probability of jellyfish injury is mainly affected by FV, DJ, JA, and OT, significantly, while SV and JV have relatively small influence. As the above variables are increased, the probability of jellyfish injury will increase.

This model provides a reference for the development of a jellyfish early warning plan and is helpful for the risk management of coastal baths. However, the model established in this chapter still needs to be analyzed based on the local hydrological characteristics and jellyfish characteristics, and the model needs to be verified according to more measured samples. In addition, in this model, the near-shore velocity (including wind speed) is set as a constant value to simplify the model calculation, which has certain practical significance in the continuous strong wind and high wave weather, but is difficult to accurately simulate in areas with large changes in water flow and wind speed. Three-dimensional simulation of this model has not been performed, which leads to opportunities and challenges for our subsequent studies on jellyfish early warning.

References

Alexandridis A, Vakalis D, Siettos C I, *et al.* 2008. A cellular automata model for forest fire spread prediction: The case of the wildfire that swept through Spetses Island in 1990. *Applied Mathematics and Computation.* **204**(1): 191–201. DOI: 10.1016/j.amc.2008.06.046.

Barona P C, Mena C. 2014. Using remote sensing and a cellular automata-Markov chains-GEOMOD model for the quantification of the future spread of an invasive plant: A case study of psidium guajava in Isabela Island, Galapagos. *International Journal of Geoinformatics.* **10**(3): 23–30.

Bricker S B, Ferreira J G, Simas T. 2003. An integrated methodology for assessment of estuarine trophic status. *Ecological Modelling.* **169**: 39–60.

Chen C S, Ji R B, Zheng L Y, *et al.* 1999. Influence of physical processes on the ecosystem in the Jinozho Bay: A coupled physical and biological model experiment. *Journal of Geophysical Research.* **104**: 29925–29949.

Cossarini G, Solidoro C. 2008. Global sensitivity analysis of a trophodynamic model of the Gulf of Trieste. *Ecological Modelling.* **212**: 16–27.

Ding D W, Shi H H, Zhang X L, *et al.* 2009. *Mechanism of Water Quality Change in Coastal Waters and Its Eco-Environmental Effects.* Ocean Publishing House, Beijing.

Dong Z, Liu D, Keesing J K. 2010. Jellyfish blooms in China-dominant species, causes and consequences. *Marine Pollution Bulletin.* **60**(7): 954–963. DOI: 10.1016/j.marpolbul.2010.04.022.

Doucet A, Pitt M K, Deligiannidis G, *et al.* 2014. Efficient implementation of Markov chain Monte Carlo when using unbiased likelihood estimator. *Biometrika.* **102**(2): 295–313.

Franks P J, Chen S C. 1996. Plankton production in tidal fronts: An model of Georges Rank in summer. *Journal of Marine Research* **54**: 631–651.

Gao M, Shi H H, Li Z Z. 2009. Chaos in a seasonally and periodically forced phytoplankton–zooplankton system. *Nonlinear Analysis: Real World Applications.* **10**: 1643–1650.

Ghermandi A, Galil B, Gowdy J, *et al.* 2015. Jellyfish outbreak impacts fare estimates from a socioeconomic pilot survey in Israel. *Ecosystem Services.* **11**: 140–147.

Graneli E, Weberg M, Salomon P S. 2008. Harmful algal blooms of allelopathic microalgal species: The role of eutrophication. *Harmful Algae.* **8**: 94–102.

Greve W. 1994. The 1989 German bight invasion of muggiaea atlantica. *ICES Journal of Marine Science: Journal du Conseil.* **51**(4): 355–358. DOI: 10.1006/jmsc.1994.1037.

Halpern B S, Walbridge S, Selkoe K A, *et al.* 2008. A global map of human impact on marine ecosystems. *Science.* **319**: 948–952.

Hou G X, Song L R , Liu J T, *et al.* 2004. Modeling of cyanobacterial blooms in hypereutrophic Lake Dianchi, China. *Journal of Freshwater Ecology.* **19**(4): 623–629.

Li H, Hou G, Feng D, *et al.* 2007. Prediction and elucidation of the population dynamics of *Microcystis* spp. in Lake Dianchi (China) by means of of artificial neural networks. *Ecological Informatics.* **2**: 184–192.

Li K, Gao Z, Ning B. 2005. Cellular automaton model for railway traffic. *Journal of Computational Physics.* **209**(1): 179–192. DOI: 10.1016/j.jcp.2005.03.016.

Lippmann J M, Fenner P J, Winkel K, *et al.* 2011. Fatal and severe box jellyfish stings, including Irukandji stings, in Malaysia, 2000–2010. *Journal of Travel Medicine.* **18**(4): 275–281.

Liu X, Ma L, Li X, *et al.* 2014. Simulating urban growth by integrating landscape expansion index (LEI) and cellular automata. *International Journal of Geographical Information Science.* **28**(1): 148–163.

Lucas C H, Gelcich S, Uye S I. 2014. *Living with Jellyfish: Management and Adaptation Strategies.* Netherlands: Springer.

Maier H R, Dandy G C. 2001. Neural network based modelling of environmental variables: A systematic approach. *Mathematical and Computer Modelling.* **33**: 669–682.

Maretic Z, Russell F E, Ladavac J. 1979. Epidemic of stings by the jellyfish Pelagia noctiluca in the Adriatic. *Natural Toxins: Proceedings of the 6th International Symposium on Animals, Plants and Microbial Toxins*, Uppsala, 77–82. DOI: 10.1016/B978-0-08-024952-0.50013-2.

Matsinos YG, Troundis AY. 2002. Modeling competition, dispersal and effects of disturbance in the dyromics of a grossland communing using a cellular automaton model. *Ecological Modelling.* **149**(1): 71–83.

Molinero J C, Buecher E, Lucic D, *et al.* 2009. Climate and Mediterranean jellyfish: Assessing the effect of temperature regimes on jellyfish outbreak dynamics. *Annales: Series historia naturalis.* **19**(2): 11–18.

Older J D. 2000. An artificial neural network approach for studying phytoplankton succession. *Hydrobiologia.* **436**(1): 131–143.

Phan K, Xie A, Kumar N, *et al.* 2014. Comparing energy for surgical ablation of atrial fibrillation: A Bayesian network meta-analysis of randomized, controlled trials. *European Journal of Cardio-Thoracic Surgery*, 1–11.

Purcell J E, Uye SI, Lo W T. 2007. Anthropogenic causes of jellyfish blooms and their direct consequences for humans: A review. *Marine Ecology Progress Series.* **350**: 153–74. DOI: 10.3354/meps07093.

Quiñones J, Monroy A, Acha E M, *et al.* 2013. Jellyfish bycatch diminishes profit in an anchovy fishery off Peru. *Fisheries Research.* **139**: 47–50.

Radwan F F, Burnett J W, Bloom D A, *et al.* 2001. A comparison of the toxicological characteristics of two Cassiopea and Aurelia species. *Toxicon.* **39**: 245–257. DOI: 10.1016/S0041-0101(00)00121-5.

Recknagel F. 1997. ANNA — Artificial neural network model for predicting species abundance and succession of blue-green algae. *Hydrobiologia.* **349**: 47–57.

Recknagel F, Kim B, Takamura N, *et al.* 2006. Unravelling and forecasting algal population dynamics in two lakes different in morphometry and eutrophication by neural and evolutionary computation. *Ecological Informatics.* **2**: 133–151.

Tome T, Oliveira M J. 2015. *Stochastic Dynamics and Irreversibility.* Springer.

van Vliet J , Hurkens J, White R, *et al.* 2012. An activity-based cellular automaton model to simulate land-dynamics. *Environment and Planning — Part B.* **39**(2): 198.

Wang H L, Feng J F 2006. *Ecological Dynamics and Prediction of Red Tide.* Tianjin University Press, Tianjin.

Wainer G A. 2014. Cellular Modeling with Cell-DEVS: A discrete-event Cellular Automata formalism. *Cellular Automata*, 6–15.

Wolfram S. 1983. Statistical mechanics of cellular automata. *Reviews of Modern Physics.* **55**(3): 601. DOI: 10.1103/RevModPhys.55.601.

Yabunaka K, Hosomi M., Murakami A. 1997. Novel application of a back-propagation artificial neural network model formulated to predict algal bloom. *Water Science and Technology.* **36**(5): 89–97.

Zheng W, Shi H H, Song X K, *et al.* 2012. Simulation of phytoplankton biomass in Quanzhou Bay using a back propagation network model and sensitivity analysis for environmental variables. *Chinese Journal of Oceanology and Limnology.* **30**(5): 843–851.

Chapter 10

Marine Ecosystem Dynamics Models Based on Statistical Dynamics Methods and Their Application

10.1 Research Background

There are mainly two traditional methods to establish ecological dynamics models: one is based on some basic physical laws (such as conservation of matter), biological laws (such as the law of least restriction), biological processes (such as photosynthesis, respiration, and biological pumps) and biological relationships (such as feeding, competition), or approximate methods to establish dynamics models (Franks *et al.*, 1996; Chen *et al.*, 2003; Gao *et al.*, 2009); another is based on statistics and observations to construct a statistics model (Wang *et al.*, 2006). Both traditional dynamics modeling and statistics modeling inevitably have researchers' subjective cognitions, such as approximations or artificial assumptions. Meanwhile, the regional characteristics of marine ecological dynamics models based on these methods are not obvious. Accuracy is often unsatisfactory in specific applications. In recent years, with the rapid development of computer technology, computer-based modeling methods, such as cellular automata and neural network methods, have also become popular. The former needs to construct the basic rules of cellular diffusion, which depends on an accurate understanding of the diffusion process

(Matsinos and Troumbis, 2002; Ding *et al.*, 2009). The neural network method uses an adaptive "black box" mode. With a sufficient sample size, high simulation accuracy can often be obtained, but it lacks explicit logical expressions of state variables and the dynamic process is not clear enough (Olden, 2000; Li *et al.*, 2007).

Therefore, when there are few observations, learning how to establish a marine ecosystem change model with local characteristics has become an important research topic. In climate prediction research, Chou (1986) proposed a dynamic-statistic inversion modeling theory. He believed that if the problem of numerical weather forecasting was set as an initial value problem, using data for a single time node was one of its relatively fundamental defects. Therefore, they used previous historical data and established inverse problem methods in numerical weather forecasting, and organically combined weather methods, statistical methods, and dynamic methods (Chou, 2007). Huang *et al.* (1991) used this theory to perform inversion experiments in the Lorenz system, and the results were consistent with the Lorenz equation. Lin *et al.* (1995) used the time series data of monthly average temperature, air pressure, and rainfall to reverse the dynamic equation of local climate in Tianjin. The successful practice of the above studies in meteorology has provided good inspiration for the inversion modeling of marine ecosystem change. In recent years, inverse problem methods have also been widely used in the field of aquatic ecosystem food web structure and dynamics (Eldridge and Jackson, 1993; Lyche *et al.*, 1996; Niquil *et al.*, 1998; Vezina and Savenkoff, 1999; Vezina and Pahlow, 2003; Kones *et al.*, 2006). However, there are few reports on the use of statistical dynamics methods in marine ecosystem modeling (Shi *et al.*, 2011).

10.2 Models Based on Statistical Dynamics

Marine ecosystems are open, dissipative, and nonlinear. Dynamic equations of complex ecological processes of each state variable can be determined with inverse modeling methods on the basis of historical observations (Shi, 2012).

If the existing observation data is regarded as a series of special solutions describing (or approximately describing) the dynamic model of the actual nonlinear dynamic system, and as long as the data can be used to solve the problem that is opposite to the numerical solution, the nonlinear dynamics model that can approximately describe the system can be reversed (Shi *et al.*, 2011).

Assuming the dynamic equation of the change of nonlinear system state variable x_i with time is:

$$\frac{dx_i}{dt} = f_i(x_1, x_2, \ldots, x_n), \quad i = 1, 2, \ldots, n \tag{10.1}$$

where n is the number of state variables.

According to the Weierstrass first theorem on R^n (Fei, 2006).

Let $f(t)$ be a continuous function defined on the compact set $F \subset R^n$, we can get $V\varepsilon > 0$, and there is a polynomial $p(\vec{t})$ of $\vec{t}$. The following can be obtained:

$$|f(\vec{t}) - p(\vec{t})| < \varepsilon. \tag{10.2}$$

It is always right with F. Here, $\vec{t}$ is an n-dimensional row vector: $t = (t_1, t_2, \ldots, t_n) \in R^n$.

Therefore, if $f_j\ (x_1, x_2, \ldots, x_n)$ in (10.1) is a nonlinear polynomial, it has universal significance as:

$$f_j(x_1, x_2, \ldots, x_n) = \sum_{i=1}^{k} a_k Q_k, \tag{10.3}$$

where a_k is the coefficient of the polynomial function, Q_k represents a term of the polynomial, and there are K term polynomials in total. Taking Eq. (10.2) into Eq. (10.1) and writing them in a different form, the following matrix is obtained:

$$D = Qa, \tag{10.4}$$

where:

$$a = \begin{bmatrix} a_1 \\ a_2 \\ \vdots \\ a_k \end{bmatrix}, \tag{10.5}$$

$$Q = \begin{bmatrix} Q_{11} & Q_{12} & \cdots & Q_{1k} \\ Q_{21} & Q_{22} & \cdots & Q_{2k} \\ \vdots & \vdots & \vdots & \vdots \\ Q_{N1} & Q_{N2} & \cdots & Q_{Nk} \end{bmatrix}, \tag{10.6}$$

$$D = \begin{bmatrix} d_1 \\ d_2 \\ \vdots \\ d_k \end{bmatrix} = \begin{bmatrix} \frac{x_j[(m+1)\Delta t]-x_j(\Delta t)}{m\Delta t} \\ \frac{x_j[(m+2)\Delta t]-x_j(2\Delta t)}{m\Delta t} \\ \vdots \\ \frac{x_j(N\Delta t)-x_j[(N-m)\Delta t]}{m\Delta t} \end{bmatrix}. \tag{10.7}$$

Here, Δt is the time interval of the time series, N is the number of sequence elements, and m is the number of moving average steps.

If matrix Q has an inverse matrix, Eq. (10.4) gives us:

$$a = Q^{-1}D. \tag{10.8}$$

Even in a more general case, if the matrix Q is a full column rank, the least square solution of a can be found as:

$$a = (Q'Q)^{-1}Q'D. \tag{10.9}$$

This formula is more universally applicable.

As long as relevant data is taken into Eqs. (10.8) or (10.9), the coefficient matrix a can be obtained theoretically to complete the inversion modeling. Taking a_k into Eqs. (10.1) and (10.2), we can get:

$$\frac{dx_j}{dt} = \sum_{k=1}^{k} Q_k a_k. \tag{10.10}$$

Thereby, the nonlinear dynamic mode of the description system is obtained.

Compared with the traditional nonlinear regression method, Eq. (10.10) has the following advantages.

(1) It has kinetic properties.
(2) In theory, the polynomial of Eq. (10.10) can contain any form with sufficiently high power (but it also depends on the richness

of historical data). Thus, it reduces human subjective factors compared with the regression method according to the artificially assumed regression function.

In specific modeling, in order to simplify the equation, some terms with very small coefficients in Eq. (10.10) can be eliminated because they contribute very little to the equation. We can use the relative contribution rate (W_k), given below, as the basis for eliminating those terms

$$W_k = \frac{|Q_k a_k|}{\sum_k |Q_k a_k|}. \tag{10.11}$$

Any item whose W_k value is less than a specified value W^* can be eliminated.

10.3 Case Study on the Change of a Typical Gulf Ecosystem

As shown in Eq. (10.10), the equation contains 20 to 30 unknown coefficients, but it is difficult to obtain positioning observations for a continuous 20 to 30 time series. Therefore, the construction of time series is critical.

In order to predict the change trend of the Jiaozhou Bay ecosystem in a typical bay, we assume in this study that the change of the ecosystem at each site in the studied area follows a uniform law. Because of this assumption, the difference between the continuous observation time series of all sites constitutes a new sample, which can meet the needs of equation inversion. The advantage of the difference is that the state quantity is replaced by the change quantity, which eliminates the influence of the time state and only focuses on the time interval. Thus, the observation data of different stations can be integrated into a new sequence. This is the key to the use of all references.

Generally, the sea surface temperature (SST) is the most basic observation factor, which is usually observed in all surveys. The following formula can also be used when the water temperature is not observed:

$$T = \overline{T} - \delta \times \cos\left[\frac{2\pi(t-90)}{365}\right]. \tag{10.12}$$

According to the nature of the cosine function, $(\overline{T}+\delta)$ and $(\overline{T}-\delta)$ are the maximum and minimum values of SST in the studied area, respectively. t represents the specific date with January 1 as the first day.

The actual observation data is used as SST in this study.

The acquisition method of light intensity data is as follows. After all daily observation data in 1995 from Jiaozhou Bay is vectorized (Wang, 2007), the annual variation of light intensity in Jiaozhou Bay is fitted with a fourth-degree polynomial, and the monthly average light intensity is represented by the mid-month data. The fitted annual variation of light intensity in Jiaozhou Bay in this study is shown in Fig. 10.1.

As an example, we set up a dynamic system of nutrients (N), phytoplankton (P), zooplankton (Z), debris (D), temperature (T), and light intensity (I) in Jiaozhou Bay, Qingdao.

Temperature (T) is included in the dynamic system because temperature is an important ecological factor for marine life. It can affect the photosynthesis and respiratory metabolism of plankton.

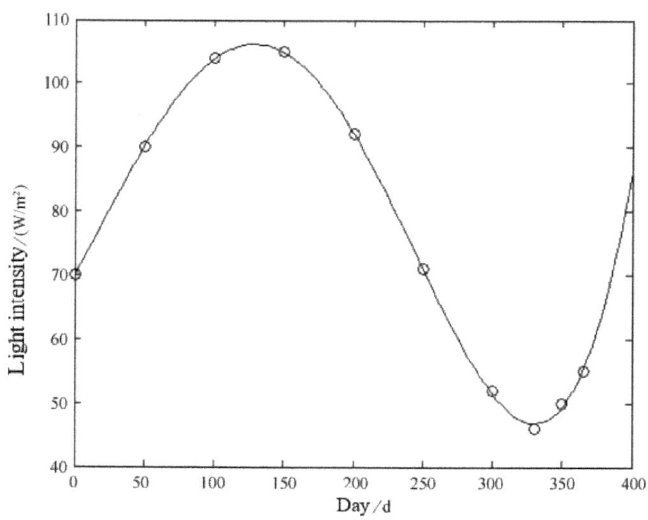

Fig. 10.1. Fitting effect of the fourth-degree polynomial on Jiaozhou Bay.

Seawater temperature not only directly affects the growth and distribution of organisms but also indirectly affects other ecological factors. For example, the concentration of dissolved oxygen in seawater can indirectly affect the growth of organisms. Temperature also directly affects the metabolism rate of biological organisms. When the temperature is increased with sufficient light, the metabolism rate is increased, and the consumption of nutrients and oxygen is also increased accordingly. The Q_{10} rule commonly used in marine ecosystem modeling quantitatively expresses the effect of temperature on phytoplankton growth and their respiration rates (Fransz *et al.*, 1985). Within the studied time scale, if the temperature change is very small, the effect of temperature can be ignored, otherwise, this effect should be considered. Because the annual temperature difference in the northern sea area of China is large, it is not appropriate to use the average value as the estimated temperature value.

Because the sunlight entering the seawater is an important energy source for the photosynthesis of phytoplankton — light intensity has a great impact on the growth and distribution of marine life, and sea surface heat flux also depends mainly on solar radiation — solar radiation intensity (I) has been included in the power system. Solar radiation directly adjusts the surface temperature of seawater, and then transfers heat to the deeper layers, even directly to the ocean floor, through the turbulent diffusion of seawater, which is true especially for shallow seas. Therefore, solar radiation plays an important role in marine ecosystems. This suggests that, if the empirical value or the average value is used as the estimated value of the light intensity (I), it will not be accurate, and the observation value in the studied sea area must be relied on.

After continuous observation data from nine stations in Jiaozhou Bay from January 2003 to January 2004 was treated with anomalization, non-dimensionalization, and smoothing, we can assume that the dynamic equation of the inversion system is:

$$\begin{aligned}\frac{dP}{dt} &= a_0 + a_1N + a_2P + a_3Z + a_4T + a_5I + a_6NP + a_7NZ \\ &\quad + a_8PZ + a_9PT + a_{10}NT + a_{11}ZT + a_{12}NI + a_{13}PI \\ &\quad + a_{14}ZI - a_{15}TI + a_{16}N^2 + a_{17}P^2 + a_{18}Z^2 \\ &\quad + a_{19}T^2 + a_{20}I^2 + a_{21}N^3 + a_{22}P^3 + a_{23}Z^3 + a_{24}T^3 + a_{25}I^3,\end{aligned} \tag{10.13}$$

$$\frac{dZ}{dt} = b_0 + b_1P + b_2Z + b_3T + b_4PZ + b_5PT + b_6ZT + b_7P^2 + b_8Z^2 + b_9T^2 + b_{10}P^3 + b_{11}Z^3 + b_{12}T^3, \tag{10.14}$$

$$\frac{dN}{dt} = c_0 + c_1N + c_2P + c_3Z + c_4T + c_5I + c_6NP + c_7NZ + c_8PZ + c_9PT + c_{10}NT + c_{11}ZT + c_{12}NI + c_{13}PI + c_{14}ZI + c_{15}TI + c_{16}N^2 + c_{17}P^2 + c_{18}Z^2 + c_{19}T^2 + c_{20}I^2c_{21}N^3 + c_{22}P^3 + c_{23}Z^3 + c_{24}T^3 + c_{24}I^3. \tag{10.15}$$

In order to reduce the impact of individual varied data and after taking the continuous impact of ecological factors on the ecosystem into account, we initialized the sample data with the moving average method. The step with the smallest absolute value of the average relative error value is selected as the optimal time step. According to the obtained data, the selected step unit in this study is a month. The relative errors are defined as follows:

$$s = 2 \times (OV - PV)/(OV + PV). \tag{10.16}$$

10.4 Results

With phytoplankton as an example, we can select different time steps to calculate the coefficients of every term on the right side of Eq. (10.13) according to the modeling program, and then calculate the average relative error, which is based on the comparison between the predicted and observed values of the phytoplankton (Table 10.1). In other states, their optimal time steps are determined according to the minimum average relative error principle. Obviously,

Table 10.1. Mean relative errors of phytoplankton dynamic equations.

Time step/Month	1	2	3	4	5	6	7	8	9
Average relative error (%)	871.1	−93.1	70.3	129.2	−43.0	28.8	−14.0	12.4	−7.4

"9" is the optimal time step for phytoplankton prediction because the growth of phytoplankton has a time-lag effect. In other words, both environmental factors and their own growth conditions within 1–9 months have impacts on their following development, rather than simple Markov performance (no post-effect).

Now, the coefficients on the right side of the phytoplankton change Eq. (10.13) can be calculated, and Eq. (10.13) can be expressed as:

$$\begin{aligned}\frac{dP}{dt} = &-187.5916 + 4.5038N - 1.3051P - 1.3789Z + 531.2373T \\ &+ 323.3452I - 2.5437NP - 4.7072NZ + 9.4569PZ \\ &- 46.3277PT - 39.1946NT - 0.3104ZT - 5.5932NI \\ &- 18.2645PI + 5.2641ZI + 13.6487TI + 58.5756N^2 \\ &+ 237.4781P^2 - 3.0751Z^2 - 903.2886T^2 - 429.0772I^2 \\ &- 46.2206N^3 - 459.4577P^3 - 1.4515Z^3 \\ &+ 524.4307T^3 + 186.9287I^3. \end{aligned} \quad (10.17)$$

In order to test the fitting of Eq. (10.17), residual analysis was done, and the resulting phytoplankton residual map is shown in Fig. 10.2. It finds that the difference between the predicted and observed values of phytoplankton is randomly distributed around 0, which indicates that the fitting of Eq. (10.17) on phytoplankton growth is good.

In order to further analyze the dynamic mechanism of phytoplankton changes, the weights of all terms on the right side of Eq. (10.13) have been calculated according to Eq. (10.11) (see Table 10.2).

Similarly, the dynamics of zooplankton and nutrient change in Jiaozhou Bay can be established as needed.

10.5 Discussion

The marine ecological dynamics model plays an important role in revealing the internal mechanism of the marine ecosystem and predicting future changes. It has become an important tool for marine ecosystem health assessment (Ding *et al.*, 2009). Most traditional ecological dynamics models are based on basic physical and biological laws and have clear dynamic and ecological significance (Franks

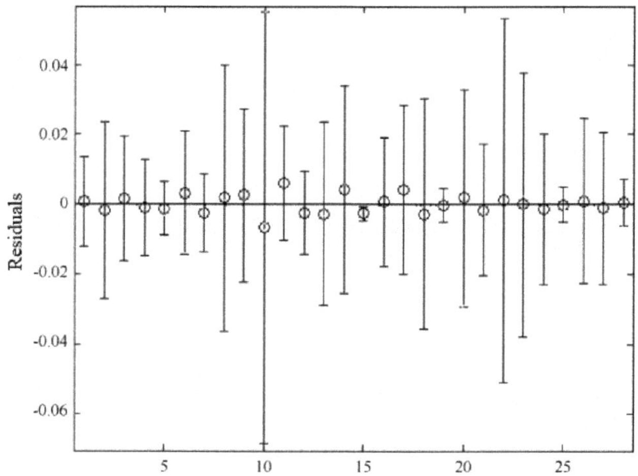

Fig. 10.2. Plot of predicted residuals of phytoplankton.

Table 10.2. Weights of items on the right side of Eq. (10.13).

Coefficients	Predicted value	Weights	Coefficients	Predicted value	Weights
a_0	−187.952	0.1191	a_{13}	−18.2645	0.0016
a_1	4.5038	0.0011	a_{14}	5.2641	0.009
a_2	−1.3051	0.0001	a_{15}	13.6487	0.004
a_3	−1.3789	0.0003	a_{16}	58.5756	0.0059
a_4	531.2373	0.1972	a_{17}	237.4781	0.0047
a_5	323.3452	0.1612	a_{18}	−3.0751	0.0002
a_6	−2.5437	0.0001	a_{19}	−903.289	0.1962
a_7	−4.7072	0.0004	a_{20}	−429.077	0.1685
a_8	9.4569	0.0003	a_{21}	−46.2206	0.0019
a_9	−46.3277	0.003	a_{22}	−459.458	0.0017
a_{10}	−39.1946	0.0058	a_{23}	−1.4515	4×10^{-5}
a_{11}	−0.3104	4×10^{-5}	a_{24}	524.4307	0.0667
a_{12}	−5.5932	0.0011	a_{25}	186.9287	0.058

and Chen, 1996; Gao *et al.*, 2009). However, due to the complexity of environmental conditions and ecological processes, these traditional models have shortcomings in the study of coastal waters, mainly due to the difficulty in obtaining model parameters and the low accuracy of models.

In this section, a new marine ecological statistical dynamics model that is based on inverse problem research ideas, integrated statistics, and differential dynamics methods, has been established. This model can be used to reveal the change of marine ecosystem structure. The basic ideas and methods for inverse problem modeling of marine ecosystems were introduced first in this section. According to the localized characteristics of the marine ecosystem in Jiaozhou Bay, based on the continuous monthly observation data from January 2003 to January 2004, the P-Z-N model has been established with sea surface temperature (SST) and light intensity as two forcing conditions. The average prediction accuracy in the case study model is over 90%. In fact, a simple model can be obtained by excluding terms with relatively small weight coefficients on the right side of the differential equation.

Compared with the traditional differential dynamic model, the statistical dynamic model proposed in this section avoids the difficulty of parameter acquisition and optimization (Chen, 2003; Gao *et al.*, 2009). At the same time, this model can make full use of existing observation data and has obvious dynamic characteristics. With a relatively mature comprehensive understanding of marine ecosystems, the model can be simplified during model construction. For example, items that are not related or have little correlation can be eliminated, which can reduce the dependence on model data. Therefore, this method provides a new way to model marine ecological dynamics.

In the modeling of the Jiaozhou Bay case, the equation difference method used in this section plays a key role in the effective utilization of observation data. With a small-scale studied area similar to Jiaozhou Bay, we can assume that the change of the ecosystems at each observation site obeys the same or similar rules, which is the prerequisite for the construction of data samples. However, with relatively large studied areas, proper processing must be done before modeling with this method (Shi *et al.*, 2011).

Acknowledgement

The reference (Shi *et al.*, 2011) was reproduced with permission from Springer Nature.

References

Chen C S. 2003. *Marine Ecosystem Dynamics and Models*. Beijing: Higher Education Press.

Chou F J. 1986. Why to combine dynamics with statistics? — And how to combine. *Plateau Meteorology*. **5**(4): 367–372.

Chou F J. 2007. An innovation road to numerical weather forecasting — From initial value problem to inverse problem. *Acata Meteorologyica Sinica*. **65**(5): 673–682.

Ding D W, Shi H H, Zhang X L, *et al.* 2009. *Water Quality Change Mechanism and Ecological Environment Effects in Coastal Waters*. Beijing: China Ocean Press.

Eldridge P, Jackson G. 1993. Benthic trophic dynamics in California coastal basin and continental slope communities analysis. *Marine Ecology Progress Series*. **99**: 115–135.

Fei L W. 2006. *Typical Problems and Methods in Mathematical Analysis*. Beijing: Higher Education Press.

Franks P J S, Chen C. 1996. Plankton production in tidal fronts: A model of Georges Bank in summer. *Journal of Marine Research*. **54**: 631–651.

Fransz H G, Verhagen J H G. 1985. Modelling research on the production cycle of phytoplankton in the southern Bight of North Sea in relation to river-borne nutrient loads. *Netherlands Journal of Sea Research*. **19**: 241–250.

Gao M, Shi H H, Li Z Z. 2009. Chaos in a seasonally and periodically forced phytoplankton–zooplankton system. *Nonlinear Analysis: Real World Applications*. **10**: 1643–1650.

Huang J P, Yi Y H. 1991. Inversion of nonlinear dynamic models with observation data. *Scientia Sinica*. (3): 331–336.

Kones J K, Soetaert K, Oevelen D, *et al.* 2006. Gaining insight into food webs reconstructed by the inverse method. *Journal of Marine Systems*. **60**: 153–166.

Lin Z S, Shi F B, Wang H. 1995. Inversion modeling and research on Tianjin local climate. *Acata Meteorologyica Sinica*. **53**(1): 115–121.

Li H, Hou G, Feng D, *et al.* 2007. Prediction and elucidation of the population dynamics of *Microcystis* spp. in Lake of artificial neural networks. *Ecological Informatics*. **2**: 184–192.

Lyche A, Andersen T, Christoffersen K, *et al.* 1996. Mesocosm tracer studies: Zooplankton the pelagic phosphorous cycle of a mesotrophic lake. *Limnology and Oceanography.* **41**: 460–474.

Matsinos Y G, Troumbis A Y. 2002. Modeling competition, dispersal and effects of disturbance in the dynamics of a grassland community using a cellular automaton model. *Ecological Modelling.* **149**: 71–83.

Niquil N, Jackson G, Legendre L, *et al.* 1998. Inverse model analysis of the planktonic food web of Takapoto Atoll (French Polynesia). *Marine Ecology Progress Series.* **165**: 17–29.

Olden J D. 2000. An artificial neural network approach for studying phytoplankton succession. *Hydrobiologia.* **436**: 131–143.

Shi H H. 2012. *Key Technologies and Applications of Coastal Zone Complex Ecosystem Evaluation, Simulation and Regulation.* Beijing: China Ocean Press.

Shi H H, Wang Z L, Fang G H. 2011. A statistical dynamics model of marine ecosystem and its application in Jiaozhou Bay. *Chinese Journal of Oceanology and Limnology.* **29**(4): 905–911.

Vezina A, Savenkoff C. 1999. Carbon and nitrogen flows in the surface layer of the NE Pacific. *Deep Sea Research Part II: Topical Studies in Oceanography.* **46**: 2909–2939.

Vezina A F, Pahlow M. 2003. Reconstruction of ecosystem flows using inverse methods: how well do they work? *Journal of Marine Systems.* **40–41**: 55–77.

Wang H L, Feng J F. 2006. *Red Tide Ecological Dynamics and Prediction.* Tianjin: Tianjin University Press.

Wang Z Y. 2007. *Modeling and Analysis of the Change of Plankton Ecosystem in Jiaozhou Bay for 40 Years.* Qingdao: Ocean University of China.

Chapter 11

Progresses on End-to-end Marine Ecosystem Dynamics Models

The end-to-end (E2E) model, also known as the whole-of-system model, emphasizes the characterization of a whole system, usually with all components included (Fulton, 2010). The end-to-end approach is adopted in marine ecosystem dynamics modeling to study the comprehensive impacts of climate change and marine fisheries on marine ecosystems, aiming to meet the needs of ecosystem fisheries management (Travers *et al.*, 2007). Marine fisheries and environmental factors affect different components and attributes of the food chain and food web in different ways. These direct effects indirectly affect the organisms via the food web. Marine fisheries show top-down effects, while environmental factors often exhibit bottom-up effects. Simple ecosystem dynamics models often only describe the components of the ecosystem that are of interest to the modeler using a single bottom-up or top-down control, and other components are considered as the driving variables, such as model inputs or closed terms (Travers *et al.*, 2007). However, these driving variables can often affect the dynamic changes of the model, yet they are usually determined by model correction rather than their ecological meanings due to oversimplification. To understand the direct and indirect effects of climate change and marine fisheries on the dynamics of marine ecosystems, the models that can cover key relationships between all components of the ecosystem from the bottom to the top of a food web are needed.

With the gradually deepened understanding of marine ecosystems and improvements of numerical computing capabilities, the end-to-end approach-based marine ecological dynamics model has become the practical model that can meet such demands. Travers *et al.* (2007) reviewed different types of related marine ecosystem dynamics models, and summarized the features of an end-to-end model: first, it characterizes the whole food web and related abiotic environments. Second, it characterizes the interactions of physical and biological processes at different scales. In addition, it couples the ecosystem components by a two-way interactive coupling approach. Last but not the least, it reveals the dynamic driving effects of climate change and human activities on multiple trophic levels. With the wide applications of adaptive management, end-to-end models have been used to study the responses and adaptation of marine ecosystems to climate changes and human activities (Plaganyi, 2007; Fulton, 2010; Steele *et al.*, 2013; Kaplan and Marshall, 2016). The development of end-to-end approaches in the modeling of marine ecosystem dynamics, and the concerns, resolutions, and pros and cons of different types of models have been reviewed in several reports (Travers *et al.*, 2007; Rose *et al.*, 2010; Moloney *et al.*, 2011; Steele *et al.*, 2013; Kaplan *et al.*, 2016).

11.1 Types of End-to-end Marine Ecosystem Dynamics Models

The simple ecosystem dynamics models can be expanded into the corresponding end-to-end models by replacing the driving variables with more detailed models. However, whether such expansion is effective depends on the structure and modeling purpose of the available models, and may also be limited by other factors, such as the discrete types, processes or differences in scale, and the feasibility of parameterization. The extended model may be invalid or lose its original advantages (Travers et al., 2007). Therefore, an end-to-end model can be realized by coupling the models that partially characterize the ecosystem or by constructing a model that can characterize the whole system.

11.1.1 *Coupling low and high levels tropic models*

(1) *One-way drive of a low trophic level model to high trophic level model*

Coupling the available models to form a hybrid model is the most direct way to construct an end-to-end marine ecosystem dynamics model. Such an approach usually mixes several models that focus on the details of different components together to characterize most of the components of a marine ecosystem and its key processes and thus is usually referred to as the minimum realistic approach. The construction of end-to-end models by this approach involves the coupling of the models of the low- and high trophic levels in the food chain (Travers *et al.*, 2007). For example, Hermann *et al.* (2001) established a model by coupling a three-dimensional hydrodynamic model, an NPZ model, and an individual model focusing on fish eggs and juveniles to evaluate the effects of climate change on the early life of fish in the western Gulf of Alaska in different spatial areas. This coupled model considers relatively high trophic level organisms, such as larval and juvenile fishes, and does not include the predation of plankton by adult fish populations. Therefore, it cannot be used to study the impacts of marine fisheries on ecosystems and is considered as the primary coupling of low and high trophic levels. In addition, the coupled model only considers the bottom-up effect of low trophic level organisms on high trophic level organisms and does not consider the feedback of juvenile fishes to the NPZ model. This one-way relationship is generally called a one-way drive, that is, the output of one model is used as the input of another model without a feedback mechanism. The first model can be regarded as an external drive of the second model. The so-called biological-physical coupling model is the "one-way driving" of the physical to biological model (Franks, 2002).

(2) *Two-way coupling of a low trophic level model with high trophic level model*

Because the biological processes of a marine ecosystem have negligible effects on the physical processes, it is feasible to connect the physical and biological models with a unidirectional drive. However, the control mechanism of a food web includes both bottom-up and top-down controls, e.g., resource control and predator control, as well as the wasp-waist control of zooplankton, which all contribute to the dynamic changes of the marine ecosystem. Therefore, the coupling of low and high trophic level models should be a two-way interaction based on the predator-prey relationship, instead of a one-way driving interaction, which is the true meaning of coupling. For example, Sourisseau (2002) simulated the annual dynamic variation of five types of copepods based on their size structure, taking the preying of copepods on phytoplankton and small zooplankton and its predation by juvenile fishes into consideration. He also pointed out that attention should be paid to the following points for studies on the effects of environmental factors on copepod populations by two-way interaction: first, the direct impacts of environmental factors on the growth and survival of copepods; second, the direct effects of environmental factors on the growth and survival of larvae and juveniles, and indirect effects on copepods via predation relationships; and third, the effects of environmental factors on the seasonal changes in biomasses of phytoplankton and small zooplankton, and the indirect effects on copepods via predation relationships. Similar to the coupled model constructed by Hermann *et al.* (2001), Sourisseau's (2002) model does not include an adult fish population, and therefore cannot be used to study the impacts of marine fisheries on an ecosystem. The OSMOSE model based on an individual-based multiple population dynamics model adopts the simple opportunistic predatory rules to simulate the trophic interactions (Shin and Cury, 2001; 2004). It can be converted into an individual-based end-to-end model if its carrying capacity parameters characterizing the low trophic level organisms are replaced with a model of low trophic level by the two-way coupling method (Travers *et al.*, 2007).

(3) *Models coupling of all components of a planktonic ecosystem with some high trophic level fish populations*

As demonstrated above, the preliminary coupling of the models of low and high trophic levels describes the interconnections of some components of the marine ecosystem. To characterize all components of the marine ecosystem and their two-way interactions, the end-to-end model can be directly constructed by coupling the biogeochemical models that mainly involve low trophic level organisms with population dynamics models of high trophic level organisms based on the strategy of expanding and coupling of available models.

The NEMURO. For Including Saury and Herring (NEMURO.FISH) model suitable for the north Pacific ecosystem is a commonly used representative model. The model is constructed by the two-way coupling of the biogeochemistry based NEMURO model with a focus on a low trophic level organisms and high trophic level fish population-bioenergetic model (Megrey *et al.*, 2002; 2007; Rose *et al.*, 2007). The fish bioenergetics model simulates the growth and spawning (as well as the entire life cycle) of the fish population with a population dynamics model that considers natural and predatory mortality. It can be applied to the herrings and sauries in the North Pacific. For example, Megrey *et al.* (2007), using the NEMURO.FISH model, simultaneously simulated the predation of zooplankton functional groups by herrings, and the nitrogen source input of particulate organic matter and ammonia nitrogen by herring excretion.

The spatial ecosystem and populations dynamics model (SEAPODYM) is another commonly used model. It can be used to study the biological-physical interactions of top predators and aquatic ecosystems, such as tuna in the Pacific Ocean (Lehodey *et al.*, 2003; Lehodey, 2005). The SEAPODYM also couples a mesotrophic box model focusing on feed species, in addition to the biogeochemical model of low-trophic organisms and the age-structure-based tuna population dynamics model. The biogeochemical model is also coupled with a hydrodynamic model to provide a spatially heterogeneous physical environment. However, SEAPODYM has high demands for the data on the dynamic changes and life cycle of tuna, and only focuses on one kind of top predator related to mesotrophic organisms. Therefore, it is not suitable for the characterization of complex coastal food webs (Travers *et al.*, 2007).

11.1.2 *Models comprehensively characterizing whole ecosystems*

(1) *Trophic dynamics based end-to-end models*

The components of a marine ecosystem range from primary producers, such as phytoplankton, to mesotrophic organisms and apex predators, such as feed species, as well as abiotic environmental factors. To achieve the complete description of a marine ecosystem, the end-to-end model not only needs to cover all the components of the ecosystem and their interactions, but also characterize the specific details of different components to a certain extent. Although it is very difficult to implement an end-to-end model, several models have been developed and widely used to characterize all trophic levels of the food web, such as the box-structure-based Ecopath with Ecosim (EwE) model. The EwE model builds a conserved network structure based on the energy flow among the various components of the ecosystem. It can be used to characterize different species and functional groups in aquatic and benthic systems, including debris, phytoplankton, as well as various feed species and top predators (Christensen and Walters, 1992). In the model, the Ecopath module characterizes the static food web, and the Ecosim module characterizes its time dynamic change. The subsequent Ecospace version was developed to describe the two-dimensional spatial distribution of each component (Walters *et al.*, 1999). The EwE model can be used to evaluate the impacts of different marine fisheries strategies on the ecosystem, but the impacts of environmental factors can only be determined indirectly with the seasonal changes in phytoplankton biomass (Christensen and Walters, 1992; Travers *et al.*, 2007). Therefore, the EwE model can be considered as a trophic dynamics-based end-to-end model.

(2) *Biogeochemistry based end-to-end models*
To directly determine the impacts of environmental factors on the food web, the biogeochemistry based end-to-end models have been developed. The European regional model (ERSEM) (Baretta *et al.*, 1995) and its enhanced version ERSEM II (Baretta-Bekker *et al.*, 1997), as well as the integrated generic bay ecosystem model (IGBEM) (Fulton *et al.*, 2004a) and its improved version bay model 2 (BM2) (Fulton *et al.*, 2004b), are the most representative ones. The ERSEM describes the biological and chemical processes of layered or mixed water bodies and the underlying systems. Its basic structure includes nutrients, bacteria, phytoplanktons, small and medium zooplanktons, and organic matters. Fish and fish predation can be further included in the model by setting predation rates or coupling population dynamics models (Baretta *et al.*, 1995; Baretta-Bekker *et al.*, 1997). The IGBEM combines the ERSEM II with a physical low-trophic level model to describe the trophic cycles and population dynamics, ranging from bacteria to large fish, at different spatial and

temporal scales. It can also be used to investigate the impacts of marine fisheries and eutrophication (Fulton *et al.*, 2004b). A disadvantage of the biogeochemistry based model is that only fishes are considered a functional group and that actual fishery management decisions are generally made based on species (Travers *et al.*, 2007). In addition, the NEMURO.FISH model can also be considered as a biogeochemistry based end-to-end model from the model coupling structure's perspective (Fulton, 2010).

(3) *End-to-end models coupling human activities*
The currently available coupled human-natural marine ecosystem dynamics models cover human components, and thus the interactive interferences and cumulative effects of various human activities on marine ecosystems are gradually included in the corresponding end-to-end models (Fulton, 2010). However, most models tend to treat human interferences as external drivers. With the development of complex adaptive systems and their management, end-to-end models gradually consider human activities as a model component for dynamic investigations to support the ecosystem-based management, by simulating each component of the adaptive management cycles and combining the evaluation and management in the framework of management strategy evaluation (MSE) or operational management procedures (OMP) (Fulton, 2010; Kaplan and Marshall, 2016). The Atlantis ecosystem model and the individual-based InVitro model are representatives of this type of end-to-end model. The Atlantis model was developed from BM2. It includes various components of physical and biogeochemical systems, as well as high trophic level organisms, and can consider human dynamics to a certain extent (Fulton *et al.*, 2005; 2007). The model can be used to study the ecological response and cumulative impacts of human activities under different assumptions, so as to evaluate the effectiveness of ecological indicators and other methods in ecosystem-based fisheries management (Fulton *et al.*, 2005; 2007). The model framework includes multiple processes, such as biophysics, industry, monitoring, evaluation, management, and socioeconomics, and thus can be used to examine the various stages of the adaptive management cycle (Fulton, 2010). The Atlantis model has also been used to investigate the impacts of human activities, such as marine fisheries, land-use change, nutrition pollution, and wind farms. Plaganyi (2007) analyzed the advantages

and disadvantages of different types of ecosystem models used in fisheries management and believed that the Atlantis model was the most effective decision support tool for ecosystem-based management. Its main advantages can be summarized as:

(1) covers the whole trophic spectra of the food web,
(2) the characterization of physiological processes is relatively simple and includes the detailed sensitivity analysis to determine the importance of included multiple processes, as well as processes not included in other models, such as mixed nutrition,
(3) simulates vertebrates, such as fish, based on the age structure,
(4) can characterize low-trophic levels biota better as compared with other most whole ecosystem models. It can also characterize high-trophic levels biota better than other biogeochemical models,
(5) decomposes the ecosystem spatially,
(6) can include multiple vertical layers,
(7) its structure allows for the replacement of different sub-models,
(8) considers the effects of different factors, such as nutrient source inputs,
(9) provides a detailed coupling mechanism of physical and biological processes, and
(10) gives a variety of optional expressions for the process description.

The code for the Atlantis model was developed by researchers at the Commonwealth Scientific and Industrial Research Organization (CSIRO), and it was subsequently applied to the studies of the coastal waters in Australia and North American (Fulton, 2010). In 2014, the Pacific Fishery Management Council's Scientific and Statistical Committee specifically reviewed the application of the Atlantis model to California Current (Kaplan *et al.*, 2016).

The agent-based model is a newer form of the end-to-end approach-based model. It is an extension of the individual-based model (IBM) with a focus on different agents. These agents are implemented by using different types of models, such as the classic individual-based model and the population models combined with difference or differential equations (Fulton, 2010). Therefore, the most effective expressions of the different components of the system are linked to form an effective model that can characterize the whole system, and the human activities and their socioeconomic drivers are coupled into the system as a subject. The InVitro model is a representative agent-based end-to-end model (Gray *et al.*, 2006). The model considers the dynamic characteristics of human activities and related assessment and management tools. It is often used to solve ecosystem-based management problems in multiple utilization modes (Fulton, 2010). The human activities involved in the model can be commercial fishing, recreational fisheries, tourism, shipping, oil and gas extraction, waste dumping and sewage, coastal development, protection, mining, fortifications, ports, regional economies, and infrastructure construction (Plaganyi, 2007). Compared with the Atlantis model, the InVitro model covers the ecosystem less (Plaganyi, 2007). However, the model exhibits advantages that can characterize the mobility of processes, such as movement, growth, trait expression, genetic inheritance, and evolution. Therefore, it includes the small-scale interactions and instability of natural systems under climate change (Fulton, 2010). Currently, the InVitro model is mainly applied to the study of the coastal waters in Australia.

11.2 Challenges in End-to-end Approach-based Ecosystem Dynamics Modeling

The components of a marine ecosystem are complex, diverse, and interactive. Therefore, it is a great challenge to characterize all components of the system with dynamics models. Travers *et al.* (2007) pointed out that the end-to-end approach-based marine ecosystem dynamics modeling faced the following difficulties: Firstly, the large number of species and functional groups, the various differences between these organisms, and the number and characteristics of interactions between different components make end-to-end modeling extremely complicated and difficult. Secondly, the spatial and temporal scales of different processes are of different orders of magnitude because these processes are related to the organisms involved in the model and the organisms can range from viruses to large marine mammals. Thirdly, marine fisheries and environmental factors may directly or indirectly affect most organisms in the ecosystem, thereby increasing the number and complexity of processes characterized by the model. Fulton (2010) also states that there are still many challenges in end-to-end modeling of marine ecosystem dynamics, especially for the long simulation period. However, the most significant difficulties may lie in the non-stationarity caused by climate change, biodiversity, ecosystem evolution, characteristics of human responses, and the uncertainty of model processing. Among them, the large amount of data demands for the large and complex end-to-end model are its biggest weakness. System imbalance is often the biggest issue that limits the application of the model, especially for end-to-end modeling at the system level. Despite these difficulties and challenges in the end-to-end approach-based marine ecosystem dynamics modeling, great progress has been made on the coupling and mixing of different models with different spatial and temporal resolutions covering different ecological components and processes, as well as the characterization of human factors and uncertainty analyses of model parameters and structures.

11.2.1 *Coupling the models with different resolutions*

(1) *Spatial and temporal resolutions*
The components in a marine ecosystem are diverse and are on significantly different spatial and temporal scales. Therefore, the implementation of the end-to-end approach should consider the integration of different spatial and temporal resolutions. Space is an important system resource and its resolution greatly impacts the simulation results (Travers *et al.*, 2007; Fulton, 2010). The spatial and temporal resolutions of high trophic level models are generally lower than those of low trophic level models. Therefore, the coupling of high trophic and low trophic level models is usually realized by the averaging method that integrates small-scale spatial information to the large scale and interpolation method that distributes large-scale spatial information to the small-scale (Travers *et al.*, 2007). For an end-to-end model with a whole system as its frame, the spatial resolutions in the horizontal and vertical directions should be determined based on the spatial heterogeneity of the study area to characterize the main features of the system, including oceanographic features, shelf slope breaks, biological distribution, and human activity source areas (Fulton, 2010). The advantages and disadvantages of the homogeneous and heterogeneous grids should be evaluated, and different spatial representations should be considered for different components. For example, the

regular grids are used to represent general terrain, polygonal patches to describe the coastal habitats, such as mangroves, and nodes to represent estuaries.

Temporal resolutions can be the tidal cycle, day, week, month, season, and year, etc., and different components should be allowed to use different temporal scales (Fulton, 2010). Two methods have been used to process temporal resolution. One is the synchronous time stepping, by which all components of the system use the same time step. The other one is adaptive time stepping, which varies the sizes of sub-steps based on the instantaneous change rate of the biome. The sub-steps accumulate and iterate until a complete time step is completed. It is commonly applied to low trophic level organisms in the biogeochemistry based end-to-end models, such as the Atlantis model (Fulton, 2010). Similar to those for spatial resolution, some models allow different biota to use different time steps. For example, the InVitro model uses the asynchronous time stepping method. It does not require all components to use the same time step, but instead, the converting time step based on the behavior of each component.

(2) *Ecological component resolution*
The various ecological components in a marine ecological dynamics model can be classified based on many factors, such as individual size, conversion rates between different variables, phylogeny, habitats and other non-trophic uses, preys and predators, the functions of ecological groups, mobile pattern, distribution, and so on (Fulton, 2010). In general, the ecological components of low trophic level models mainly belong to the functional groups, and the high trophic level models are dominated by species. The species may include substructures of age, growth stage, and individual size. The components in most whole-of-system models, such as the Atlantis, ERSEM, EwE, and InVitro models, belong to the functional groups, and some of them are species-divided components (Travers *et al.*, 2007; Fulton, 2010). The substructures included in the high trophic level model can be used to characterize the transformations and delays of individuals in the system, which ensures the multi-stable states of the end-to-end model to certain extents and thereby helps to study the behaviors of the system under different environmental conditions and stress levels (Fulton, 2010). Currently, end-to-end modeling platforms, such as EwE,

Atlantis, and InVitro, are also gradually combining various characterizations, such as component-based individual characterization, age structure characterization, subpopulation characterization, biomass pool characterization, and so on. For the integration of multiple ecological components with different resolutions in the end-to-end model, attention should be paid to the conversions between different components, such as the flow of biogenic elements between functional groups and the changes in the biomass conversions between species. Such conversions are generally characterized by the conversion rates, such as the Redfield ratio. Sensitivity analysis is usually required to determine the impacts of different conversion rates on model performance (Travers *et al.*, 2007).

(3) *Process resolution and external driver*

The physical environment components are usually coupled into the marine ecosystem dynamics models as an external driving function to provide the initial driving force for the system. The coupling of human activity components can also adopt the external driving method. However, the currently available end-to-end models of the whole system framework, such as Atlantis and InVitro, gradually use the two-way coupling method to clearly characterize the dynamic changes. Because the end-to-end models generally require the explicit characterizations of feedback and system delays, most ecological components are coupled by the dynamic two-way coupling approach (Fulton, 2010). The ecological components can be coupled through preying-predation, excretion, biomass loss as debris and bacterial ponds, and growth of fish zooplankton to larger individuals (Travers *et al.*, 2007). The studies on the impacts of marine fisheries and climate change on the food web should consider opportunistic predation. The predators and prey should appear in the same space-time, and the size of the prey is suitable for predation to make the simulated food web flexible and adaptable (Travers *et al.*, 2007).

Studies have shown that end-to-end models do not need to describe all processes in detail in order to improve modeling efficiency. The models should be appropriately simplified according to the purpose of the modeling (Fulton, 2010). For example, a model for system-level resource utilization problems needs only describe the simple growth and consumption processes without characterizing detailed physiological processes. The implementation of currently available models is subject to multiple simplifications. For example, the InVitro model only describes in detail the feeding and movement behaviors of rare species to characterize most ecological populations with ensemble populations. In addition, even if the key processes cannot be clearly described, their impacts can be characterized with statistical models or experience curves (Fulton, 2010). The process resolutions should match the spatial and temporal resolutions of the ecological components. Meanwhile, the sensitivity analysis of the model structure should be used to examine the characterizations of different processes and the rationality of the simplification to select relatively suitable process resolutions.

(4) *Resolution of human activities*

There are a variety of socioeconomic models that focuses only on human activities, and those models have been gradually coupled with biophysical models in recent years. Currently, the resolutions of human components involved in most end-to-end models are often lower than those of biophysical components. The EwE, Atlantis, and InVitro models are end-to-end models examining the human factors in detail. The EwE model includes clear economic and social parameterization, product chain tracking, and fleet behaviors, etc. The Atlantis and InVitro models give different options to characterize detailed socioeconomic driving behaviors (Fulton, 2010). Experience shows that behavioral uncertainty, that is, unpredictable responses to resource changes and management decisions, plays a key role in the long-term dynamic characteristics of the affected system. However, it is often more difficult to obtain the dynamic characteristics of the human components of a system than to characterize its biophysical components. On one hand, the end-to-end modeling approach that couples human activities is relatively immature. On the other hand, the interdisciplinary research of social sciences and natural sciences is still in the slow integration state. Fulton (2010) pointed out that with the progress of cross-disciplinary integration, it is only a matter of time before similar resolutions are adopted to couple human components with biophysical components, and in the current transition period, the multi-scenario settings of human activities with different behaviors, patterns, and intensities might be a workable method.

11.2.2 *Uncertainty of end-to-end model*

(1) *Uncertainties of model parameters*
An end-to-end marine ecosystem dynamics model involves diverse and complex components, making it necessary to determine model uncertainties, which is a great challenge. In general, a variety of data sets are established to limit model parameters in the modeling, which can improve the accuracy of simulation results. However, overfocusing on the accuracy of modeling results without thoroughly considering the overall model and interaction relationships makes the sensitivity analyses of the large number of parameters involved in the end-to-end model and feedback mechanism impractical. Therefore, new mathematical methods applicable to end-to-end models need to be developed by examining the model structure (Fulton, 2010). Uncertainty analysis methods for large feedback models are already available in other fields, such as the functional filter method that can be used to determine the most sensitive parameter subsets or system components. The shortcomings of the incomplete coverage of this method can be mitigated by combining it with Bayesian hierarchical modeling (Pantus, 2007). Currently, two sensitivity analysis methods have been successfully applied to end-to-end models—the perturbation analysis that examines the effects of artificial perturbations of different components, such as reducing, increasing, and systematically changing, on all other components of the system to provide information for model sensitivity analysis and potential system functions and channels (Pantus, 2007), and the bounded parameterization method, which conducts parameterization using a set of bounded sets and explores possible variation ranges of dynamic characteristics of the system with the worst, neutral and best scenarios of each major dimension of the system, such as productivity, vulnerability, external drivers, human impact levels, and so on (Fulton, 2010). The bounded parameterization method is very simple and easy to operate for the analysis of parameter uncertainty. Successful cases include the application of the InVitro model to the tropical waters of Northwest Australia (Little *et al.*, 2006) and the application of the Atlantis model to Southeast Australia (Fulton *et al.*, 2007) for the potential management effectiveness in different scenarios of a series of system states, management combinations, and human activities.

(2) *Uncertainty of the model structure*
Because of the high complexity of end-to-end modeling caused by the diversity of system components, more attention should be paid to the model structural uncertainty than uncertainties of model parameters, e.g., how the model is characterized and what processes, scales, and system components are included (Fulton, 2010). Loop analysis is an effective structural uncertainty analysis method that is generally used to qualitatively examine the structure of a system, as well as the potential structural sensitivity of the model based on the trophic level or aggregation degree of processes (Puccia and Levins, 1985). For example, Metcalf *et al.* (2008) investigated the aggregation errors caused by three simplification methods: Euclidean distance, Bray-Curtis similarity, and regular equivalence, and found that the last method gave the smallest aggregation errors and retained the most details of the food web. Model structural uncertainty and parameter uncertainty often correlate with each other, and the methods for parameter uncertainty analysis can generally be applied to structural uncertainty analysis (Fulton, 2010). For example, the robustness analysis of the bounded parameterization method largely considers the structural uncertainty. Models are usually a simplification of reality. As our understanding of marine ecosystem dynamics is still limited, there is no such thing as a correct model. To reduce the uncertainty caused by the model structure, the modeler should carefully consider the complexity of the model, build a variety of optional models, and select a relatively effective structure by evaluating and comparing the performances of the models.

References

Baretta J W, Ebenhoh W, Ruardij P. 1995. The European regional seas ecosystem model, a complex marine ecosystem model. *Netherlands Journal of Sea Research.* **33**: 233–246.

Baretta-Bekker J G, Baretta J W. 1997. Special issue: European regional seas ecosystem model II. *Journal of Sea Research.* **38**: 169–438.

Christensen V, Walters V. 1992. Ecopath with Ecosim: Methods, capabilities and limitations. *Ecological Modeling.* **172**: 109–139.

Franks P J. 2002. NPZ models of plankeon dynamics their conservation, coupling to physics, and application. *Journal of Oceanography.* **58**(2): 379–287.

Fulton E A. 2010. Approaches to end-to-end ecosystem models. *Journal of Marine Systems*. **81**(1–2): 171–183.

Fulton E A, Smith A D M, Smith D C. 2007. Alternative management strategies for Southeast Australian wealth fisheries: Stage 2: Quantitative management strategy evaluation. Australian Fisheries Management Authority Report. 378.

Fulton E A, Smith A D M, Johnson C R. 2004a. Biogeochemical marine ecosystem models I: IGBEM-a model of marine bay ecosystems. *Ecological Modelling*. **174**: 267–307.

Fulton E A, Parslow J S, Smith A D M, *et al.* 2004b. Biogeochemical marine ecosystem models II: The effect of physiological detail on model performance. *Ecological Modelling*. **173**: 371–406.

Fulton E A, Fuller M, Smith A D M, *et al.* 2005. Ecological indicators of the ecosystem effects of fishing: Final report. Australian Fisheries Management Authority Report, R99/1546.

Gray R, Fulton E A, Little L R, *et al.* 2006. Operating model specification within an agent based framework. North West Shelf Joint Environmental Management Study Technical Report, 16. CSIRO, Hobart, Tasmania. 127.

Hermann A, Hinckley S, Megrey B, *et al.* 2001. Applied and theoretical considerations for constructing spatially explicit individual-based models of marine larval fish that include multiple trophic levels. *ICES Journal of Marine Science*. **58**: 1030–1041.

Kaplan I C, Marshall K N. 2016, A guinea pig's tale: Learning to review end-to-end marine ecosystem models for management applications. *ICES Journal of Marine Science*. DOI: 10.1093/icesjms/fsw047.

Lehodey P. 2005. Reference manual for the spatial ecosystem and populations dynamics model — SEAPODYM. WCPFC-SCI, ME IP-1.

Lehodey P, Chai F, Hampton J. 2003. Modelling climate-related variability of tuna populations from a coupled oceanbiogeochemical-populations dynamics model. *Fisheries Oceanography*. 12: 483–494.

Little L R, Fulton E A, Gray R, *et al.* 2006. Multiple use management strategy evaluation for the north west shelf — results and discussion. North West Shelf Joint Environmental Management Study Technical Report, 18. CSIRO, Ho-bart, Tasmania.

Megrey B, Rose K, Klumb R, *et al.* 2007. A bioenergetic/population dynamics model of Pacific herring (*Clupea harengus pallasi*) coupled to a lower trophic level nutrient-phytoplankton-zooplankton model: Dynamics, description, validation and sensitivity analysis. *Ecological Modelling*. **202**: 144–164.

Megrey B A, Kishi M J. 2002. Model/REX workshop to develop a marine ecosystem model of the north pacific ocean including pelagic fishes. *PICES Scientific Report*. **20**: 77–176.

Metcalf S J, Dambacher J M, Hobday A J, *et al.* 2008. Importance of trophic information, simplification and aggregation error in ecosystem models. *Marine Ecology Progress Series.* **360**: 25–36.

Moloney C L, John M A S, Denman K L, *et al.* 2011. Weaving marine food webs from end to end under global change. *Journal of Marine Systems.* **84**(3–4): 106–116.

Pantus F J. 2007 Sensitivity analysis for complex ecosystem models. PhD Thesis, School of Physical Sciences, The University of Queensland, Brisbane.

Plaganyi E E. 2007. Models for an ecosystem approach to fisheries. FAO Fisheries Technical Paper. No. 477. Rome, FAO. 108.

Puccia C J, Levins R. 1985. *Qualitative Modeling of Complex Systems.* Cambridge: Harvard University Press.

Rose K A, Allen J I, Artioli Y, *et al.* 2010, End-to-end models for the analysis of marine ecosystems: Challenges, issues and next steps. *Marine & Coastal Fisheries Dynamics Management & Ecosystem Science.* **2**(1): 115–130.

Rose K A, Werner F, Megrey B A, *et al.* 2007. Simulated herring growth responses in the Northeastern Pacific to historic temperature and zooplankton conditions generated by the 3-dimensional NEMURO nutrient-phytoplankton-zooplankton model. *Ecological Modelling.* **202**: 184–195.

Shin Y J, Cury P. 2001. Exploring fish community dynamics through size-dependent trophic interactions using a spatialized individual-based model. *Aquatic Living Resources.* **14**: 65–80.

Shin Y J, Cury P. 2004. Using an individual-based model of fish assemblages to study the response of size spectra to changes in fishing. *Canadian Journal of Fisheries and Aquatic Sciences.* **61**: 414–431.

Sourisseau M. 2002. Etude de la structure de taille de la communaute des copepodes par l analyse de spectres comptuer optique et par la modelisation de la dynamique des populations. PhD Thesis Universite Piet Marie Curie-Paris VI.

Steele J H, Aydin K, Gifford D J, *et al.* 2013. Construction kits or virtual worlds: Management applications of E2E models. *Journal of Marine Systems.* **109–110**: 103–108.

Travers M, Shin Y J, Jennings S, *et al.* 2007. Towards end-to-end models for investigating the effects of climate and fishing in marine ecosystems. *Progress in Oceanography.* **75**(4): 751–770.

Walters C, Pauly D, Christensen V. 1999. Ecospace: Prediction of mesoscale spatial patterns in trophic relationships of exploited ecosystems, with emphasis on the impacts of marine protected areas. *Ecosystems.* **2**: 539–554.

Chapter 12

Introduction to Land-sea Coordination-based Coastal Ecosystem Dynamics and Prospects for Marine Ecosystem Dynamics Models

We are conducting a research titled "Sea-land coordination based synergistic effects of multiple human activities on Shandong coastal ecosystem and the adaptive management" funded by NSFC. The goals of the research are, by analyzing the temporal and spatial change characteristics of ecological pressure and its driving factors, to identify long-term and cumulative effects on the key ecological processes of coastal waters, establish a dynamic model of the coastal ecosystem under the influences by multiple human activities, simulate the scenarios of the change and responses of the coastal ecosystem under different pressures, reveal the change mechanism of the coastal ecosystem under the synergistic effects of multiple human activities in the "watershed-estuary-coastal waters" system, analyze the ecosystem health maintenance mechanism of the coastal ecosystem under the influence of multiple human activities, and eventually propose land-sea-coordination-based adaptive management strategies for the coastal ecosystem according to the spatial heterogeneity of the health pressure source (land) and carrier (sea) of the coastal ecosystem. This project will promote the construction of a bridge between natural science research and ocean management and provide important scientific support for the environmental protection of coastal areas and the construction of a marine ecological civilization.

12.1 Research Background and Significance

With rapid economic development in coastal areas, accelerated urbanization, and increased shortage of land resources, scientific development of marine resources and marine ecosystem protection based on the concept of land-sea coordination have become important ways for the sustainable development of human society. Coastal waters, a typical ocean ecosystem, have long been affected by high-intensity human activities, such as reclamation, land-sourced pollutant discharge, and mariculture, and undergone natural feature alternation, degradation of key habitats, increased eutrophication, frequent ecological disasters, and functional damages or losses of the ecosystem. Most of the coastal waters ecosystems in China have long been in sub-healthy or unhealthy conditions (Wang *et al.*, 2006; Su *et al.*, 2011).

At present, the impact of human activities on coastal ecosystems has been studied, yet the results are insufficient for understanding the ecosystem's change and predicting its development as a result of human activities due to the multi-factor, multi-disciplinary, and complexity properties of this scientific problem (MA, 2005; Halpern *et al.*, 2008, 2012; Duarte *et al.*, 2009; Lester *et al.*, 2018; Bevilacqua *et al.*, 2018). The pressure imposed by multiple human activities, such as land-sourced pollutant discharge, reclamation, and mariculture, on the coastal ecosystem is a dynamic and multi-directional synergistic effect, rather than a simple linear superposition. The process, mechanism, and effects are very complex and urgently need to be further studied.

We noticed that the main pressure of coastal ecosystem degradation coming from human activities and social development on land causes the regional dislocation of the ecological pressure carrier (coastal waters) and pressure source (coast and watershed), and the current degradation of the coastal ecosystem results from the long-term effects of multiple human activities. Studying the marine and terrestrial ecosystems separately would cause the separation of the pressure sources from the carrier of the marine ecosystem, and result in the difficult implementation of regulation measures due to the separation of the marine environmental management and terrestrial environmental management. Therefore, studies should be conducted with the "watershed-estuary-coastal waters" as a whole, in accordance with the concept of land-sea coordination, to put forward adaptive management strategies for ocean health development.

There are more than 20 bays of different sizes along the coast of Shandong Province, among which Laizhou Bay, Jiaozhou Bay, and Sanggou Bay are relatively large ones. Human activities have imposed significant effects on these bays. For example, Laizhou Bay receives large amounts of pollutants from more than a dozen rivers along the coast, causing increased pollution in the coastal water (Ma *et al.*, 2014). The sea area in Jiaozhou Bay has decreased rapidly in the past 80 years due to reclamation (Wu *et al.*, 2011). The structure of the planktonic ecosystem in Sanggou Bay has undergone major changes under the influence of large-scale mariculture activities (Song *et al.*, 2007; Shi *et al.*, 2013). The long-term observational study shows that the structures and functions of the ecosystems of those bays have undergone great changes, and human activities are the major influencing factor. Most related studies have focused on the impact analysis of a single type of human activity, which ignores the synergistic effects of multiple human activities (Liu *et al.*, 2013). The main research contents of this project are to identify the influences of different human activities from other environmental factors that drive the changes of the coastal ecosystem, reveal the synergistic mechanism of multiple human activities on the ecosystem, and then propose adaptive management strategies for the coastal ecosystem based on the concept of land-sea coordination.

Maintaining the sustainable and healthy development of the coastal ecosystem under multiple human disturbances is a great challenge for coastal environmental protection and management. Currently, the study on the dynamic mechanism of typical coastal ecosystems under the influences of high-intensity human activities is still in its infancy, and in-depth explorations on the theoretical system and technical methods of a coastal ecosystem are highly urgent. A dynamics model of the marine ecosystem for the impacts of human activities in the "watershed-estuary-coastal waters" system can provide an important means to analyze the synergistic effects of multiple human activities on a coastal ecosystem (Bonan and Doney, 2018). The construction of an ecosystem dynamics model requires accurate parameters that can be obtained by the combination of historical data, on-site supplementary surveys, and controlled experiments. The study of the impacts of human activities on the key ecological processes of Shandong Coastal waters will also help improve the model framework and quantify the relationships of the parameters.

The project "Land-sea coordination based synergistic effects of multiple human activities on Shandong coastal ecosystem and the adaptive management" aims to (1) analyze the temporal and spatial change characteristics of the ecological pressure caused by typical human activities in the past 40 years in Laizhou Bay, Jiaozhou Bay, and Sanggou Bay along the Shandong coast based on the marine ecosystem dynamics model and the pressure-state-response (PSR) framework, (2) systematically investigate the synergistic influence mechanism of typical human activities, such as land-sourced pollution, reclamation and mariculture, on the bay ecosystems, and (3), in accordance with the concept of land-sea coordination, put forward pressure mitigation strategies including reducing the amount of pollutants entering the sea, reclamation control, and optimization of mariculture methods, revealing the health maintenance mechanism of typical coastal ecosystem, clarifying the main stress factors, and proposing adaptive management strategies for coastal ecosystem health. The project will also provide a decision-making reference for the sustainable development of coastal areas and strong support for environmental protection and marine ecological civilization construction in coastal areas of China.

12.2 Research Progress

12.2.1 *Impacts of typical human activities on coastal ecosystem*

(1) *Land-sourced pollutant discharge*

Land-sourced pollutant causes the eutrophication of a coastal water body and thus affects the corresponding ecological environment health. Nutrients are a major land-sourced pollutant affecting the ecological environment of coastal waters and causes the above said eutrophication of the coastal water body (Huang *et al.*, 2005; Yin *et al.*, 2011; Huang *et al.*, 2015). According to the *Bulletin of Marine Ecology and Environment Status of China, 2016*, 72,000 km^2 of coastal waters in China are eutrophicated and many bay areas are heavily eutrophicated. This eutrophication has led to a series of complex ecological effects, such as imbalanced nutrient structure, alterations of the species compositions and distribution patterns of phytoplankton, zooplankton, and benthic organisms, and poor output quality of ecosystems (Li *et al.*, 2006; Yu *et al.*, 2015; Zhang *et al.*, 2015).

With the increasing frequencies of human activities along the coast, not only have the pollutant concentrations in coastal waters increased significantly, but the composition structure of pollutants has also changed significantly. Organic nutrients have gradually become the main pollutants (Dafner *et al.*, 2007). For example, the nutrient and restrictive biogenesis of the Dokai Bay in Japan has evolved (Hamada *et al.*, 2012). The biogeochemical process of phytoplankton growth-limiting nutrients has become complex, and the changes in nutrient flux and structure may alter marine biogeochemical processes (Trommer *et al.*, 2013). Studies have shown that a large amount of ammonia nitrogen discharged from land can cause different effects on the nutrition absorption of phytoplankton. For example, a large amount of ammonia nitrogen discharged into the Suisun Bay in the United States (US) inhibits the absorption of nitrate nitrogen and the primary production of phytoplankton (Parker *et al.*, 2012), while that of the Northport-Huntington Bay, US, enhances the nutrient absorption of toxic algae and stimulates its toxicity (Hattenrath *et al.*, 2010).

Although great progress has been made domestically and internationally in the change process and mechanism of the coastal ecosystem under the influence of land-sourced pollution, systematic studies are still needed. Efforts and in-depth studies should be made to reveal the impact mechanism of land-sourced pollution in the "watershed-estuary-coastal waters" system on the coastal ecosystem from a longer time scale and larger spatial scope.

(2) *Reclamation*

Reclamation degrades key habitats, changes the hydrodynamic environment, and increases the pollution degree of coastal waters. The direct degradation of bay habitats by reclamation is violent, devastating, and irreversible (Zhang *et al.*, 2014). The marine soil extraction, reclamation, and burial processes drastically change the living conditions of marine organisms, reduce the number of benthic organisms, change community structure, decrease biodiversity, and greatly impact benthic organisms (Shen *et al.*, 2016a). Reclamation directly changes the natural features of the coastal water and even has devastating effects on the organisms in the reclaimed area (Zhang *et al.*, 2014). Coastal engineering constructions cut, block, and thus fragment the habitats, which reduces the space for biological activities. It also blocks the genetic information communication of the population, and causes unfavorable border effects (Lin *et al.*, 2016). Reclamation also reduces the tidal-bearing sea area in coastal waters, which greatly affects the water exchange and self-purification capacity in that area, and increases pollution pressure (Wang *et al.*, 2015).

A coastal ecosystem is an open and complex giant system that exhibits comprehensive symptoms under the pressure of multiple human activities. Identifying the healthy response characteristics of the ecosystem to reclamation is key to the selection and weight function of the health evaluation index of coastal ecosystems (Zheng *et al.*, 2011a; Shen *et al.*, 2016a). Most studies have focused on the impacts of a single reclamation project. There is no systematic study on the superimposed effects and nonlinear characteristics of the impacts of reclamation at different shores during different periods on the key ecological processes of coastal waters. Multi-level, multi-scale, and multi-perspective analyses are needed.

(3) *Mariculture*

Mariculture aggravates the eutrophication of a coastal ecosystem. It increases the nutrient load in mariculture waters (Lu *et al.*, 2015), and directly leaves large amounts of excess fodders, and produces biological excrement that increases the organic nutrients in the water body. Feces, fodder residue, secretions and sheddings, dead bodies, etc., from mariculture result in nitrogen (N) and phosphorus (P) pollution. The changes in the forms and concentrations of N and P pollutants in seawater significantly affect the biogeochemical cycle of the mariculture ecosystem. Compared with other pollutants discharged into the ocean by human activities, the total amount of pollutants produced by mariculture is small, but can impose extremely strong impacts on the local sea area (Lester *et al.*, 2018).

Mariculture affects the structure and function of the coastal ecosystem in many aspects (Shi *et al.*, 2013). First, the increases in nutrient amount caused by excess fodder promote the growth of phytoplankton, while the decreases in light transparency caused by the deterioration of water quality inhibit the growth of phytoplankton. Second, the feeding pressure of filter-feeding shellfish affects the population structure of phytoplankton and zooplankton and reduces biodiversity (Dong and Wang, 1999). In addition, the excretion of residual fodder and metabolites in cage mariculture increases the concentrations of organic matter, and solid and dissolved nutrients, which eventually changes the benthic community in the culture area (Xu and Qian, 2004). Last but not least, mariculture activities not only accelerate the nutrient circulation rate but also fluidize the sediment (Zheng *et al.*, 2011b).

Currently, domestic and international studies on the impacts of mariculture have mainly focused on monitoring process changes. The effects of mariculture activities on the internal ecosystem structure transformation mechanism and steady-state changes have been rarely reported.

12.2.2 *Marine ecosystem dynamics model under the influence of human activities*

The dynamics model is an important means to reveal the change mechanism of an ecosystem and predict its changes. Accuracy is the main factor limiting the application of the dynamic model to the marine ecosystem. The model's accuracy can be improved by optimizing the model parameters (Shen *et al.*, 2016c). Data assimilation based on parameter sensitivity analysis is an effective method for parameter optimization (Fan and Lv, 2009; Shi *et al.*, 2014). At present, most studies have focused on the simulation of hydrodynamics and its influences on the temporal and spatial distributions of chemical and biological factors. Further exploration of the dynamics model of the marine ecosystem under the influence of human activities is urgently needed to provide sufficient support for marine management and decision-making (Bevilacqua *et al.*, 2018; Hiroshi *et al.*, 2018).

With the increasing complexity of marine environmental issues and the popularization of the concept of sustainable development, the marine ecosystem dynamic model gradually evolves to the characterization of all components of a marine ecosystem, and mutually couples with socioeconomic models to study the impacts of the marine ecosystem on climate and the response and adaptation to human activities (Travers *et al.*, 2007; Plaganyi, 2007; Fulton, 2010). Taking human activities as an internal component, this type of model simulates dynamic interactions between human activities and the marine ecosystem and supports the ecosystem-based marine management by simulating various components of the adaptive management cycle (Fulton, 2010; Kaplan and Marshall, 2016). The Atlantis ecosystem model and the individual-based InVitro model are representative coupling models. Expressing the dynamic characteristics of human activities is often more difficult as compared with the characterization of system biophysical components. However, multi-scenario simulations of different behaviors, patterns, and intensities of human activities can provide important technical references for regulating human activities in the coastal area (Fulton, 2010).

12.2.3 *Health assessment methods and regulation strategies of marine ecosystem*

The concept of ecosystem health was originally proposed and developed by Rapport *et al.* (1998) and Costanza (1992). It is defined as the ability of an ecosystem to maintain its structure and function for a period of time under external pressures. The health of a marine ecosystem is affected by many factors, which not only change with different natural habitat conditions, but also have been increasingly affected by human disturbances in recent years (Halpern *et al.*, 2008). Comprehensively and quantitatively characterizing the spatial distribution of ecosystem health and the impacts by human disturbance can provide important technical support for maintaining the balance of the marine ecosystem and the rational development and utilization of the ocean. Yet, direct measurement of ecosystem health remains challenging. In coastal ecosystems, influencing factors are becoming more and more complex. Therefore, the evaluation indicators of ecosystem health have gradually evolved from physical, chemical, and biological characteristics to ecosystem services and functions. The indicator composition evolves from a single indicator to multi-indicators (Halpern *et al.*, 2008, 2012; Chen *et al.*, 2010; Rombouts *et al.*, 2013; Shi *et al.*, 2017).

Currently, most studies have focused on the overall health status of an ecosystem, and the health spatial distribution characteristics are rarely involved. A few studies on the spatial distribution of ecosystem health are mostly at the national or global level at low spatial resolutions. Due to the spatial differences in natural conditions and human activities, the health of a marine ecosystem displays obvious spatial differentiation characteristics (Shen *et al.*, 2016a; 2016b). Though the health of coastal ecosystem health has received some attention, most of these studies intend to evaluate its health only, rarely provide further management and regulation strategies based on the evaluation results, especially for the maintenance mechanism of coastal ecosystem health under the synergistic effect of multiple human activities, and thus cannot provide a scientific basis for the ecosystem-based adaptive management of coastal waters.

12.3 Research Questions

The rapid economic development and strong demand for land have triggered high-intensity marine development activities, which puts unprecedented pressures on the health of coastal ecosystems. Land-sourced pollution, reclamation, and mariculture are typical human activities that affect the ecosystem's health. The driving mechanism of these human activities to the change of the structure and function of the coastal ecosystem is still unclear. The impacts of a single type of human activity have been studied, though no systematic investigation has been conducted on the synergistic influence mechanism of multiple human activities on a coastal ecosystem (Lotze *et al.*, 2006; Shen *et al.*, 2016a; Bonan and Doney, 2018).

In all, current studies on marine ecosystem health assessment are insufficient to support marine management decision-making. Further studies should be conducted based on the concept of sea-land coordination and simulation of different stress scenarios of different human activities by the dynamics model to determine the health stress factors of the ecosystem, reveal the adaptive regulation mechanism of the ecosystem, and eventually propose corresponding regulation strategies, which would be of great theoretical significance and high application value.

12.4 Research Objective, Contents, and Key Scientific Issues to be Solved

12.4.1 *Overall objective*

The main research objective of this project is to determine the temporal and spatial changes of land-sourced pollution, reclamation, mariculture, and other human activities and their influence mechanism, especially long-time cumulative impacts on the key coastal ecological processes by studying the change trend of human activity pressure on the typical bay ecosystems in Shandong Coastal waters and its driving mechanism, establishing a coastal ecosystem dynamics model under the influence of multiple human activities in the "watershed-estuary-coastal waters" system, exploring the synergistic influence mechanism of multiple human activities on the bay ecosystems, and provide support for clarifying the degradation mechanism of the bay ecosystems. This project will also identify the main stress factors for the health of typical coastal ecosystems and propose adaptive regulation strategies to maintain coastal ecosystem health, such as reducing pollutants discharge into the sea, limiting reclamation, optimizing mariculture techniques, and ecological restoration, based on the concept of land-sea coordination. These countermeasures can provide decision-making references for the sustainable development of the coastal area, and strong support for environmental protection and marine ecological civilization construction in coastal areas.

12.4.2 *Research contents*

The project intends to follow these four aspects: (1) change of typical human activity stresses in the coastal area and its driving mechanism, (2) key impact processes and parameters of human activities on coastal ecosystem, (3) coastal ecosystem dynamic model under the influences of multiple human activities, and (4) ecosystem health

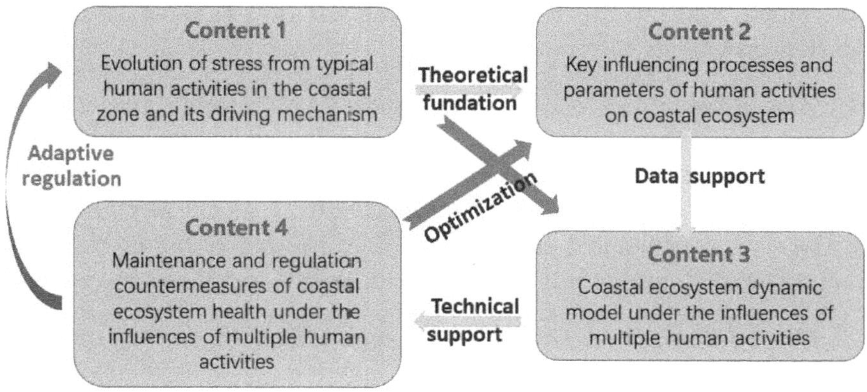

Fig. 12.1. Relations between the research contents.

maintenance mechanism and regulation countermeasures under the influence of multiple human activities. Among them, the first two aspects will reveal the characteristics of human activity stress and its impact on the main ecological processes of the coast, and provide the key parameters for the ecosystem dynamic model and a theoretical foundation and data support for "Content 3" (see Fig. 12.1), which aims to identify main stress factors affecting coastal ecosystem health and provide technical support for "Content 4" by constructing a marine ecosystem dynamic model under multiple human activity impact scenarios. Based on the results of "Content 3", "Content 4" will conduct the scenario simulations with different combinations of the stress factors using the model constructed in "Content 3", clarify the health of the coastal ecosystem and its regulation mechanism, and propose adaptive regulation measures for the coastal ecosystem based on the land-sea coordination concept, as well as the regulation strategies of human activity behaviors based on their impacts identified in "Content 1" and "Content 2".

(1) *Change of stress from typical human activities in the coastal zone and its driving mechanism*

In this section, we analyze the temporal and spatial change of the stresses from typical human activities in Shandong Coastal waters and discuss the driving mechanism of the stresses based on data from the past 40 years of major human activities, including land-sourced pollution, reclamation, and mariculture in typical bays of

Shandong, and urbanization level, economic development, and population growth in the coastal areas.

(2) *Key influencing processes and parameters of human activities on coastal ecosystem*

In this section, the impacts of human activities including land-sourced pollution, reclamation, and mariculture on the key processes of the coastal ecological environment, and the key parameters of the marine ecosystem dynamic model are identified and then determined.

(3) *Coastal ecosystem dynamic model under the influences of multiple human activities*

A marine ecosystem dynamic model that contains multiple nutrients, phytoplankton, and zooplankton under the influence of multiple human activities will be established. The model parameters are optimized with historical data, ecological simulations, and conventional observation data, and the model is further verified by observation data. The main stress factors of the coastal ecosystem change will be identified by a large number of scenario simulations (of Laizhou Bay) using the established marine ecosystem dynamics model, which will reveal the synergistic effect of multiple human activities on the change of the coastal ecosystem and provide a scientific foundation for choosing the optimal ecosystem health assessment indicators.

(4) *Maintenance and regulation countermeasures of coastal ecosystem health under the influences of multiple human activities*

This section aims to identify the main stress factors causing the change of the coastal ecosystem, construct a coastal ecosystem health assessment indicator system and model, characterize the temporal and spatial distributions of the health of a typical coastal ecosystem in the past 40 years and its response mechanism to the stress changes of human activities, and propose land-sea-coordination-based adaptive regulatory countermeasures of the coastal ecosystem.

12.4.3 *Key scientific issues to be solved*

(1) The coastal ecosystem is located in the land-sea interlaced zone and thus is synergistically affected by various human activities. Land-sourced pollution, reclamation, and mariculture are the main sources of ecological stress in a coastal ecosystem. However, they have long

coexisted with each other. The questions to be answered are: What is the mechanism of their synergistic influences on the coastal ecosystem? How will the structure and function of the coastal ecosystem change in different influencing scenarios of different human activity structures and intensities?

(2) The main pressures of coastal ecosystem degradation come from economic and social development on land, which causes the regional dislocation of the ecological pressure carrier (coastal waters) and the pressure source (watershed). The uncertainty of future pressures of different human activities and their combinations leads to many impact scenarios on the health of the coastal ecosystem. Revealing the maintenance mechanism and stress factors of a healthy coastal ecosystem based on the technical framework of "stress (human activities)-state (function of a coastal ecosystem)-response (ecosystem change)" and the concept of land-sea coordination is the key scientific issue of this project, as well as the foundation for studying and proposing the adaptive management strategies for a coastal ecosystem.

12.5 General Research Ideas

The key goal of this project is to reveal the in-depth changes in the intensities of multiple human activities under the background of global changes and their synergistic influence mechanisms on the coastal ecosystem. Centering on the dynamic model of the coastal ecosystem, the technical route of this project follows the background order of multiple human activity stresses, data acquisition, impacts on coastal ecological processes, ecological dynamics model, and then regulation and control countermeasures (Fig. 12.2).

According to the concept of land-sea coordination, the land "funnel", the catchment area of rivers entering the sea (watershed), is considered as the land-related scope, and the sea "funnel", the coastal water, belongs to the sea-related scope. A typical "watershed-estuary-coastal waters" research conceptual framework can then be established as shown in Fig. 12.3 for studying the impacts of human activities on the coastal area. The influence mechanism of land-sourced pollution on key ecological processes in the coastal waters can be discussed by scientifically estimating the pollution load of

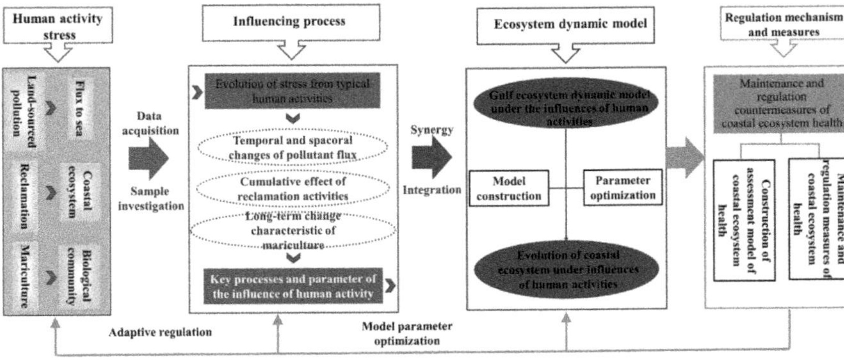

Fig. 12.2. Technical route of this project.

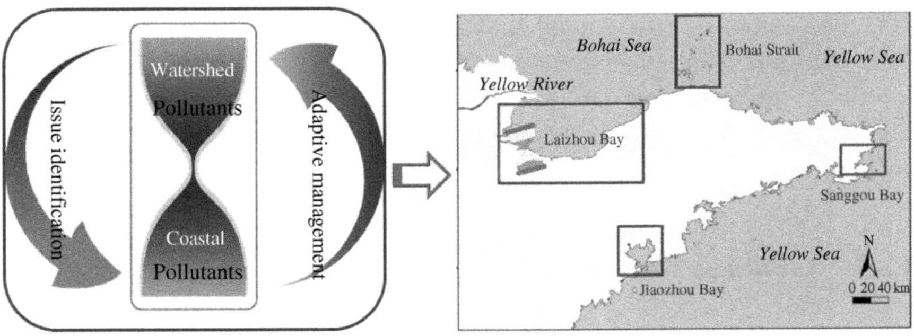

Fig. 12.3. Schematic diagram of land-sea coordination of the "watershed-estuary-coastal waters" system.

the land funnel to the sea funnel through the estuary, and analyzing the temporal and spatial changes of land-sourced pollutant fluxes into the sea. Also, the cumulative impact of reclamation activities on the coastal ecosystem will be explored by analyzing the temporal and spatial change characteristics of typical coastal reclamation and its driving mechanism. The study will also analyze the influencing process and mechanism of typical mariculture activities on the dynamics of the coastal plankton community. Based on the above-mentioned work, a dynamic model for the coastal ecosystem influenced by multiple human activities will be established to reveal the synergistic influence mechanism of the multiple human activities in the "watershed-estuary-coastal waters" system on the coastal ecosystem, identify the main stressors of the ecological dynamics model by the sensitivity analysis, establish a health assessment model of the coastal ecosystem, clarify the adaptive regulation mechanism of the coastal ecosystem, and propose land-sea-coordination-based adaptive management countermeasures.

12.6 Prospects for Global Optimization of Marine Ecosystem Dynamics Models throughout Entire Modeling Process

Currently, the voices for ecosystem-based coastal zone management and natural solutions in ocean management become increasingly louder, and marine ecosystem dynamics is in the ascendant. However, there is still room for improvement in the systematic theory and comprehensive performance of the model. Therefore, the next step is to further deepen the understanding of marine ecosystem change and its driving mechanism, optimize and improve the key technologies of marine ecosystem simulation and prediction, and then promote the study of marine ecosystem dynamics to make new progress.

In order to achieve the above goals, we believe that it is imperative to improve the comprehensive performance of coastal ecosystem models. Thus, from the perspective of systems engineering, we propose a technical framework to carry out global optimization of marine ecosystem dynamics models throughout entire modeling process (GOMEM) such aspects as the initial value problem of ecological variables, the boundary value problem of simulation area, model parameter optimization, modeling method, efficient calculation strategy, and model application & feedback. By integrating artificial intelligence, big data, nonlinear optimization, supercomputing, and other technologies, marine ecosystem simulation and prediction capabilities and comprehensive performance will be improved. Specifically, the initial value problem of ecological variables refers to the initial value of each state variable of the marine ecological dynamic model (such as phytoplankton, zooplankton, and nutrients) that is the discrete value of scattered observation stations and how to form a relatively real initial value of continuous spatial distribution through spatial interpolation. The boundary value problem of the simulation area refers to the processing of boundary conditions, such as land boundary, sea boundary, air-sea interface, and seabed interface in the simulation area, and also includes how to incorporate climate change and typical human disturbance into the overall model of marine ecological dynamics. In terms of the model parameter optimization, besides the development of mathematical methods of parameter optimization, it is necessary to further update and enrich the research of marine ecological indicators data and innovate the scientific cognition of key coastal ecological processes and parameters through the comprehensive use of remote sensing and GIS, genomics, molecular biology and traditional survey experiments. Different modeling methods have different applicability to the study of coastal ecological dynamics models. It is necessary to attach importance to the application of artificial intelligence and big data analysis technology in marine ecological modeling, analyze the application scope and advantages and disadvantages of different modeling methods, and integrate different modeling methods to carry out research and application of coastal ecological dynamics model. Aiming at the problems of complex model, large amount of calculation, and strong requirement of computing power, efficient calculation method and simulation experiment strategy are studied. Model application & feedback refers to carrying out simulation experiments and prediction of specific cases based on the constructed model, accumulating experience for the improvement of the model based on the application effect, forming a feedback loop of "construction → application → improvement → application", and promoting the continuous improvement of the comprehensive performance of the model.

References

Bevilacqua S, Guarnieri G, Farella G, *et al.* 2018. A regional assessment of cumulative impact mapping on Mediterranean coralligenous outcrops. *Scientific Reports.* **8**: 1757. DOI: 10.1038/s41598-018-20297-1.

Bonan G B, Doney S C. 2018. Climate, ecosystems, and planetary futures: The challenge to predict life in Earth system models. *Science.* **359**: 6375. DOI: 10.1126/science.aam8328.

Chen X, Gao H, Yao X, *et al.* 2010. Ecosystem-based assessment indices of restoration for Daya Bay near a nuclear power plant in South China. *Environmental Science & Technology.* **44**: 7589–7595.

Costanza R. 1992. Toward an operational definition of ecosystem health. In: *Ecosystem Health: New Goals for Environmental Management.* Washington, D C: Island Press.

Dafner E V, Mallin M A, Souza J J, *et al.* 2007. Nitrogen and phosphorus species in the coastal and shelf waters of Southeastern North Carolina, Mid-Atlantic US coast. *Marine Chemistry.* **103**(3–4): 289–303.

Dong S, Wang F. 1999. Effects of bay scallop on plankton and water quality of mariculture pond. *Acta Oceanologica Sinica.* **6**: 138–144.

Duarte C M, Conley D J, Carstensen J, *et al.* 2009. Return to Neverland: Shifting baselines affect eutrophication restoration targets. *Estuaries and Coasts.* **32**(1): 29–36.

Fan W, Lv X. 2009. Data assimilation in a simple marine ecosystem model based on spatial biological parameterizations. *Ecological Modelling.* **220**: 1997–2008.

Fulton E A. 2010. Approaches to end-to-end ecosystem models. *Journal of Marine Systems.* **81**(1–2): 171–183.

Halpern B S, Longo C, Hardy D, *et al.* 2012. An index to assess the health and benefits of the global ocean. *Nature.* **488**: 615–620.

Halpern B S, Walbridge S, Selkoe K A, *et al.* 2008. A global map of human impact on marine ecosystems. *Science.* **319**: 948–952.

Hamada K, Ueda N, Yamada M, *et al.* 2012. Decrease in anthropogenic nutrients and its effect on the C/N/P molar ratio of suspended particulate matter in hypertrophic Dokai Bay (Japan) in summer. *Journal of Oceanography.* **68**(1): 173–182

Hattenrath T K, Anderson D M, Gobler C J. 2010. The influence of anthropogenic nitrogen loading and meteorological conditions on the dynamics and toxicity of Alexandrium fundyense blooms in a New York (USA) estuary. *Harmful Algae.* **9**(4): 402–412.

Hiroshi K, Akinori T, Yuichi H, *et al.* 2018. Numerical experiments based on a coupled physical–biochemical ocean model to study the Kuroshio-induced nutrient supply on the shelf-slope region off the southwestern coast of Japan. *Journal of Marine Systems.* **179**: 38–54.

Huang X, Zhang J, Jiang Z. 2015. Eco-environmental effects of nutrients input caused by human activities on the semi-enclosed bay and its management strategy. *Advances in Earth Science.* **30**(9): 961–969.

Kaplan I C, Marshall K N. 2016. A guinea pig's tale: Learning to review end-to-end marine ecosystem models for management applications. *ICES Journal of Marine Science.* DOI: 10.1093/icesjms/fsw047.

Lester S E, Stevens J M, Gentry R R, *et al.* 2018. Marine spatial planning makes room for offshore aquaculture in crowded coastal waters. *Nature Communications.* **9**: 945. DOI: 10.1038/s41467-018-03249-1.

Li K Z, Yin J Q, Huang L M, *et al.* 2006. Spatial and temporal variations of mesozooplankton in the Pearl River estuary, China. *Estuarine Coastal & Shelf Science.* **67**(4): 543–552.

Lin L, Liu D, Liu Z, *et al.* 2016. Impact of land reclamation on marine hydrodynamice and ecological environment. *Acta Oceanologica Sinica.* **38**(8): 1–11.

Liu D, Shen X, Di B, *et al.* 2013. Palaeoecological analysis of phytoplankton regime shifts in response to coastal eutrophication. *Marine Ecology Progress Series.* **475**: 1–14.

Lotze H K, Lenihan H S, Bourque B J, *et al.* 2006. Depletion, degradation, and recovery potential of estuaries and coastal seas. *Science.* **312**(5781): 1806–1809.

Lu J, Huang L, Luo Y, *et al.* 2015. Effects of freshwater input and mariculture (bivalves and macroalgae) on spatial distribution of nanoflagellates in Sungo Bay, China. *Aquaculture Environment Interactions.* **6**(2): 191–203.

Ma D M, Shi H H, Feng A P. 2014. Estimation of agricultural non-point source pollution based on watershed unit: A case study of Laizhou Bay. *Acta Ecologica Sinica.* **34**(1): 173–181.

Millennium Ecosystem Assessment (MA). 2005. *Ecosystems and Human Well-Being: Synthesis.* Washington D C: Island Press.

Parker A E, Wilkerson F P, Dugdale R C. 2012. Elevated ammonium concentrations from wastewater discharge depress primary productivity in the Sacramento River and the northern San Francisco Estuary. *Marine Pollution Bulletin.* **64**: 574–586.

Plaganyi E E. 2007. Models for an ecosystem approach to fisheries. FAO Fisheries Technical Paper. No. 477. Rome, FAO.

Rapport D J, Costanza R, McMichael A J. 1998. Assessing ecosystem health. *Trends in Ecology & Evolution.* 13: 397–402.

Rombouts I, Beaugrand G, Artigas L F, *et al.* 2013. Evaluating marine ecosystem health: Case studies of indicators using direct observations and modelling methods. *Ecological Indicators.* **24**: 353–365.

Shen C, Shi H, Zheng W, *et al.* 2016a. Study on the cumulative impact of reclamation activities on ecosystem health in coastal waters. *Marine Pollution Bulletin.* **103**(1–2): 144–150.

Shen C, Shi H, Zheng W, *et al.* 2016b. Spatial heterogeneity of ecosystem health and its sensitivity to pressure in the waters of nearshore archipelago. *Ecological Indicators.* **61**(2): 822–832.

Shen C, Shi H, Liu Y, *et al.* 2016c. Discussion of skill improvement in marine ecosystem dynamic models based on parameter optimization and skill assessment. *Chinese Journal of Oceanology and Limnology.* **34**(4): 683–696.

Shi, H, Ding D W, Huo W Z, *et al.* 2017. *Research on China's Marine Ecological Civilization Construction Strategy Based on the Integration of Sea and Land — Theoretical Basis and Typical Case Study.* Beijing: Ocean Press

Shi H, Zheng W, Zhang X, *et al.* 2013. Ecological–economic assessment of monoculture and integrated multi-trophic aquaculture in Sanggou Bay of China. *Aquaculture.* **410**(11): 172–178.

Shi H H, Shen C C, Li F, *et al.* 2014. Parameter sensitivity analysis of a coupled biological-physical model in Jiaozhou Bay. *Acta Ecologica Sinica.* **34**(1): 41–49.

Song H-J, Li R, Wang Z-L, *et al.* 2007. Interannual variations in phytoplankton diversity in the Sanggou Bay. *Advances in Marine Science.* **25**(3): 332–339.

Su J, Wang Y, Qin Y, *et al.* 2011. Some scientific problems and suggestions for countermeasures in China's reclamation projects. *Bulletin of Chinese Academy of Sciences.* **26**(2): 171–173.

Travers M, Shin Y J, Jennings S, *et al.* 2007. Towards end-to-end models for investigating the effects of climate and fishing in marine ecosystems. *Progress in Oceanography.* **75**(4): 751–770.

Trommer G, Leynaert A, Klein C, *et al.* 2013. Phytoplankton phosphorus limitation in a North Atlantic coastal ecosystem not predicted by nutrient load. *Journal of Plankton Research.* **35**(6): 1207–1219.

Wang X-L, Li K-Q, Shi X-Y. 2006. *Marine Environmental Capacity of Major Chemical Pollutants in Jiaozhou Bay.* Beijing: Science Press.

Wang Y-Z, Wu D, Shi H-H, *et al.* 2015. Impact of reclamation on water exchange in Bohai Bay in recent decade. *Oceanologia Et Limnologia Sinica.* **46**(3): 471–480.

Wu S Y, Wang W H, Feng A P. 2011. *China's Bay Development Activities and Their Environmental Effects.* Beijing: Ocean Press.

Xu Y, Qian L. 2004. Impacts of cage culture on marine environment. *Chinese Journal Of Applied Ecology.* **15**(3): 532–536.

Yin K, Harrison P J, Broom M. *et al.* 2011. Ratio of nitrogen to phosphorus in the Pearl River and effects on the estuarine coastal waters: Nutrient management strategy in Hong Kong. *Chemical Geology.* **36**(9–11): 411–419.

Yu H, Yu Z, Song X, *et al.* 2015. Seasonal variations in the nitrogen isotopic composition of dissolved nitrate in the Changjiang River estuary, China. *Estuarine Coastal & Shelf Science.* **155**: 148–155.

Zhang H, Wang D, Xie Z, *et al.* 2015. Comparative proteomics reveals highly and differentially expressed proteins in field-collected and laboratory-cultured blooming cells of the diatom Skeletonema costatum. *Environmental Microbiology.* **17**(10): 3976.

Zhang J M, Song W P, Gao S, *et al.* 2014. *Research and Application of Marine Environmental Impact Assessment Techniques for Intensive Sea Use on the Bohai Sea.* Beijing: Ocean Press.

Zheng B, Zhao X, Liu L, *et al.* 2011a. Effects of hydrodynamics on the distribution of trace persistent organic pollutants and macrobenthic communities in Bohai Bay. *Chemosphere.* **84**(3): 336–341.

Zheng W, Wang Z, Shi H. 2011b. *Assessment of Impacts of Typical Human Activities on Marine Ecosystem Services and Ecological Compensation Study.* Beijing: Ocean Press.

Index

Y

Z

www.ingramcontent.com/pod-product-compliance
Ingram Content Group UK Ltd.
Pitfield, Milton Keynes, MK11 3LW, UK
UKHW020350150926
8624IPUK00004B/90

9 789811 264207